Monographs on
Theoretical and Applied Genetics 8

Edited by
R. Frankel (Coordinating Editor), Bet-Dagan
G. A. E. Gall, Davis · M. Grossman, Urbana
H. F. Linskens, Nijmegen · R. Riley, London

Y. Y. Gleba K. M. Sytnik

Protoplast Fusion

Genetic Engineering in Higher Plants

Edited by R. Shoeman

With 62 Figures

Springer-Verlag
Berlin Heidelberg New York Tokyo 1984

Dr. YURY Y. GLEBA
Dr. KONSTANTIN M. SYTNIK

Institute of Botany
Academy of Science
Ukrainian SSR
Repina 2
Kiev 252601, USSR

Editor:

Dr. ROBERT L. SHOEMAN
Max-Planck-Institut für Zellbiologie
D-6802 Ladenburg
Fed. Rep. of Germany

Present address:

Department of Biochemistry
Roche Institute of Molecular Biology
Nutley, NJ 07110, USA

ISBN-13: 978-3-642-82247-6 e-ISBN-13: 978-3-642-82245-2
DOI: 10.1007/ 978-3-642-82245-2

Library of Congress Catalogıng in Publicatıon Data. Gleba, ÎU. ÎU. (ÎUriĭ ÎUr'evıch) Protoplast fusion, genetıc engıneerıng in higher plants. (Monographs on theoretıcal and applıed genetics, 8) Includes ındex. Bıblıography: p. 1. Plant genetic engıneering. 2. Plant Protoplasts. 3. Somatic hybrıds. I. Sytnik, K. M. (Konstantın Merkur'evich) II. Shoeman, R. III. Title. IV. Serıes. QK981.5.G5413 1984 581.87'3 84-5315

Softcover reprint of the hardcover 1st edition 1984

Typesettıng. Bruhlsche Unıversıtatsdruckereı, Gıessen
2131/3130-543210

Preface

Although only about 3 years have passed since the preparation of the original manuscript of this book for the Russian edition, the number of successful experiments on somatic hybridization of higher plants has doubled. Although the main inferences of the first edition still remain in force, most of them have received conclusive experimental support and, moreover, some new conclusions have been drawn. It can be expected that these inferences and conclusions will constitute a more or less durable foundation for somatic cell genetics of higher plants. We thus hope this book will also remain useful over the next years, in spite of the rapid progress of experiments and the increase in the number of scientific reports in this field.

Though it might appear strange to an uninvolved observer, the principal progress in hybridization of somatic cells of higher plants has been due to plant physiologists (who entered the field by elaborating methods and techniques for plant cell and, later, for isolated protoplast culture) rather than plant geneticists. However, further qualitative improvement in this field is inconceivable without the instillation of genetic ideology and the strict logic of genetic experiments. The main purpose of this book is the attempt to organize the available experimental data in terms and categories of genetic analysis. For this reason, this book lays no claim on being a comprehensive treatise on somatic hybridization. The consideration of the principally genetic aspects of somatic hybridization seemed indispensible to us, thus, numerous important questions of physiology of protoplasts and hybrid cells are not discussed. Although we tried to include all of the significant studies in this field, not all of the available literature has been cited. Unfortunately, the unnecessary publishing of the same results several times (in a scientific journal, then in proceedings of congresses or symposia, and then in combination with other data, etc.) is very widespread. Therefore, we have attempted to cite the same data only once.

We are aware that this book is subjective to a certain extent and we apologize beforehand to our colleagues whose works are cited perhaps less frequently than they merit. In describing the experimental data, we have given preference to the works of those colleagues whose investigations were known to us not only from published reports. Similarly, our own data have been reported in more detail. It is easier to discuss and critically analyze data which are better known. As was mentioned earlier, this book represents a geneticist's point of view and the conclusions herein may therefore appear to be at odds with the view of those dealing with somatic cell hybridization in higher plants (mostly physiologists or sometimes "culturists").

Certainly, we bear complete responsibility for the possible successes and, first and foremost, for all failures of this book. However, it should be acknowledged that the appearance of this book is the result of the work of many people. This book is rather profusely illustrated and contains the illustrations from the works of numerous investigators. We wish to thank Prof. G. Melchers, Dr. K. N. Kao, Dr. K. Glimelius, Dr. L. R. Wetter, Dr. L. C. Fowke, Dr. F. Constabel, Dr. O. Schieder, Dr.L. Menczel, Dr. V. A. Sidorov, and Dr. F. Nagy for their kind permission to incorporate their illustrations into this book and for providing the originals of these illustrations.

Dr. V. A. Sidorov kindly read the manuscript and we are thankful to him for a number of valuable remarks.

This book would not have appeared without the work of the editor of Springer-Verlag, Dr. R. L. Shoeman, who revised and completely rewrote the English text.

We would also like to express our gratitude to our colleagues and collaborators for their tremendous help in the preparations of the manuscript and especially to M. N. Moskalenko, I. F. Kanevsky, and V. P. Momot.

Y. Y. Gleba
K. M. Sytnik

Contents

Still no forbidden fruits on the tree of knowledge: one of the pioneers in somatic hybridization research,

Professor GEORG MELCHERS

with his creation – potato (x) tomato somatic hybrid plant.

1 Introduction

The research described in this book elaborates the production of higher plant hybrids by using the experimental technique of parasexual hybridization by means of protoplast fusion. Additionally, investigations into the genetic novelty of the resulting hybrid plants will be described.

The main distinction owned by this new hybridization method is that somatic plant cells, rather than sexual ones (gametes), are used as the parental cells. Treatment of these cells with specific enzymes results in the removal of their rigid polysaccharide cell walls: thus, "naked" plant cells (i.e., isolated protoplasts) are obtained. Certain, specific experimental treatment of these protoplasts results in cell fusion. Hybrid cells produced in this way may, in many cases, be subsequently regenerated into entire hybrid plants. This parasexual hybridization via protoplast fusion has been made possible by the elaboration of two experimental approaches in plant physiology: (1) methods of plant cell and tissue culture and (2) techniques of production and manipulation of isolated protoplasts. The avenues opened by the first of these methods may be summarized as follows. Special nutrient media have been developed and conditions have been described in which individual cells or isolated groups of cell dedifferentiate (i.e., they lose the characters of the initial tissue) and begin dividing as independent unicelluar organisms. This sort of unorganized growth and cell reproduction can be practically maintained in vitro for tens of years. The organized growth of the cells can be restored and, as a result, the entire plant can be produced de novo by altering the culture conditions, in particular by modifying the hormonal content of the media. As matter of fact, these techniques enable one to step from the organism level down to the cellular level, making feasible experiments on millions of cells in a test tube that would otherwise be prohibitive or impossible if attempted on a similar number of plants. At the same time, these techniques permit the regeneration of the entire plant at the end of an experiment on isolated cells, thus allowing the visualization of the manipulation at the organism level as well.

The techniques of production and manipulation of isolated protoplasts that have been described in the last 10 years depend upon the enzymatic hydrolysis of the cell walls, making possible not only the fusion of the resulting naked cells (protoplasts), but also enabling the cells to take up macromolecules (proteins, nucleic acids, etc.) and particles (isolated cell organelles, microorganisms, etc.) from the surrounding solution. Cultivation of protoplasts under defined conditions results in synthesis of a new cell wall. The resulting cells behave like "normal" cells cultured in vitro and, in particular, are capable of dividing and forming cell colonies and/or even entire plants. Thus, to reiterate a crucial point, these two methods permit the experimenter to easily change from the organism level to the cellular level and back again, with the added bonus that the contents of the plant cell may be manipulated at the isolated protoplast stage. The combination of these techniques clears the way for hybridization of higher plants without sexual crossing, i.e., parasexual hybridization. Substantial progress has been made up to now in only one of the above discussed realms of parasexual hybridization, namely, in hybrid production by fusion of isolated protoplasts. The full realization of the other aspects, such as the transplantation of organelles or the introduction of foreign DNA into protoplasts, depends largely on the refinement and application of existing methods.

The subject of this book, the hybridization of plant somatic cells by fusion of protoplasts, is a method similar to the previously elaborated method of hybridization of animal somatic cells. There is, however, one fundamental distinction between these two methods–hybridization of animal somatic cells enables the production of *hybrid cells*, whereas the fusion of plant somatic cell protoplasts results, in many cases, in the production of *hybrid plants*. Thus, parasexual hybridization is not only a novel pathway for genetic analysis, but may also develop into a powerful tool for industrial plant breeding in the future. Up to now, both the geneticist, investigating the mechanics of plant cell inheritance, and the breeder, attempting to feed mankind by creating more productive plant varieties, possessed only sexual crossing as the sole tool for the engineering of novel plants. Our purpose is to equip them with a new method to achieve their diverse goals.

The necessity of elaboration of a new method of hybridization is also dictated by the limitations of the old method: sexual crossing is a strictly determined system where only a defined group of organisms in restricted combinations may be used as parental forms. The resulting progeny from such a sexual cross possess defined and limited genotypes. A short review of the general characteristics of sexual crossing will illustrate the restrictions imposed on genotype engineering by conventional methods of hybridization and also serve to point out the potential advantages offered by parasexual hybridization.

Highly specialized plant cells, gametes, are involved in sexual crossing. Gameteogenesis involves meiotic reduction and segregation of chromosomal material. Only plants which are capable of normal gametogenesis can be sexually crossed. The zygote is a product of the fusion of two gametes. This process results in nuclear fusion and in the restoration of the sporophytic chromosome number. Nuclear genetic determinants are inherited biparentally. Sexual crossing is symmetrical in the sense that both the male and female gametes contribute equally and supply a gametophytic set of nuclear genetic material to the zygote. The segregation of nuclear characters is in accordance with the Mendelian laws of inheritance. Extranuclear genetic determinants are inherited strictly uniparentally and maternally in most higher plants. Sexual crosses are limited to only a few genotype combinations out of many theoretically possible combinations, because of inherent systems of incompatibility. Sexual crosses are limited to phylogenetically closely related species.

Plant hybridization by protoplast fusion may therefore be of interest if it can be used:

1. to perform hybridization between species which are too distantly related to be crossed sexually;
2. to obtain different assymetric hybrids, in which the progeny would have all the chromosomal material from one of the parents, but only a few chromosomes, a few genes, or only the cytoplasm and organelles from the other parent;
3. to devise hybridization systems involving simultaneous fusion of more than two parental cells;
4. to obtain hybrids genetically representing a sum of parental idiotypes;
5. to obtain plants heterozygous for extranuclear genes;
6. to bypass restrictions on hybridization imposed by inherent systems of incompability;
7. to produce hybrids between plants that are not sexually crossable because of abnormalities in the morphogenesis (gametogenesis) of the parents;
8. to obtain hybrids between cells containing different epigenetic programs.

The most important distinction of parasexual hybridization as compared to the usual sexual crossing is the completely artificial nature of this new method. Theoretically, it may be assumed that many restrictions inherent in sexual crossing (for example, the restrictions due to abnormal gametogenesis in the parent or due to reactions of inherent incompatibility) are not a problem in parasexual hybridization, since gametes and their exchange are not involved in this new method. Moreover, since parasexual hybridization is a complex process consisting of a number of events (cell fusion, fusion/nonfusion of cell nuclei, fusion/nonfusion of organelles, etc.), each of which may lead to genetic alterations and since the genotype of the progeny is a function of all these events, it is quite evident that by defining and standardizing some parameters of the artificial process of parasexual hybridization, one can predetermine, to a certain extent, the genetic results. The great importance of this presumable control of the parasexual hybridization process is stressed by the fact that we deliberately chose to include the term "genetic engineering of plants" rather than "plant hybridization" in the title of this book. We believe the term genetic engineering emphasizes a creative human presence and accentuates the synthetic rather than the analytic nature of the scientific methodology utilized in these investigations of plant inheritance. Thus, parasexual hybridization is not only a way of producing novel plant forms, but is also a new synthetic method for investigations of biological systems.

Not all methodological problems of plant cell genetic analysis can be solved by using the techniques of parasexual hybridization. We would like to point out two of the major limitations to the use of parasexual hybridization. First, the technique of parasexual hybridization, although dealing with isolated plant cells rather than entire plants, is nonetheless relatively time-consuming since plant cells divide more slowly than animal cells and microorganisms (the duration of the cell cycle in plant cells in vitro is typically 25–40 h); moreover, the regeneration of entire plants takes several months in most cases. For this reason, one experiment for production of *Nicotiana* hybrids by protoplast fusion and subsequent analysis of their progeny takes 1–1½ years and no foreseeable improvements in experimental technique are likely to reduce the time necessary for this process. Second, the plant cell is very labile and responds readily to external signals, producing the undesirable consequence of a great genetic heterogeneity of the plant cells via external induction. This heterogeneity complicates the proper analysis of genetic events caused by hybridization, since the spectrum of induced genetic alterations are superimposed upon those resulting from parasexual hybridization. This heterogeneity may also become a grave obstacle to practical work using protoplast fusion to achieve the goal of production of material for plant breeding.

2 Techniques of Parasexual Hybridization

2.1 Culture Techniques

2.1.1 Culture of Isolated Plant Protoplasts

Methods of hybridization based on the phenomenon of induced fusion of somatic cells have been used successfully in animal cell genetics for over 20 years (see reviews of Ephrussy 1972; Harris 1974; Ringertz and Savage 1976). Until recently, the principal obstacle to the introduction of analogous methods in plant cell genetics was the cellulose-pectin cell wall which completely excluded the possibility of plant cell fusion. Although naked plant cells (isolated protoplasts) were mechanically obtained as early as 1892 (Klercker 1892), the improvement of the method of producing isolated protoplasts became possible only after the elaboration of the techniques of enzymatic isolation of a large quantity of naked protoplasts (Cocking 1960). Over 2000 papers concerning isolated protoplasts have been published on a world scale. The present state of the art and the potentials of this method have been reported in detail in the proceedings of several symposia and congresses. The latest of these have been: The Fourth Cell and Tissue Culture International Congress, August 20–25, 1978, Calgary, Canada (Frontiers of Plant Tissue Culture 1978, 1978); Fifth International Protoplast Symposium, July 9–14, 1979, Szeged, Hungary (Abstracts of the Fifth International Protoplast Symposium, 1979); Second International Congress on Cell Biology, August 31-September 5, 1980, Berlin (West); and the Fifth International Congress of Plant Tissue and Cell Culture, July 11–16, 1982, Tokyo, Japan. The more or less modern data are reported in the following books (full bibliographic data are contained in the references): *Cell genetics in higher plants,* 1976; *Microbial* and *plant protoplasts,* 1976; *Plant cell and tissue culture*, 1979; *Plant cell, tissue and organ culture*, 1977; *Advances in Protoplast Research*, 1980; and *International Cell Biology* 1980–1981.

This chapter contains a brief enumeration of the techniques, methods, and conditions which determine the modern trends in the development of genetic engineering of higher plants.

A list of the plant species for which the techniques of protoplast culturing and cell colony or entire plant regeneration have already been described is given in Table 1. In most experiments, leaf mesophyll or callus protoplasts were used as the initial material. The regeneration of tobacco mesophyll protoplasts and *Arabidopsis* callus protoplasts are illustrated in Fig. 1–4. As can be seem from the data contained in Table 1, techniques of protoplast culture and plant regeneration are available for a great number of species.

Table 1. Plant species for which techniques of protoplast culturing and cell colony or entire plant regeneration have been described

Species	Source[a]	Product[b]	Ref.
Ammi visnaga	Callus	Callus	Gamborg et al. (1972)
Antirrhinum majus	Leaf	Callus	Poirier-Hamon et al. (1974)
Antirrhinum majus	Shoot tip	Callus, root	Binding et al. (1981)
Arabidopsis thaliana	Callus	Callus	Gresshoff (1976)
Arabidopsis thaliana	Callus	Plant	Gleba et al. (1978d); Le Thi-Xuan and Menczel (1980)
Arabidopsis thaliana	Shoot tip	Plant	Binding et al. (1981)
Asparagus officinalis	Leaf	Plant	Bui-Dang-Ha and Mackenzie (1973); Bui-Dang-Ha et al. (1975)
Atropa belladonna	Leaf	Plant	Gosch et al. (1975b); Bajaj et al. (1978), Lörz and Potrykus (1979)
Beta vulgaris	Callus	Callus	Evans et al. (1980b); Smolenskaya and Raldugina (1981)
Brassica alba	Root	Callus	Xu et al. (1982b)
Brassica campestris	Leaf	Callus, root	Schenck and Hoffmann (1979); Kumar et al. (1981)
Brassica campestris	Root	Root	Xu et al. (1982b)
Brassica campestris	Cotyledon	Root	Lu et al. (1982)
Brassica napus	Leaf	Plant	Katha et al. (1974); Thomas et al. (1976)
Brassica napus	Root	Plant	Xu et al. (1982b)
Brassica napus	Stem embryo	Plant	Kohlenbach et al. (1982)
Brassica napus	Cotyledon	Root	Lu et al. (1982)
Brassica nigra	Leaf	Callus	Schenck and Hoffmann (1979)
Brassica oleracea	Leaf	Callus, root	Gatenby and Cocking (1977), Schenck and Hoffmann (1979)
Brassica oleracea	Shoot tip	Callus, root	Binding et al. (1981)
Brassica oleracea	Root	Plant	Xu et al. (1982b)
Brassica oleracea	Cotyledon	Plant	Lu et al. (1982)
Bromus inermis	Callus	Embryo	Kao et al. (1973)
Bryum erythocarpum	Protonema	Protonema	Binding (1966)
Calendula arvensis	Shoot tip	Callus	Binding et al. (1981)
Capsicum annuum	Leaf	Plant	Saxena et al. (1981b)
Caragana arborescens	Leaf	Callus	Constabel et al. (1975)
Catharantus roseus			Koblitz (1975)
Catharantus roseus	Leaf	Callus	Constabel et al. (1981)
Chrysanthemum morifolium	Leaf	Callus	Schum and Preil (1981)
Chrysanthemum segetum	Shoot tip	Callus, root	Binding, et al. (1981)
Cicer arietinum	Leaf	Callus	Gamborg et al. (1974)
Cichorium intybus	Shoot tip	Plant	Binding et al. (1981)
Cichorium intybus	Leaf	Plant	Crepy et al. (1982)
Citrus aurantium	Nucellar callus	Plant	Vardi et al. (1982)
Citrus limon	Nuc. callus	Plant	Vardi et al. (1982)
Citrus paradisi	Nuc. callus	Plant	Vardi et al. (1982)

[a] Initial cell type used for protoplast production
[b] Product obtained after culturing of the protoplasts

Table 1 (continued)

Species	Source[a]	Product[b]	Ref.
Citrus reticulata	Nuc. callus	Plant	Vardi et al. (1982)
Citrus sinensis	Callus	Embryo	Vardi et al. (1975, 1982)
Coffea arabica	Callus	Callus	Sondahl et al. (1980)
Crepis capillaris	Shoot tip	Callus, root	Binding et al. (1981)
Cucumis sativus	Leaf	Callus	Coutts and Wood (1975)
Cucumis melo	Callus	Callus	Mareno et al. (1980)
Datura innoxia	Leaf	Plant	Schieder (1975 b)
Datura innoxia	Cotyledon	Plant	Lu et al. (1982)
Datura innoxia	Callus	Plant	Furner et al. (1978)
Datura metaloides	Leaf	Plant	Schieder (1977 b)
Daucus carota	Callus	Plant	Grambow et al. (1972); Dudits et al.(1976)
Daucus carota	Shoot tip	Plant	Binding et al. (1981)
Digitalis lanata	Leaf	Plant	Xiang-hui (1981)
Digitalis lanata	Callus	Callus	Diettrich et al. (1982)
Digitalis purpurea	Callus	Callus	Diettrich et al. (1980)
Funaria hygrometrica	Protonema	Protonema	Binding (1966, 1974); Gwozdz and Waliszewska (1979); Bastra and Abel (1981)
Gaillardia grandiflora	Shoot tip	Plant	Binding et al. (1981)
Glycine max	Callus	Callus	Kao et al. (1970)
Glycine max	Pod	Callus, root	Zieg and Outka (1980)
Glycine max	Shoot tip	Callus, root	Binding et al. (1981)
Glycine max	Root	Callus	Xu et al. (1982 a)
Gossypium hirsutum	Callus	Callus	Bhojwani et al. (1977)
Gossypium hirsutum	Leaf	Callus	Khasanov and Butenko (1979)
Gossypium klotschianum	Callus	Callus	Finer and Smith (1982)
Haplopappus gracilis	Callus	Callus	Kao et al. (1971)
Helianthus annuus	Shoot tip	Plant	Binding et al. (1981)
Helianthus tuberosus	Shoot tip	Callus, root	Binding et al. (1981)
Hordeum vulgare			Koblitz (1976)
Hyosciamus albus	Leaf	Root	Lörz et al. (1979)
Hyosciamus muticus	Leaf, callus	Plant	Lörz et al. (1979); Wernicke et al. (1979); Wernicke and Thomas (1980)
Hyosciamus niger	Leaf	Callus	Kohlenbach and Bohnke (1975)
Hyosciamus niger	Cotyledon	Callus	Lu et al. (1982)
Kentranthus ruber	Shoot tip	Callus, root	Binding et al. (1981)
Lathyrus odoratus	Leaf	Callus	Razdan et al. (1980)
Lilium speciosum × *henryi*	Callus	Callus	Simmonds et al. (1979)
Limnophyla chinensis	Leaf	Callus	Sangwan and Harada (1978)
Linum usitatissimum	Leaf	Callus	Gamborg and Shyluk (1976)

Table 1 (continued)

Species	Source[a]	Product[b]	Ref.
Linum usitatissimum	Shoot tip	Callus, shoot	Binding et al. (1981)
Lycopersicon esculentum	Leaf	Shoot	Zapata et al. (1977); Zapata and Sink (1981)
Lycopersicon esculentum	Leaf	Callus	Mülbach (1980)
Lycopersicon hirsutum	Leaf	Callus	Zapata et al. (1981)
Lycopersicon peruvianum	Leaf	Callus, shoot	Zapata et al. (1977)
Lycopersicon peruvianum	Leaf	Callus	Mülbach (1980)
Lycopersicon pimpinellifolium	Leaf	Callus	Zapata et al. (1981)
Lygodium japonicum	Prothallia	Gametophyte	Maeda and Ito (1981)
Manihot esculenta	Leaf	Plant	Shahin and Shepard (1980)
Marchantia polymorpha	Cell culture	Thalli	Ono et al. (1979)
Matricaria chamomilla	Shoot tip	Callus	Binding et al. (1981)
Medicago coerulea	Leaf, callus	Embryo	Arcioni et al. (1982)
Medicago glutinosa	Leaf, callus	Embryo	Arcioni et al. (1982)
Medicago sativa	Leaf	Plant	Mesentsev (1978); Kao and Michayluk (1980); dos Santos et al. (1980); Johnson et al. (1981)
Medicago sativa	Cotyledon	Plant	Lu et al. (1982)
Medicago sativa	Root	Plant	Xu et al. (1982c)
Melilotus alba	Leaf	Callus	Constabel et al. (1975)
Mollugo nudicaulis	Leaf	Callus, root	Chellappan et al. (1980)
Nicotiana accuminata	Leaf	Plant	Chupeau et al. (1974); Bourgin et al. (1979)
Nicotiana alata	Leaf	Plant	Chupeau et al. (1974); Bourgin and Missonier (1978a, b); Bourgin et al. (1979)
Nicotiana alata	Leaf	Plant	Passiatore and Sink (1981)
Nicotiana alata	Flower petal	Plant	Flick and Evans (1983)
Nicotiana alata (×) *N. sanderae*	Flower petal	Plant	Flick and Evans (1983)
Nicotiana debneyi	Leaf	Plant	Scowcroft and Larkin (1980); Piven (1981)
Nicotiana forgetiana	Leaf	Plant	Passiatore and Sink (1981)
Nicotiana glauca	Leaf	Plant	Chupeau et al. (1974); Bourgin et al. (1979)
Nicotiana glutinosa	Leaf	Callus	Chupeau et al. (1974); Bourgin et al. (1979)
Nicotiana langsdorffii	Leaf	Plant	Chupeau et al. (1974); Bourgin et al. (1979)
Nicotiana langsdorffii	Leaf	Plant	Evans (1979)
Nicotiana langsdorffii × *N. glauca*	Leaf	Plant	Chupeau et al. (1974)
Nicotiana longiflora	Leaf	Plant	Chupeau et al. (1974); Bourgin et al. (1979)
Nicotiana nesophila	Leaf	Plant	Evans (1979)
Nicotiana noctiflora	Leaf	Plant	Chupeau et al. (1974); Bourgin et al. (1979)

Table 1 (continued)

Species	Source[a]	Product[b]	Ref.
Nicotiana otophora	Leaf	Plant	Banks and Evans (1976); Bourgin et al. (1979)
Nicotiana paniculata	Leaf	Plant	Chupeau et al. (1974); Bourgin et al. (1979)
Nicotiana plumbaginifolia	Leaf	Plant	Bourgin et al. (1979); Magnien et al. (1980); Negrutiu (1981)
Nicotiana repanda	Leaf	Plant	Evans (1979)
Nicotiana rustica	Leaf	Plant	Gill et al. (1979)
Nicotiana sanderae	Leaf	Plant	Passiatore and Sink (1981)
Nicotiana stocknii	Leaf	Plant	Evans (1979)
Nicotiana suaveolens	Leaf	Plant	Chupeau et al. (1974); Bourgin et al. (1979)
Nicotiana sylvestris	Leaf	Plant	Chupeau et al. (1974); Bourgin et al. (1976, 1979); Banks and Evans (1976); Nagy and Maliga (1976); Durand (1979); Facciotti and Pilet (1979); Negrutiu and Mousseau (1980, 1981); Negrutiu and Muller (1981)
Nicotiana sylvestris × *N. otophora*	Leaf	Plant	Bank and Evans (1976)
Nicotiana tabacum	Leaf	Plant	Tabeke et al. (1971); Nagata and Tabeke (1971); Nitsch and Ohyama (1971); Gleba et al. (1974a); Vasil and Vasil (1974); Binding (1975); Raveh and Galun (1975); Uschimiya and Murashige (1976); Lörz et al. (1977); Gleba (1978a); Gamborg et al. (1979); Landgren and Bonnett (1979); Caboche (1980); Flick et al. (1981)
Nicotiana tabacum	Callus	Plant	Chi-Kuei et al. (1977)
Nicotiana tabacum	Leaf epidermis	Plant	Davey et al. (1974a)
Nicotiana tabacum	Cotyledon	Plant	Gleba et al. (1974a); Lu et al. (1982)
Nicotiana tabacum × *N. otophora*	Leaf	Plant	Bank and Evans (1976)
Nigella arvensis	Shoot tip	Plant	Binding et al. (1981)
Nigella damascena	Shoot tip	Callus, root	Binding et al. (1981)
Nigella sativa	Shoot tip	Callus	Binding et al. (1981)
Oryza sativa	Callus	Callus	Anonymous (1975); Deka and Sen (1976)
Panicum maximum	Callus	Plant	Lu et al. (1981)
Parthenocissus tricuspidata	Callus	Callus	Anonymous (1975)
Pelargonium hortorum	Callus	Callus	Abo El-Nil and Hildebrandt (1976)
Pelargonium zonale	Leaf	Plant	Kameya (1975b)
Pennisetum americanum	Callus	Plant	Vasil and Vasil (1980, 1981)

Table 1 (continued)

Species	Source[a]	Product[b]	Ref.
Pennisetum americanum × *Pennisetum purpureum*	Callus	Plant	Vasil and Vasil (1981)
Pennisetum mexicanum	Callus	Callus	Vasil and Vasil (1979)
Petunia axiliaris	Leaf	Shoot	Power et al. (1976a)
Petunia hybrida	Leaf	Plant	Durand et al. (1973); Frearson et al. (1973); Binding (1974a, b); Vasil and Vasil (1974); Power et al. (1976a)
Petunia hybrida (×) *Petunia parodii*	Leaf	Plant	Power et al. (1976b)
Petunia inflata	Leaf	Root, shoot	Power et al. (1976a)
Petunia parodii	Leaf, leaf, callus	Plant, shoot	Hayward and Power (1975); Power et al. (1976a)
Petunia parviflora	Leaf	Plant	Sink and Power (1977)
Petunia violacea			Power et al. (1976a)
Pharbitis nil	Leaf	Callus	Messerschmidt (1974)
Phaseolus aureus	Root	Callus	Xu et al. (1981)
Phaseolus vulgaris	Leaf	Callus	Pelcher et al. (1974)
Phaseolus vulgaris	Callus	Callus	Gamborg et al. (1974)
Physalis alkekengi	Leaf	Callus	Bapat and Schieder (1981)
Physalis ixocarpa	Leaf	Callus	Bapat and Schieder (1981)
Physalis minima	Leaf	Callus	Bapat and Schieder (1981)
Physalis peruviana	Leaf	Callus	Bapat and Schieder (1981)
Physalis pruinosa	Leaf	Callus	Bapat and Schieder (1981)
Physcomitrella patens	Protonema	Protonema	Stumm et al. (1975); Grimsley et al. (1977a); Ashton and Cove (1977); Burgess and Linstead (1981)
Physcomitrium eurystomum	Protonema	Protonema	Binding (1964, 1966)
Physcomitrium piriforme	Protonema	Protonema	Binding (1964, 1966)
Pinus pinaster	Cotyledon	Callus	David and David (1979)
Pisum sativum	Leaf, shoot	Callus	Constabel et al. (1973); Gamborg et al. (1975); Nandi and Eriksson (1977)
Pisum sativum	Root	Callus	Landgren and Torrey (1973a, b); Landgren (1976a, b)
Pisum sativum	Callus	Callus	Gamborg et al. (1974)
Plagiomnium vesicatum	Protonemata	Protonemata	Maeda (1979)
Polytrichum juniperinum	Prothalli	Gemetophyte	Gay (1976)
Pottia intermedia	Protonema	Protonema	Ripetsky (1978)
Pteridium aquilinum	Prothalli	Prothalli	Partanen et al. (1980)
Ranunculus sceleratus	Leaf	Plant	Dorion et al. (1975)
Reseda luteola	Shoot tip	Callus, shoot	Binding et al. (1981)
Rosa	Callus	Callus	Krishnamurthy et al. (1979)
Saccharum officinarum	Callus	Callus	Maretzky and Nickell (1973)
Salpiglossis sinuta	Callus	Plant	Boyes et al. (1980)
Senecio jacobaea	Shoot tip	Plant	Binding et al. (1981)

Table 1 (continued)

Species	Source[a]	Product[b]	Ref.
Senecio silvaticus	Shoot tip	Plant	Binding et al. (1981)
Senecio vernalis	Shoot tip	Callus, shoot	Binding et al. (1981)
Senecio viscosus	Shoot tip	Plant	Binding et al. (1981)
Senecio vulgaris	Shoot tip	Plant	Binding et al. (1981)
Senecio vulgaris	Leaf	Plant	Binding and Nehls (1980)
Sinapis arvensis	Shoot tip	Callus, root	Binding et al. (1981)
Solanum chacoense	Leaf	Callus	Kuchko and Butenko (1976); Butenko et al. (1977)
Solanum dulcamara	Shoot tip	Plant	Binding et al. (1981)
Solanum dulcamara	Leaf	Plant	Binding and Nehls (1977)
Solanum incanum	Shoot tip	Callus	Binding et al. (1981)
Solanum luteum	Shoot tip	Plant	Binding et al. (1981)
Solanum melongena	Leaf	Plant	Jing-fen Jia and Potrykus (1981); Saxena et al. (1981 a)
Solanum nigrum	Leaf	Plant	Nehls (1978 b)
Solanum nigrum	Shoot tip	Plant	Binding et al. (1981)
Solanum tuberosum	Leaf	Plant	Shepard and Totten (1977); Butenko et al. (1977); Binding et al. (1978); Gunn and Shepard (1981); Uhrig (1981)
Solanum tuberosum	Callus	Callus	Kuchko and Butenko (1976); Butenko et al. (1977)
Solanum tuberosum	Shoot tip	Plant	Binding et al. (1981)
Solanum tuberosum	Leaf	Plant	Shepard (1982)
Sorghum bicolor	Callus	Callus	Brar et al. (1979, 1980)
Sphaerocarpos donnellii	Protonema	Protonema	Wenzel and Schieder (1973)
Trifolium arvense	Callus	Callus	White and Bhojwani (1981)
Trifolium repens	Callus	Plant	Gresshoff (1980)
Trifolium repens	Leaf	Root, shoot	Bhojwani and White (1982)
Trigonella foenum-graecum	Root	Root	Xu et al. (1982c)
Triticum monococcum	Callus	Callus	Dudits (1976)
Vicia faba	Callus	Callus	Kao et al. (1970); Röper (1981)
Vicia faba	Leaf, shoot apex	Callus	Binding and Nehls (1978)
Vicia faba	Shoot tip	Callus, root	Binding et al. (1981)
Vicia narbonensis	Leaf	Callus	Donn (1978)
Vicia hajastana	Callus	Callus	Kao et al. (1973); Kao and Michayluk (1975)
Vigna sinensis	Callus	Callus	Davey et al. (1974); Gamborg et al. (1974 b)
Vitis vinifera	Callus	Callus	Skene (1975)
Zea mays	Stem	Callus	Potrykus et al. (1977 b)
Zea mays	Callus	Callus	Potrykus et al. (1979 c); Choury and Zurawsky (1981)
Zea mays	Leaf	Callus	Invantsov and Achmetov (1978)

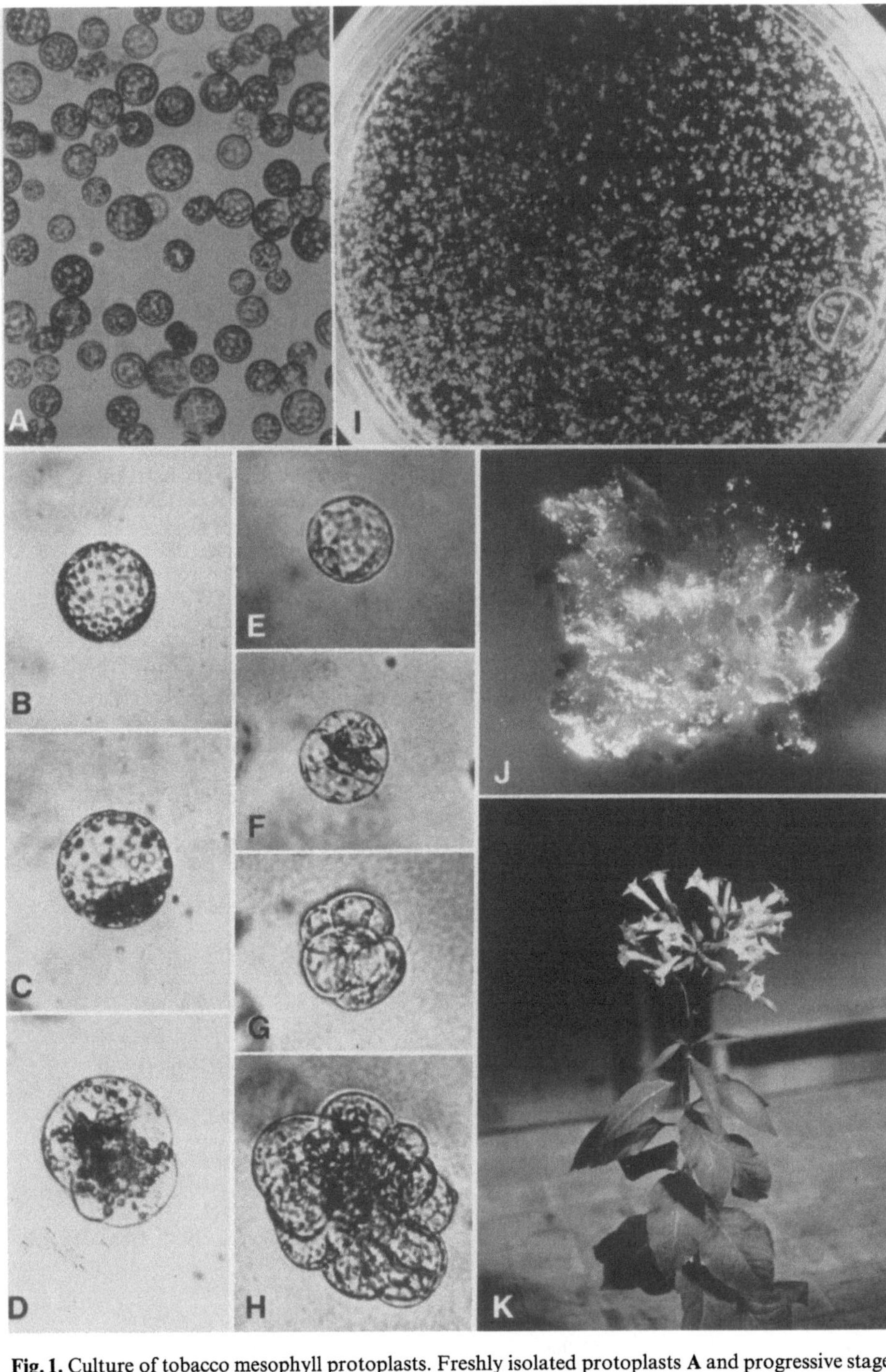

Fig. 1. Culture of tobacco mesophyll protoplasts. Freshly isolated protoplasts **A** and progressive stages of cell colony development from a normal tobacco protoplast **B–D** and from a chlorophyll deficient plastome mutant of tobacco **E–H**, × 400. A petri dish **I** containing numerous colonies regenerated from

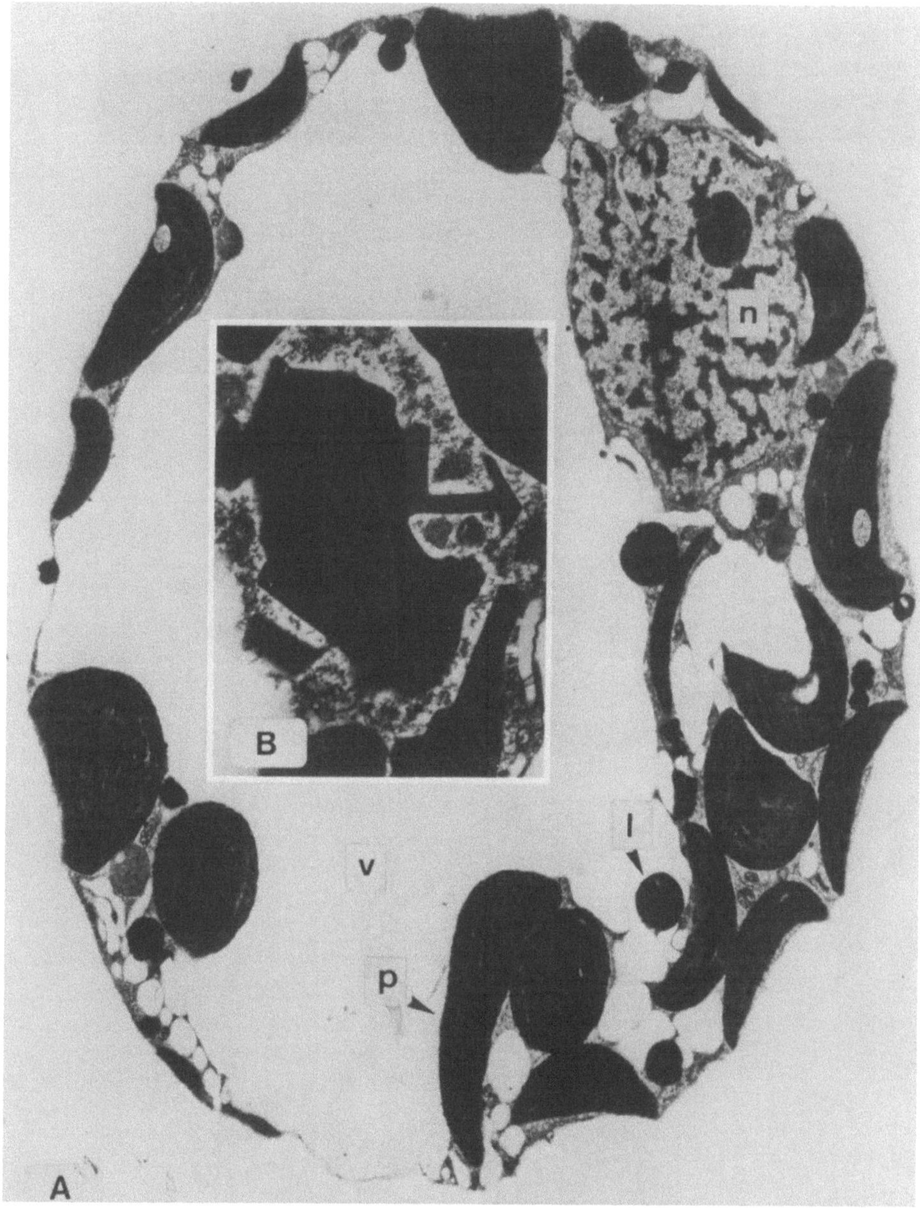

Fig. 2. Electron micrograph of an isolated tobacco mesophyll protoplast, var. Dubeck 44. **A** Cross-section of the entire protoplast. Note the conspicuous absence of a cell wall; × 10 000. **B** Enlargement of a chloroplast containing crystalline inclusions of ribulose-1,5-bisphosphate carboxylase (Fraction I protein), × 14 000. The nucleus *n*, plastids *p*, vacuole *v*, and lipid globules *l* are labelled. (Belitser et al. 1977)

isolated protoplasts after 21 days of culture. Shoot organogenesis in calli **J** regenerated from protoplasts. A flowering tobacco plant **K** regenerated from a mesophyll protoplast. (Gleba et al. 1974; Gleba 1980)

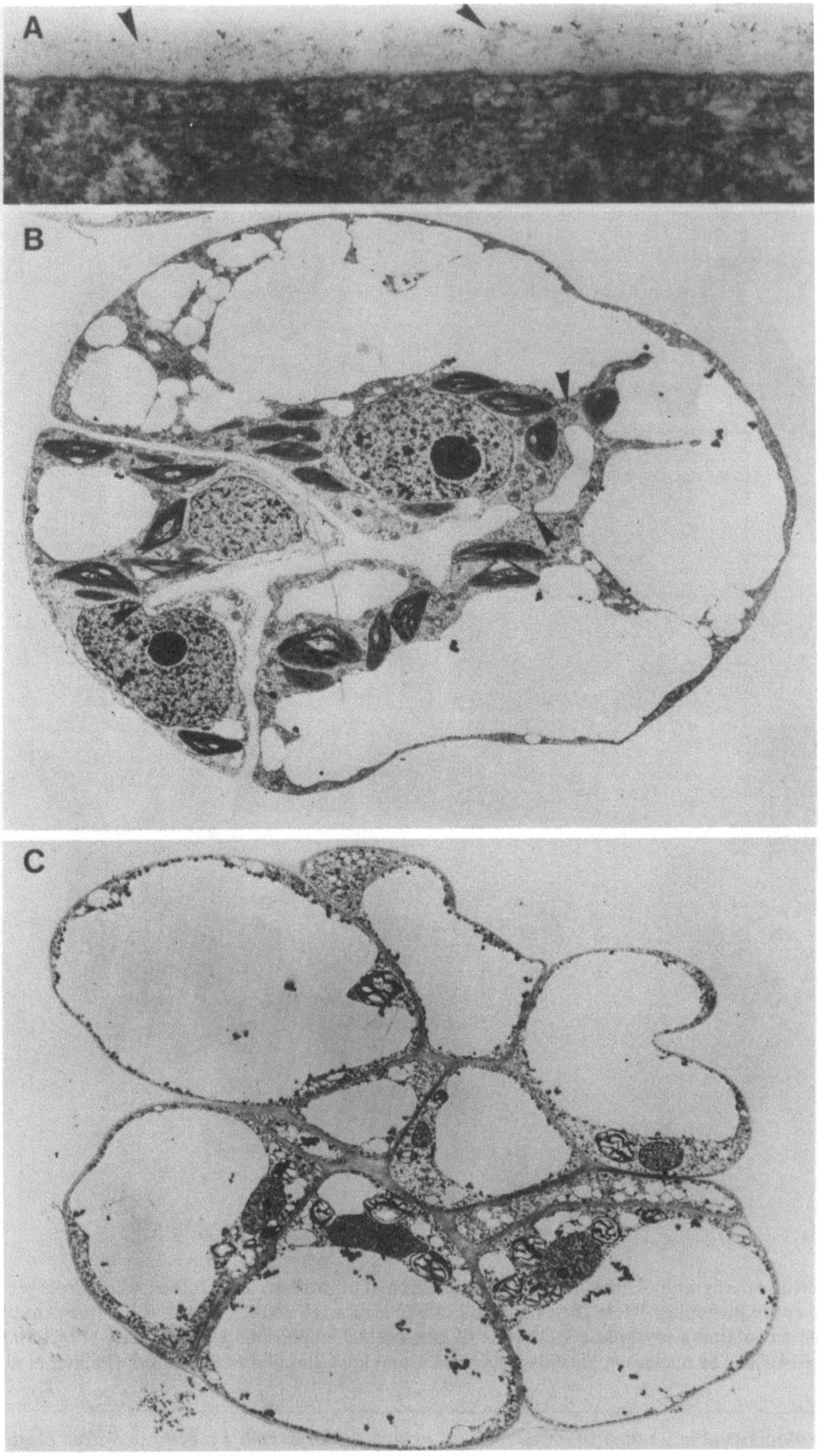

Fig. 3. Electron micrographs of cultured tobacco mesophyll protoplasts. **A** Resynthesis of the polysaccharide cell wall (*arrows*) on the surface of the protoplast after 48 h of culture, × 50 000. **B** Cross-section through the site of the initial cell division of a cell regenerated from a protoplast after 5 days of culture, × 4 000. The abnormalities of cytokinesis of this first cell division are indicated by *arrows*. **C** Cross-section through a microcolony regenerated from a protoplast after 15 days of culture, × 2 000. (Belitser et al. 1977; Sidorov et al. 1977)

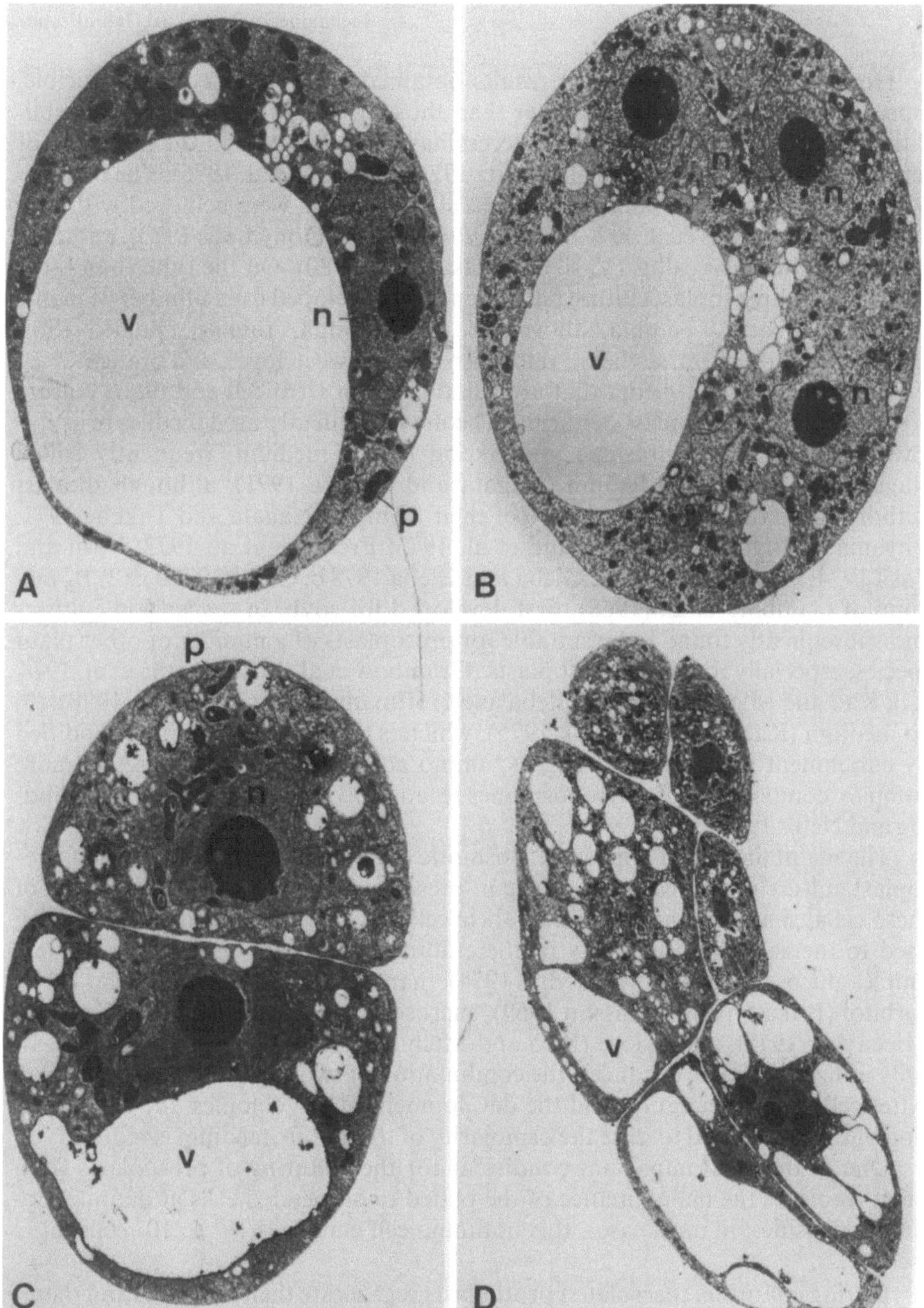

Fig. 4. Electron micrographs of isolated callus protoplasts from *Arabidopsis thaliana* (Heyhn L.), race Enkheim. The nucleus (*n*), vacuole (*v*), and plastids (*p*) are clearly visible. **A** A freshly isolated protoplast, ×3100. **B** A freshly isolated multinuclear protoplast, ×2000. **C** The first cell division after 3 days of culture, ×4300. **D** A microcolony of cells regenerated from a callus protoplast after 5 days of culture, ×2200. (Sidorov et al. 1978)

However, in many cases, the results obtained are not always reproducible, especially for races or varieties other than those used by the authors of a publication. Moreover, no appreciable success has been achieved in the culturing of cereal protoplasts (see reviews of Koblitz 1975; Potrykus et al. 1976). The relatively few positive results obtained with cereal protoplasts were achieved with *Zea mays* stem (Potrykus et al. 1977), *Oryza sativa* callus (anonymous 1975), and *Pennisetum americanum* callus (Vasil and Vasil 1979, 1980). On the other hand, the conditions for protoplast culture have been fully developed for a number of plants including tobacco, petunia, thornapple, belladonna, tomato, potato (Solanaceae), pelargonium, soybean, vetch, asparagus, sweetclover, and orange.

The same nutrient media that are suitable for in vitro cell and tissue culture are also used for protoplast culturing. The most frequently used media are as follows: (1) modified Murashige and Skoog (1962) medium, frequently called Nagata and Takebe B Medium (Nagata and Takebe 1971), although the last authors stress the secondary nature of their formula (Nagata and Takebe 1971; Ohyama and Nitsch 1972; Durand et al. 1973; Frearson et al. 1973; Vasil and Vasil 1974; Gleba et al. 1974a; Gleba and Gleba 1978b; Gleba 1978a); (2) B_5 medium of Gamborg et al. (1968), first developed for soybean suspension cultures and subsequently found to be suitable for protoplasts of a number of other plant species, especially for callus protoplasts (Grambow et al. 1972; Kartha et al. 1974 a, b; Kao and Michayluk 1975; Gleba and Hoffmann 1978; Gleba et al. 1978); (3) 8p medium (Kao and Michayluk 1975), which is Gamborg B_5 medium modified by enrichment with vitamins, sugars, amino acids, and, in some cases, a more complex composition of phytohormones (Kao 1977; Melchers et al. 1978; Binding and Nehls 1978a, b).

The maintainance of appropriate osmotic pressure at the initial stages of protoplast culture is of great importance in preserving and ensuring the stability of these cells, since they are devoid of their cell walls. Different sugars have been used to increase the osmolarity of the culture medium for protoplasts; for example, glucose (Kao and Michayluk 1974), mannitol (Nagata and Takebe 1971), sorbitol (Eriksson and Jonasson 1969), sucrose (Mackenzie et al. 1973; Kao and Michayluk 1975), and xylose (Kao and Michayluk 1975) have all been successfully utilized. In some instances, the combination of two sugars may be beneficial. After cell wall regeneration and the development of cell colonies, the concentration of the solute used to raise the osmolarity of the culture medium is reduced.

One of the most important conditions for the culturing of protoplasts from most species is the maintainance of the plated or suspended cells at definite, optimum densities; in most cases, this optimum cell density is 10^4 to 10^5 cells ml^{-1} (see 2.1.2).

During culturing, the isolated protoplasts regenerate their cell walls and thereafter resemble normal in vitro cultured cells (Fig. 1–4). The description of conditions for the in vitro culture of plant cells has been the object of numerous investigations over the last years, resulting in the accumulation of a large wealth of data in this area (Butenko 1964, *Plant Cell, Tissue and Organ Culture*, 1977, *Plant Cell and Tissue Culture*, 1979). The regeneration of an entire plant from a cultured cell has not been successful in a number of species, particularly the cereals, due to insufficient knowledge of the conditions for induction of morphogenesis in these species (see Table 1). Unfortunately, this embarrassing situation has arisen and still persists, not because of peculiarities of these species, but because the investigation of the regulation of growth and morphogenesis in cultured plant cells is on the whole purely empirical.

Therefore, we believe it is critically important to not only elaborate the conditions for protoplast cultivation and plant regeneration from callus tissue for the various plant species and varieties, but to also develop new methods and components of cell biotechnology to facilitate the search for suitable culture conditions. In our opinion, one such valuable approach is the method of protoplast cultivation in serial droplets proposed by Potrykus et al. (1979a). It enables the testing of thousands of variations in nutrient media composition without inordinate expenditures of time and materials. The protoplasts are cultured in hanging droplets containing 40 μl of medium, the components of which are varied along a gradient. The droplets are ordered into geometric arrays, which enable the testing of a complete matrix of combinations of the various factors. The method allows two laboratory assistants to test 5000 combinations of media, including simultaneous variations in two or three factors, within 2 days using only 2×10^6 cells. The efficacy of this method has been confirmed by the work on the optimization of a nutrient medium for tobacco (Potrykus et al. 1979b). A further refinement of this approach might involve the addition of computer scanning and evaluation for automation of the screening of the variations of the culture conditions. In spite of the advantages offered by the method of culturing in serial test droplets, the same authors were not able to establish suitable conditions for culturing of protoplasts from Gramineae using this method (Potrykus et al. 1977). This failure has nothing to do with the efficacy of the method, but rather once more demonstrates our ignorance of the real factors involved in cell differentiation and proliferation.

2.1.2 Culture of Single Cells and Protoplasts

As was previously mentioned, most of the techniques for plant cell and isolated protoplast culture have been developed for mass culturing of large populations of cells (macrocultures). Only a few reports have appeared on attempts to grow individual plant cells or protoplasts. Culturing of individual cells as microcultures, although a potentially promising tool, has been generally unsuccessful because of the limited ability of plant cells to grow at low population densities. Some investigators feel that this limited ability is not associated with any specific cellular interactions, but rather a result of the loss of some essential metabolites from the individual cells (Kao and Michayluk 1975). Thus, there have been attempts to provide for the conditioning of cells in dilute suspension by either the use of a "feeder layer" of irradiated cells which have lost their capacity for dividing, but are able to maintain the growth of other cells (Raveh and Galun 1975), or by using "conditioned" media containing either extracts of undefined composition (co-

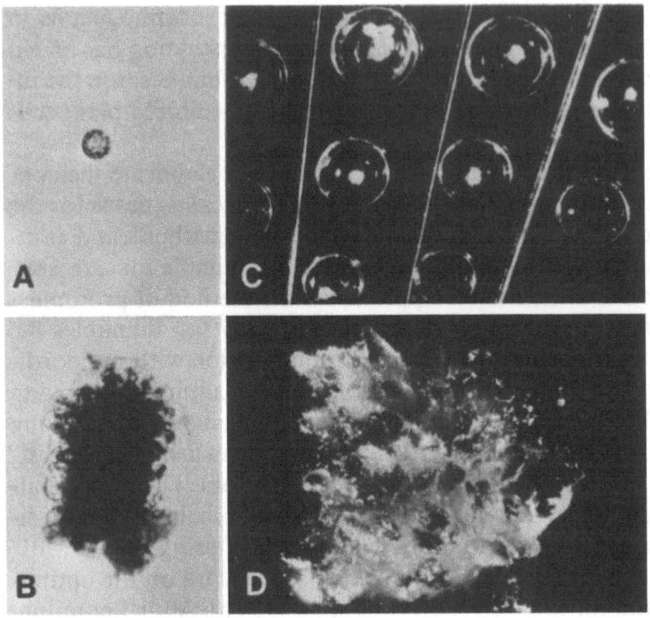

Fig. 5. Microdroplet culture of individual tobacco mesophyll protoplasts and subsequent plant regeneration. **A** A single haploid protoplast of *Nicotiana tabacum*, var. Samsun in an inner chamber of a Cuprak dish, × 200. **B** This cell colony regenerated from the protoplast seen in **A** after 21 days of culture. **C** After 21 days of culture, several colonies have formed in the individual chambers of this Cuprack dish, × 3. **D** Shoot organogenesis is evident in this cell colony obtained from a single protoplast. (Gleba 1978 a)

conut milk or caseine hydrolysate) or supplements of different amino acids, vitamins, and sugars (Kao and Michayluk 1975).

Galun's utilization of a feeder layer of irradiated cells has found almost universal application although, in our experience, it is technically difficult to consistently reproduce. The dose of irradiation resulting in inactivation, but not in death of the feeder cells, is difficult to determine, since it varies depending on the physiological state of the cells used. The enriched media method of Kao is a perfect solution to the problem. Unfortunately, it is presently completely suitable for only two or three plant species, such as *Vicia faba* and *Glycine max* (Kao and Wetter 1977). In all the other cases tested, 8p medium (Kao and Michayluk 1975), though found appropriate for macrocultures (Melchers et al. 1978; Binding and Nehls 1978 a, b), is nonetheless inappropriate for culturing of individual protoplasts.

A different approach to this problem is the reduction of the volume of the culture to a minimum. We have been able to demonstrate that single amphidiploid and amphihaploid mesophyll protoplasts of tobacco can be regenerated into cell colonies and even entire plants after initial culture of them in volumes of medium smaller than 1 μl (Gleba 1978 a) (Fig. 5). By using improved media and/or by preculturing the cells at high densities prior to transfer into microculture, it is possible to grow individual cells in microdroplets with efficiencies comparable to those obtained with macrocultures.

Mesophyll protoplasts isolated from aseptically grown plants of *Nicotiana tabacum* L., var. Samsun, were used in the following experiments. We first determined the minimum culture volume necessary for growing protoplasts. Nagata and Takebe medium (Nagata and Takebe 1971) containing 0.4 *M* rather than 0.7 *M* mannitol and protoplasts at a normal density of 4×10^4 cells ml^{-1} were used. Cells were cultured in droplets of various volumes in Cuprak dishes (Kao 1977) or as hanging droplets (Potrykus et al. 1976). These experiments indicated that the division rates in protoplast cultures in microdroplets as small as 0.25–0.5 μl did not differ significantly from division rates in macrovolumes. Further reduction of the culture volume was prevented by technical difficulties. The ratio of cell to culture volume in microdroplets of such volume, provided that only one cell is contained in the droplet, is the same as in macrosuspensions with a cell density of 2 to 4×10^3 cells ml^{-1}. The smallest culture volumes previously reported were 25 μl (Kao 1977) and 40 μl (Potrykus et al. 1976), which are two orders of magnitude more than used for our studies.

Single haploid and diploid protoplasts, when plated in 0.25–0.5 μl microdroplets of Nagata and Takebe medium supplemented with 20 ml l^{-1} of coconut milk, 300 mg l^{-1} of caseine hydrolysate, and 100 mg l^{-1} of glutamine, regenerated within 2–3 weeks into cell colonies with a 5%–40% plating efficiency. These colonies may be subcultured further in droplets of 25–50 μl of the same medium and after another 2–3 weeks may be transferred to regeneration medium (Nagata and Takebe 1971). Complete plants have been obtained from individually cultured haploid protoplasts (Fig. 5). Preculturing of the protoplasts for 1–3 days in suspensions at higher than normal cell densities (i.e., 10^4 to 10^5 cells ml^{-1}) can significantly improve the plating efficiency of these cells in individual culture. This preculturing may alleviate the need for enriched medium.

Another new approach to the problem of growing protoplasts at low plating density or as single cells was proposed by Caboche (1980a). It was demonstrated that tobacco mesophyll protoplasts require high concentrations of auxin and high plating densities only for the first few days of culture; thereafter, they may be diluted to a cell concentration as low as one to two cells ml^{-1}, provided that the concentration of exogenous auxin is simultaneously lowered.

The technique of cell cultivation at low plating density has already been successfully applied for studies on the frequency of mutations for valine resistance in tobacco following experimental mutagenesis (Caboche 1980b). White and Bhojwani (1981) have used the same method with good results for the culturing of protoplasts from *Trifolium arvense* callus tissue. They obtained plating efficiencies of about 60% by using the following schema of cultivation. The protoplasts were first grown at a density of 5×10^5 cells ml^{-1} in a medium with a high concentration of auxin [2,4-dichlorophenoxyacetic acid, (2,4D) at 2 mg l^{-1}] for 4 days. Subsequently, the protoplasts were transferred to a medium with a lower concentration of auxin (2,4D at 0.1 mg l^{-1}) and cultured at very low cell densities (in some experiments not more than 15 cells ml^{-1}). It should also be mentioned that the cell colonies obtained in this way turned out to be autotrophic for auxin, in contradistinction to the initial tissue.

2.2 Isolated Protoplast Stage and Plant Cell Engineering

2.2.1 Protoplast Fusion

This brief survey of methods of higher plant cell hybridization by fusion of isolated protoplasts must be prefaced by the following remarks. Much speculation exists concerning the mechanism of the induction of fusion caused by various substances that influence either the protoplast on the whole or, in particular, the plasmalemma of the protoplasts (Fig. 6). One can even state that a special branch of science exists that may be called fusology (for example, see Lucy 1970).

Our present task is to become acquainted with the techniques of fusion induction as a component of hybridization technology. We shall not attempt a rigorous analysis of the delicate processes accompanying this sort of fusion since the present state-of-the-art in the fusology field is still inadequate to explain the relationship between the utilization of a specific "fusogen" (a fusion inducing agent) and the peculiarities of *genetic* processes in the systems created by fusion. Therefore, we shall mainly dwell on the question of how to obtain the maximum yield of hybrid cells while minimizing or eliminating damage to the cells involved in fusion.

The spontaneous fusion of mechanically isolated protoplasts was first observed by Küster as early as 1909 (Küster 1909). The first induced protoplast fusion was produced by Cocking and collaborators using sodium nitrate as the fusogen (Power et al. 1970). However, the efficiency of this technique was found to be low. During subsequent searches for a more suitable fusogen, treatment with gelatine (Kameya 1973), concanavalin (Hartmann et al. 1973), different salt solutions (Eriksson 1971; Kameya and Takahashi 1972), and dextran sulfate (Kameya 1975, 1979), were tried. Additionally, a solution with a high pH (pH 9–11) and a high Ca^{2+} concentration (100–300 m*M*) has been successfully used as fusogen (Keller and Melchers 1973). The protoplasts are usually agglutinated prior to fusion by the addition of concentrated polyethylene glycol solutions (PEG, 1500–6000 mol. wt.) (Kao and Michayluk 1974; Eriksson et al. 1974) or by centrifugation (Schieder 1978a) (Fig. 7). The suitability of a combination of polyethylene glycol-high pH-high Ca^{2+} concentration (Kao 1977) has been generally recognized and modifications of this method are now applied in the majority of laboratories (Fig. 8). This method results in the fusion of 10%–50% of the cells (Kao 1977; Maliga et al. 1977). The principal shortcomings of this method are the relatively high lethality of the treatment (the more efficient the fusion, the more the protoplasts are damaged), as well as the high frequency of formation of multinuclear fusion products, which are not viable in most cases. Moreover, the unfused cells remain agglutinated after the treatment and thereby hamper the analysis of hybridization events.

The frequency of protoplast fusion seems to be independent of the species from which the protoplasts are derived. The fusion process is affected to a much greater degree by external factors and by the peculiarities of the protoplast ultrastructure (i.e., tissue affiliation). For example, meristem and callus protoplasts can be easily fused, whereas the fusion between protoplasts of vacuolized cells or of cells with developed chloroplasts is significantly less efficient (the fusion of me-

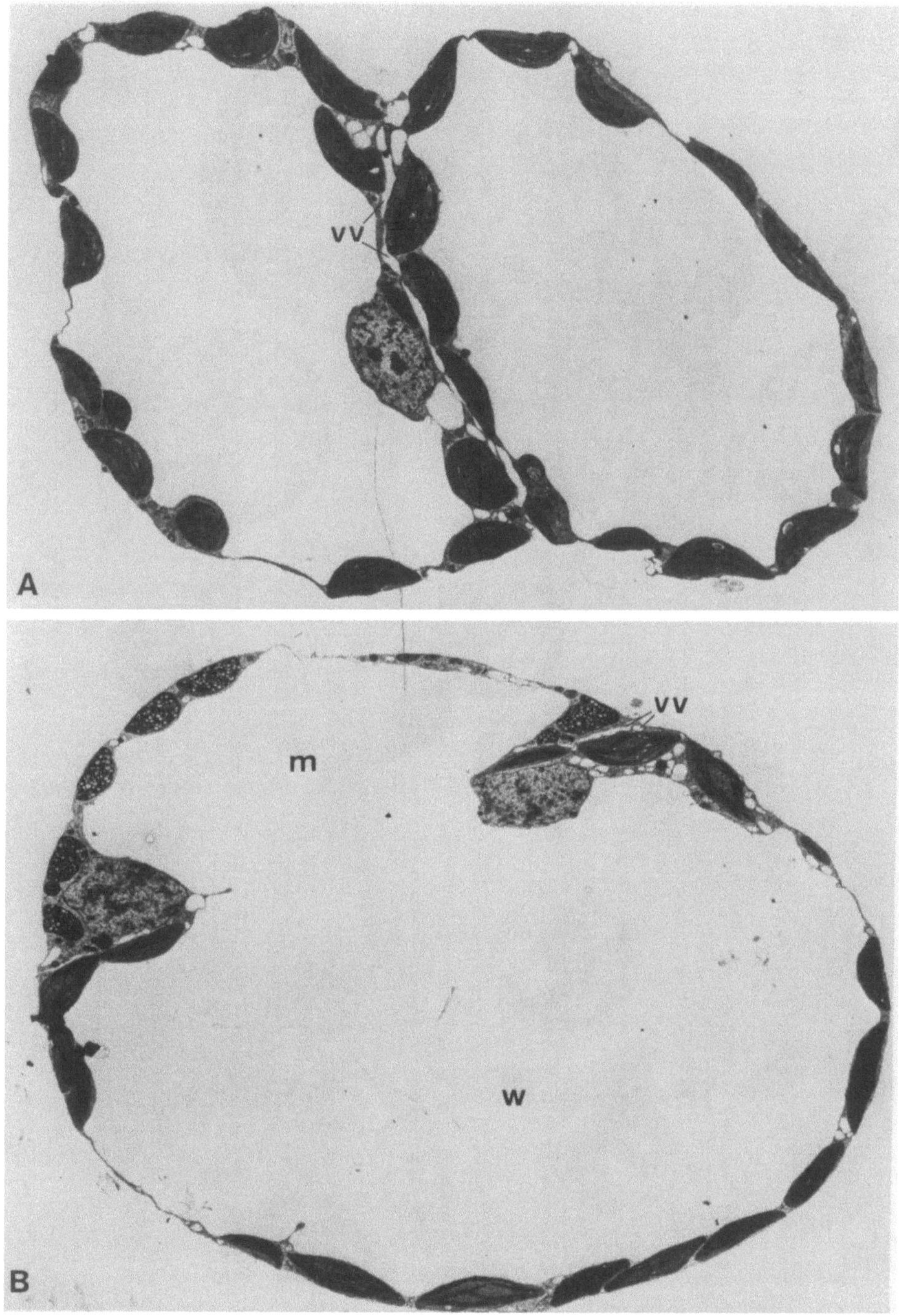

Fig. 6. Electron micrographs of tobacco mesophyll protoplasts, var. Samsun, fused by polyethylene glycol treatment. **A** A homoplasmic fusion product formed by the fusion of two wild type cells, × 3 500. **B** A heteroplasmic fusion product formed by the fusion of a wild type *w* with a plastome mutant *m* cell, × 3 400. The fusion of the central vacuoles can be clearly seen, as can the difference in plastids. Vacuolar vesicles *vv*, formed as a result of fusion of the plasmalemmas, are apparent in both fusion products

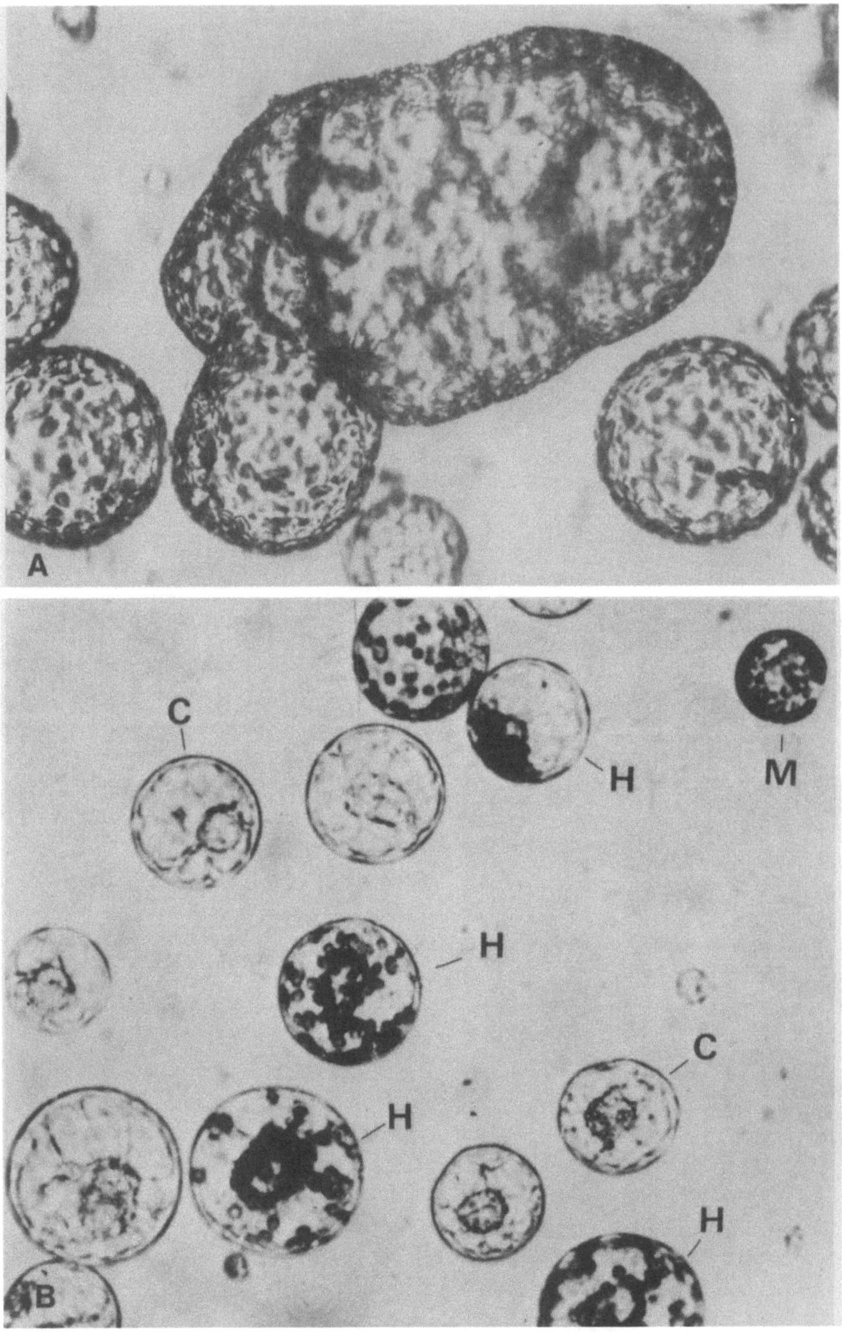

Fig. 7. Fusion of protoplasts. **A** Homoplasmic fusion products in preparations of tobacco mesophyll protoplasts ×400. **B** Fusion of tobacco mesophyll and *N. plumbaginifolia* callus protoplasts, ×400 (*h* heteroplasmic fusion products, *c* callus protoplasts, *m* mesophyll protoplasts)

Fig. 8. Ultrastructural analysis of tobacco mesophyll protoplasts fused with callus protoplasts of *Arabidopsis*. **A** Protoplast agglutination products, × 1 800. **B** The region of fusion of two protoplasts. The area of fusion is indicated by a *long arrow* and the vesicles formed by fusion of the plasmalemmas are

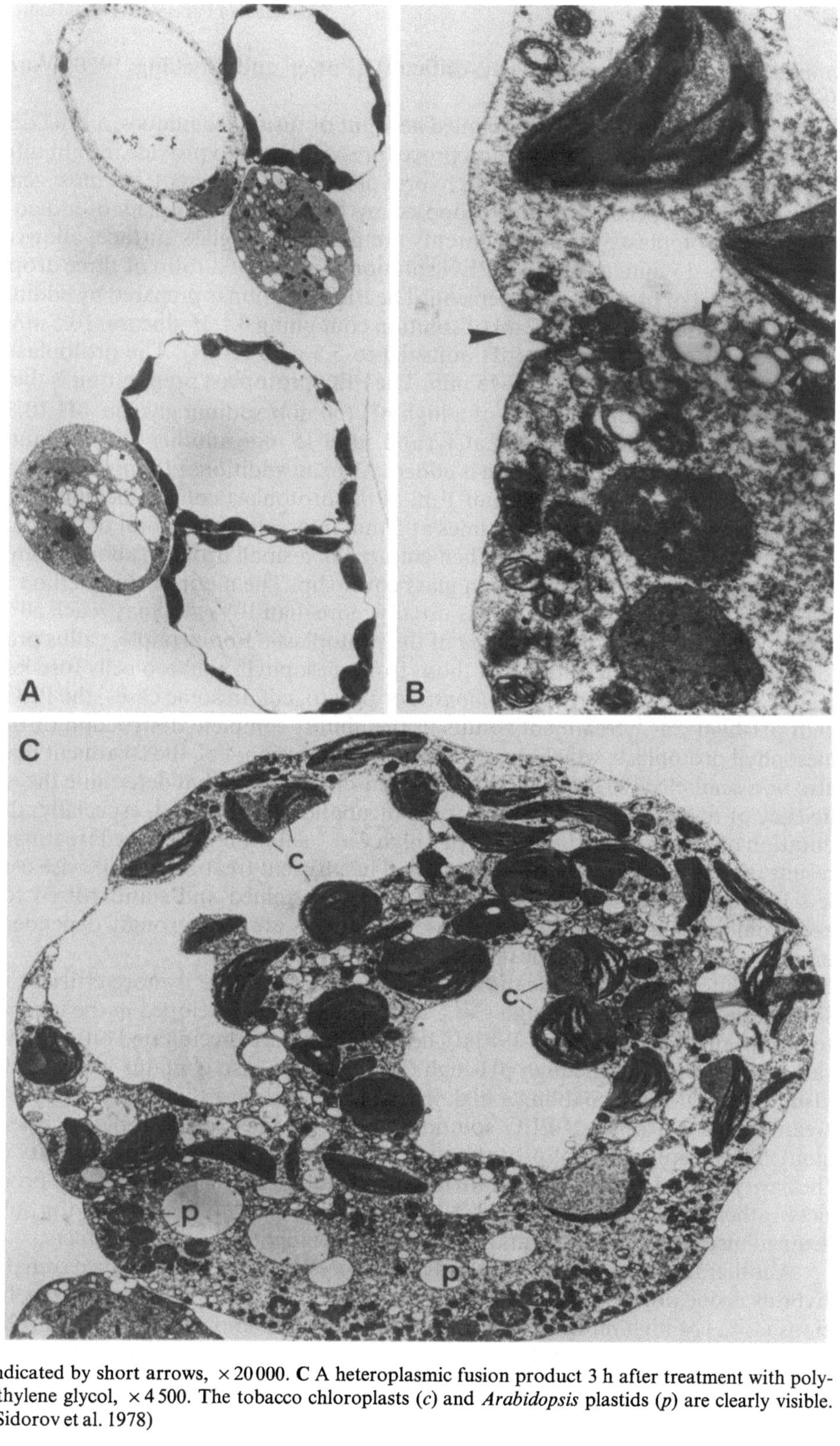

indicated by short arrows, ×20000. **C** A heteroplasmic fusion product 3 h after treatment with polyethylene glycol, ×4500. The tobacco chloroplasts (*c*) and *Arabidopsis* plastids (*p*) are clearly visible. (Sidorov et al. 1978) sophyll protoplasts is particularly difficult) (Power and Cocking 1970; Vasil 1976).

Although our aim is not the detailed account of fusion techniques, a brief description of one of the most common procedures will serve to provide insight into the methodology of experiments with hybrid plant cells. At present, the most generally accepted procedure is that proposed by Kao (1977). Droplets of a dense mixture of protoplasts from two parents are placed on a glass surface, allowed to settle for 5–15 min, and then a PEG solution is added at a ratio of three droplets per droplet of protoplast suspension. The PEG solution is prepared by adding 1 g of PEG (mol. wt. 1540) to 2 ml of solution containing 0.1 *M* glucose, 10.5 m*M* $CaCl_2$, and 0.7 m*M* KH_2PO_4 (pH adjusted to 5.5 with KOH). The protoplasts are incubated in the PEG for 15–45 min. The PEG protoplast preparation is then diluted by the addition of 0.5 ml of a high pH (50 m*M* sodium glycine, pH 10.5) and high Ca^{2+} solution (50 m*M* $CaCl_2$) and, after 15 min, another 0.5 ml aliquot of this high pH-high Ca^{2+} solution is added. After an additional 15 min, the preparation is diluted by the addition of 1 ml of the protoplast culture medium. The protoplasts are then washed five times at 5 min intervals with a total of 20 ml of this medium. The protoplasts are then cultured in a small droplet (about 0.5 ml) of protoplast culture medium on an glass cover slip. The merit of this method is the high efficiency of fusion, which is usually more than 10% and may reach 50% of the cells, depending on the source of the protoplasts. For example, callus protoplasts fuse more efficiently than those from mesophyll; tobacco cells fuse less readily than those of carrot, *Arabidopsis*, or potato, etc. In some cases, the PEG-high pH-high Ca^{2+} treatment results in the almost complete destruction of the mesophyll protoplasts which were not involved in fusion, i.e., the treatment may also be a semiselective procedure (Fig. 9). The critical steps that determine the efficiency of fusion are the duration of the incubation in PEG and, especially, the duration of treatment with the high pH-high Ca^{2+} solution: protracted treatment results in destruction of the protoplasts and insufficient treatment is likewise useless for fusion. Optimal parameters must be determined and standardized for each combination of protoplasts. These parameters are also strongly dependent on external factors, such as the temperature of the medium.

A highly popular modification, which in our experience is more efficient in some cases, of the high pH-high Ca^{2+} method has been developed in the laboratory of Maliga (Menczel et al. 1981). The protoplasts are precipitated onto plastic rather than glass and the high pH-high Ca^{2+} solution also contains 9% dimethylsulfoxide (DMSO), enabling a high frequency of fusion with a less drastic PEG treatment (one droplet of PEG solution per two droplets of protoplast suspension). The agglutination of protoplasts is reduced and as a result the majority of the newly formed heterokaryons are found in solution as freefloating single products rather than large clumps. The last point is especially important for subsequent mechanical isolation and culturing of the individual fusion products.

Another fusion technique that has been successfully used to produce somatic hybrids is one proposed by Kameya (1975a, 1979). It utilizes concentrated solutions (15%) of high molecular weight dextran (500000 mol. wt.) in a NaCl solution to induce cell agglutination and fusion is subsequently induced by the addi-

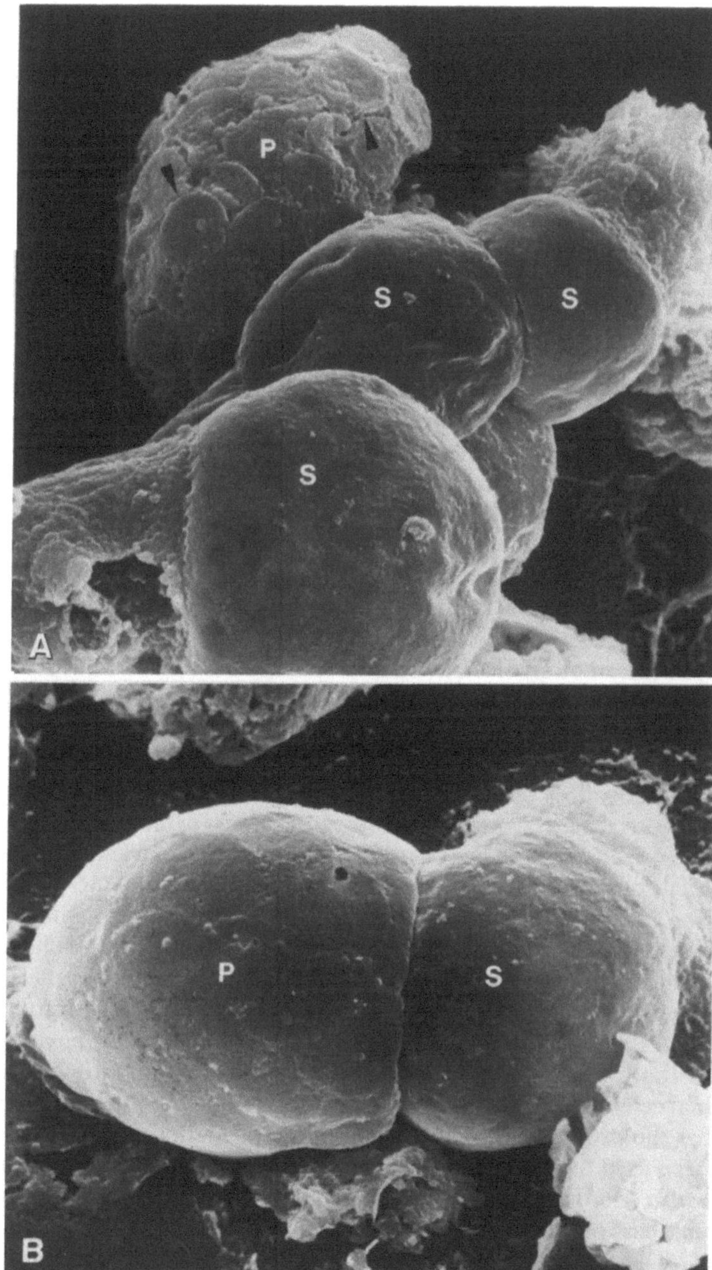

Fig. 9. Fine structure of fusion products from soybean cell culture and pea leaf protoplasts. **A** Scanning electron micrograph showing agglutinated soybean (*s*) protoplasts. The pea (*p*) protoplast has lost its plasma membrane revealing numerous chloroplasts (arrows). × 1 800. **B** Scanning electron micrograph showing agglutinated pea (*p*) and soybean (*s*) protoplasts. Note the outlines of the chloroplasts under the plasma membrane of the pea protoplast. × 1 800. (Fowke et al. 1977)

tion of 0.1 N NaOH (a recent variation of this techniques has been described by Kameya et al. 1981).

There are a few other fusion techniques that may be promising. For example, phospholipids, such as lysolecithin, have been found to be efficient fusogens for animal cells (Lucy 1970). However, according to our observations, lysolecithin leads to lysis rather than fusion of higher plant protoplasts. A similar conclusion concerning lysolecithin was also reached by Nagata et al. (1979); nonetheless, these same authors discovered that the synthetic phospholipid 1,2-O-dipentadecylmethylidene-glycerol-3-phosphoryl-(N-ethylamino)-ethanolamine is an efficient fusogen. The best results were obtained by incubation of the protoplasts in a solution containing 1 m*M* phospholipid, 50 m*M* $CaCl_2$, and 0.7 *M* mannitol for 20–30 min at 30 °C. Ten to 30% of the protoplasts derived from tobacco diploid mesophyll fuse under these conditions. The phospholipid did not have an obvious harmful effect on the protoplasts, although the viability was reduced to 70% after treatment as compared to 90% in control experiments. The availability of the dioxolane ring and the positive charge of this synthetic phospholipid are thought to be responsible for the high efficiency of fusion. Unfortunately, this work has not been further developed and this fusogen has not yet been utilized for the production of somatic hybrids in higher plants.

In recent publications, Zimmermann and Scheurich (1981 a, b; see review by Zimmermann 1982) describe a completely new approach to fusion, namely, the application of an electric field for protoplast agglutination and fusion. The protoplasts are placed between two closely spaced electrodes (with a gap of 200 μm) between which a high frequency alternating electric field is applied (5–10 V peak-to-peak sine wave with a frequency of 500 kHz). Under these conditions, the protoplasts collect by dielectrophoresis and form bridges consisting of several cells between the electrodes. The bridge or chain formation and protoplast agglutination are stable only during application of the electric field. The simultaneous application of an additional single impulse (rectangular, 15 V, 50 μs) leads to the fusion of the protoplasts. A high efficiency of fusion is achieved at room temperature and physiological pH values without the use of any chemical stimulators. It is presumed that the single additional impulse results in dielectric rupture of the protoplast plasmalemmas at the point of contact, thus initiating the fusion process. Protoplasts from different tissues and species have been fused via this method, which has also been utilized for the production of viable hybrids of animal cells and of yeasts. There are still no reports of the application of this method for the production of somatic hybrids of higher plants. This technique has also been used to release individual chloroplasts from mesophyll protoplasts of *Avena sativa* (Zimmermann et al. 1982) and may prove suitable for the isolation of small numbers of pure plastids (see 2.2.3).

The fusion frequency is technically difficult to determine and is, therefore, rarely measured. However, the efficiency of fusion, defined as the difference, in percent, of the number of nuclei contained in fusion products relative to the total number of nuclei before and after the treatment, is one of the criteria by which the suitability of a fusion method must be judged. The following formula may be used for determining the percentage of hybrid cells in a population (H):

$$H = \frac{\text{(number of heteroplasmic fusion products)}}{\text{(total number of living cells)}} \times 100 .$$

On the whole, one may conclude that the existing techniques of protoplast fusion are suitably efficient and appropriate for most applications in parasexual plant hybridization.

2.2.2 Cell Reconstruction Involving Fusion of Subprotoplasts

Several groups of investigators have explored the possibilities for plant cell reconstruction by parasexual hybridization using subprotoplasts (miniprotoplasts, nucleoplasts, or cytoplasts) as one or both of the parents (Wallin et al. 1979; Binding 1979; Maliga et al. 1982). The major obstacles encountered in this area are the absence of suitable techniques for cell fractionation and difficulties in the isolation of large populations of homogeneous, intact cell fragments.

As a rule, subprotoplasts arise during isolation of protoplasts; however, they represent only a small fraction (up to a few percent) of the total number of cells and subcellular particles in most cases. The formation of subprotoplasts (cytoplasts and nucleoplasts) may be stimulated to some degree by alterations of some of the parameters of the protoplast isolation procedure, such as by increasing the osmotic pressure with *Brassica oleracea* protoplasts (Vatsya and Bhaskaran 1981), by treating the isolated protoplasts with cytochalasin B (Wallin et al. 1978; Lin and Ma Cheng 1981), and/or by forcing the protoplasts through gradient solutions by ultracentrifugation (Wallin et al. 1978; Lörz et al. 1981). Subprotoplasts can be separated from protoplasts by filtration through a sieve with 10 μm pores (Landgreen, personal communication) and cytoplasts and nucleoplasts may be separated by ultracentrifugation (see below). Unfortunately, the existing techniques do not yield subcellular fractions that are completely free of contamination by cells or other cell fragments.

Wallin et al. (1978) produced miniprotoplasts (subprotoplasts containing a nucleus and a small amount of cytoplasm) by treating the cells with cytochalasin B at 50 μg ml^{-1} and centrifuging them at 37 °C and 40000 × g through a layer of 1.5 *M* sorbitol containing cytochalasin B. Wallin et al. (1979) have demonstrated the possibility of production of viable hybrids as a result of fusion of miniprotoplasts derived from two mutant lines of tobacco defective for nitrate reductase (lines nia-63 and cnx-68; Müller and Grafe 1978). As has been observed in experiments with the fusion of entire protoplasts of these mutants (Glimelius et al. 1978a), the investigators were able to obtain autotrophic cell lines which could only have originated as the result of hybridization. Unfortunately, the authors did not present any evidence that the preparations of miniprotoplasts were free of intact protoplasts.

Lörz et al. (1981) have developed a method for the production of large populations of intact cytoplasts (enucleate subprotoplasts) which consists of ultracentrifugation of the protoplasts through an isodsmotic colloidal silica (Percoll) gradient in the presence of mannitol and calcium cloride. The cytoplast preparations obtained are at least 96% pure; the remaining 4% consist of contaminating protoplasts and/or nucleoplasts. Cytoplasts from tobacco, maize, and *Hyosciamus muticus* have been produced by this technique.

To date, only one research group has succeeded in transferring plastids from one species to another using isolated cytoplasts (Maliga et al. 1982). Similar work in Maliga's research group on the transfer of cytoplasmic (organellar) genes by fusing normal protoplasts with protoplasts inactivated by iodoacetate (Medgyesy et al. 1980) or irradiation (Menczel et al. 1982) or by fusing protoplasts inactivated by iodoacetate with irradiated protoplasts (Sidorov et al. 1981 a) will be discussed later (see 4.4). In all these experiments, the screening of the cybrids has been based on the transfer of resistance to streptomycin. Maliga et al. (1982) have reported plastid transfer via a system in which protoplasts of *Nicotiana plumbaginifolia* were fused with cytoplasts derived from mesophyll protoplasts of tobacco plastome mutant SR1, which is resistant to streptomycin. The populations of cytoplasts used in this experiment were contaminated with 2%–7% protoplasts. Seventy-five lines resistant to the antibiotic were found among the colonies regenerated from the PEG-induced fusion products. However, the experimenters were successful in producing plants from only 17 of the colonies and it was found that 8 of the plants (10% of the number of supposed recombinants) contained *N. plumbaginifolia* nuclei and *N. tabacum* plastids. This conclusion is based not only on the resistance of the plants to streptomycin, but also on analysis of the restriction endonuclease digests of their chloroplast DNA. In the author's opinion, these eight plants are the product of fusion of protoplasts with cytoplasts. The remaining nine lines of regenerants turned out to be sensitive to the antibiotic. The authors explain this fact as a persistance of heterozygosity for plastome during cell culturing on streptomycin-containing medium. The analysis of esterase isoenzymes (multiple molecular forms) in the 56 cell lines incapable of morphogenesis has revealed their hybrid character for nuclei. Two hybrids with tobacco nuclei and *N. plumbaginifolia* cytoplasm were also observed. In toto, these results demonstrate that the protoplasts contaminating cytoplast preparations, even though present in small quantities, are more stable or efficient in the process of fusion. Hence, this system of cytoplasm transfer is less efficient than fusion of intact protoplasts and, strictly speaking, the conclusion that the eight lines of plants containing the nuclei of *N. plumbaginifolia* and the plastids of tobacco orginated from fusion of a cytoplast with a protoplast, as opposed to fusion of a protoplast with a protoplast with subsequent nuclear segregation, is not quite convincing. The 10% efficiency of plastid transfer (or cybrid formation) observed in these experiments surpasses that of hybridization systems utilizing iodoacetate, which have an efficiency of 0%–3% (Menczel et al. 1982). There are at least two principal distinctions between these two experimental systems that may account for the discrepancy in efficiency: the use of iodoacetate in the one case and the use of cytoplasts contaminated with protoplasts in the other.

The unequivocal analysis of this and other analogous experiments is hampered by the application of statistical approaches and estimations. A satisfactory solution to this problem might be the production, cultivation, and analysis of individual fusion products of a cytoplast and a protoplast, rather than analysis of the fusion products arising in mass cultures.

2.2.3 Cell Modification Through Uptake of Isolated Cell Organelles by Plant Protoplasts

Isolated protoplasts are able to take up macromolecules and particles from the environment (Cocking 1970, 1973; Potrykus and Lörz 1976; Giles 1977). However, no concensus has been reached on the mechanism(s) of this process. According to Cocking (1970, 1973), the uptake is a result of endocytosis, a natural process of the plant cell which can be enhanced by different experimental manipulations. Other investigators favor the possibility of passive infiltration (at least for macromolecules and viruses) through the ruptures of the plasmalemma which are always extant on the surface of newly prepared protoplasts (Takebe 1975).

The transplantation of isolated cell organelles into protoplasts represents a popular trend in investigations of genetic reconstruction of novel plant cells via the introduction of foreign particles. Unfortunately, progress in this field has been largely limited to the development of suitable methods. In particular, several methods for the induction of organelle uptake by protoplasts have been elaborated, namely: the simultaneous application of organelles and chemical modifiers of membranes such as lysozyme or sodium nitrate in combination with centrifugation (Potrykus 1973; Potrykus and Hoffmann 1973); and agglutination of protoplasts and isolated organelles with polyethylene glycol along with centrifugation (Bonnett and Eriksson 1974; Bonnett 1976). The efficiency of the uptake, evaluated by microscopy, may be quite high and depends on the type of protoplasts and organelles utilized. The data of Potrykus showed that 0.5% of the protoplasts participated in the absorption and that one to five nuclei and one to 20 chloroplasts were taken up. The results of Bonnett show a higher efficiency of absorption in that 8%–24% of the plastids were taken up by the protoplasts. However, electron microscopy of the cells reconstructed by this method has demonstrated that the external membrane of most of the absorbed chloroplasts is absent, indicating that the transfer of the chloroplasts into the protoplasts seems to be the result of fusion of the protoplast plasmalemma with the external membrane of the chloroplast (Davey et al. 1976; Bonnett and Banks 1977). The few intact chloroplasts revealed in the recipient cells have apparently gotten into them by fusion of a protoplast with a cytoplast rather than by uptake of chloroplasts. Once again, the preparations of the isolated organelles are heterogeneous in all cases and contain both cytoplasts and damaged organelles as well as the intact organelles.

All attempts to obtain genetic evidence for organelle transfer by the use of the techniques mentioned have been unsuccessful up to now. For example, in the experiments of Lörz (Lörz and Potrykus 1978), 5.5×10^7 protoplasts of a light-sensitive tobacco mutant homozygous for the recessive gene causing chlorophyll deficiency were treated with nuclei isolated from a phenotypically similar tobacco mutant that is nonallelic to the receptor protoplasts. Despite a high efficiency of 5% for the absorption of the nuclei by the protoplasts, no wild type plants have been found among those plants regenerated. The experiments of Carlson aiming at the transfer of isolated chloroplasts from wild type tobacco plants into the protoplasts from a tobacco plastome mutant were unconvincing (Carlson 1973); the data from the cytological studies reported by the author cannot be seriously discussed (see Potrykus 1973), and the biochemical analysis of the Fraction I protein of the sole variegated plant regenerated in these experiments has shown the presence of both the plastome and the genome of the recipient (Kung et al. 1975). Negative results in the transfer of chloroplasts have also been obtained by other investigators (Gleba 1974; Uchimiya 1978; Landgren and Bonnett 1979).

Good progress has been achieved in experiments on the transfer of isolated metaphase chromosomes in animal cells (see the review of Willecke 1978). Application of this method to higher plant protoplasts is difficult because populations of plant cells are not easily synchronized and this synchronization is a prerequisite for the isolation of large quantities of chromosomes. Temporary starvation (absence of auxin, cytokinin, or phosphate) or temporary suppression of DNA synthesis using inhibitors are the routine methods of plant cell synchronization; however, they achieve only a partial synchronization of about 30% according to the mitotic index (King 1980). Nonetheless, there are several promising experiments concerning production of preparations of metaphase chromosomes (Malmberg and Griesbach 1980; Szabados et al. 1981; Griesbach et al. 1982) and even their utilization for chromosome transfer into protoplasts (Szabados et al. 1981). Unfortunately, the last work contains only cytological evidence obtained by light microscopy for the absorption of the chromosomes by the protoplast; thus, rigorous and unequivocal demonstration is lacking.

Future development of techniques of organelle transplantation is obviously dependent upon the solution of several problems. First, judging from the data discussed, the isolated organelles are easily damaged or destroyed by the treatments inducing the uptake; in particular, the plastids and perhaps other organelles may lose their structural integrity. Therefore, it is necessary to find conditions that protect the organelles, perhaps by enclosing them into liposomes, as has been performed with animal cells using macromolecules for transformation experiments. Second, efficient methods for screening the reconstructed cells in genetic experiments are still lacking. These methods might be based on the use of extranuclear mutations for antibiotic resistance, in the case of plastid transfer, although such mutants are still rare (Maliga et al. 1973, 1975; Dix et al. 1977). Third, in spite of recent progress, the methods of isolation and fractionation of plant cell organelles are still far from being perfect. The best methods involve protoplast production as a step prior to further fractionation (for example: Blaschek et al. 1974; Edwards et al. 1978; Malmberg and Griesbach 1980). A promising development in plastid isolation is the application of an electric field to plant protoplasts (see 2.2.1 for fusion of protoplasts via this method). Under defined conditions, one or more chloroplasts may be expelled from a protoplast of *Avena sativa* (Zimmermann et al. 1982). Although this method has not yet been fully characterized and exploited, it is conceivable that one or a few chloroplasts could be isolated in this manner and then immediately fused with another protoplast or subprotoplast to give rise to a cybrid under rigorously controlled and defined conditions. The existence of refined techniques for the culturing of single cells (see 2.1.2) makes feasible the use of this method and thereby circumvents many of the problems encountered with the techniques involving mass cultures.

2.3 Problems of Genetic Variability

2.3.1 Genetic Variability Associated with in Vitro Manipulation of Plant Cells

The utilization of in vitro cultured plant cells for genetic studies is greatly hampered because of the genetic heterogeneity of the great majority of cell lines. An analysis of many reports produces the impression that a plant cell cultured in vitro is not capable of normal reproduction and that the very conditions of culturing are a source of signals that induce different genetic changes.

The majority of the data available on genetic variability deal with chromosome variability of cultured cells and may be summarized as follows (according to Dulieu 1972):

1. aneuploid and polyploid cells constitute the major part of populations of long-term culture lines (Sacristan and Melchers 1969; Zagorska et al. 1971 a, b);
2. aneuploid, polyploid, and diploid cells make up newly introduced cell lines (Mitra et al. 1960; Murashige and Nakano 1966, 1967; Kao et al. 1970 b), just as is found upon induction of the primary divisions in tissue explants (Shimada and Tabata 1967; Tabata et al. 1978; D'Amato et al. 1978);
3. the process of culturing is accompanied by a stable, more (during the first passages) or less (afterwards) rapid qualitative change in the spectrum of chromosome numbers in the populations of cultured tissue lines (Braun 1959; Sacristan and Melchers 1969, 1977; Kunakh 1975);
4. the process of culturing is accompanied by changes in chromosome morphology (Sacristan 1971; Zosimovich and Kunakh 1975);
5. plants obtained by regeneration of the cultured cells may be diploid (Das et al. 1956; Braun 1959), polyploid (Murashige and Nakano 1966), and/or aneuploid (Sacristan and Melchers 1969, 1977; Bennici and D'Amato 1978);
6. upon plant regeneration, a selection for cells bearing a reduced set of chromosomes takes place (Tabata et al. 1968; Bennici et al. 1968; Sacristan and Melchers 1969, 1977).

The observed heterogeneity of in vitro cultured plant cells may be partly explained by the heterogeneity (heteroeuploidy) of the initial explants. A DNA endoreduplication takes place in differentiating plant cells of different tissues, in particular in those tissues frequently used for primary explants such as stem pith parenchyma or the extension zone of root tissue (Das et al. 1956) or in cells of shoot parenchyma (Shimada and Tabata 1967). In contradistinction to in vivo genome changes, the chromosome variations and rearrangements, to the best of our knowledge, arise mainly or only during the in vitro culturing stage. It has been shown in a number of experiments that the proportion of polyploid cells in a culture depends on the balance of phytohormones. The changes in phytohormone balance greatly affect the ratio of cells with different chromosome sets in the population and, in some cases, these changes cannot be explained only by selection processes (Das et al. 1956). The chromosome variability is a function of genotype and differs in different plant species and varieties. Under certain conditions, the cultured cells of *Haplopappus* remain strictly diploid over many years and are therefore an exception in this aspect (Eriksson 1965, 1967; chromosome rearrangements were not studied in the experiments). The overwhelming majority of in vitro cultured cells of tobacco, petunia, soybean, and many other species are aneuploid (however, see Evans and Gamborg 1982).

Genetic variability is not limited to only genome changes, changes in chromosome number, or chromosome rearrangements and mutations. It also includes the formation of polytene chromsosomes and amplification of certain chromosomal regions. All of these processes occur in differentiating plant somatic cells (Kunach 1978; Nagl 1979; Larkin and Scowcroft 1981) and these cells certainly bring the entire pool of genetic diversity into the system after being introduced into in vitro culture. To date, no experiments exploring these aspects have been published. Likewise, for all practical purposes, the rate of point mutations in plant tissue culture is essentially unexplored.

As should be clear from the foregoing discussion, the term cell "clone" in in vitro plant tissue culture, while retaining its physical meaning of origination from one cell, loses its genetic sense because of the high rate of chromosome variability, which immediately results in chromosome heterogeneity within the clone. Any investigations on the genetics and physiology of cultured plant cells must take into account the initial genetic heterogeneity as well as the rate and direction of variability of the populations of the clones utilized. This means that the experimental and control variants of such experiments must be performed not only with cells of the same cell line and from the same clone but also, in most cases, with cells from the same passage. Finally, if a return to the intact plant is the main or ultimate goal of an experiment, the duration of unorganized growth of the cells must be reduced to a minimum (Gleba and Gleba 1978a). These problems have been reviewed by Sunderland (1973), D'Amato (1978), and Larkin and Scowcroft (1981).

2.3.2 Genetic Changes in Cells Induced During the Stage of Isolated Protoplasts

This problem still has not been the subject of detailed studies. Most investigators assume that the plants regenerated from isolated protoplasts are normal and undergo no divergence from the parental forms. The increase in ploidy is the most frequent abnormality observed in plants regenerated from the protoplasts of certain species (Takebe et al. 1971; Bourgin and Missonier 1978a, b; Wagner and Hess 1974; Banks and Evans 1976; Bajaj et al. 1978). Several authors have noted the high percentage of polyploids among the regenerated plants (Takebe et al. 1971; Bajaj et al. 1978; Bourgin and Missonier 1979). The inference has been made that the polyploidy was not produced at the stage of unorganized growth in vitro of the cells obtained from protoplasts, but resulted from the regeneration of plants from products of spontaneous protoplast fusion. However, it is significant that the proportion of polyploid plants is also large in those cases where there is a low level of spontaneous protoplast fusion (Bourgin and Missonier 1979; Gleba et al. 1978c).

Our electron microscopic studies and cytophotometric measurement of nuclear DNA content of populations of isolated haploid and diploid tobacco mesophyll protoplasts during the process of culturing have resulted in the formulation of a new explanation for the process of polyploidization that accompanies the stage of isolated protoplasts. In this investigation, the changes in the ploidy spectra were also studied after treatment with the mutagen N-nitroso-N-methylurea and the fusogen polyethylene glycol. Mesophyll protoplasts were isolated from haploid and diploid tobacco plants. The protoplasts were cultured in liquid B medium (Nagata and Takebe 1971), modified according to Gleba et al. (1974 a), at an initial plating density of 4.4×10^4 cells ml^{-1} and 6.3×10^4 cells ml^{-1} for diploid and haploid protoplasts, respectively. In the initial preparations, the rate of spontaneous protoplast fusion, determined from the number of multinuclear cells, averaged 0.3%–0.5% after 14 days of culturing. During this period, most of the cells underwent 5.7 cell divisions and formed colonies of 30–100 cells at a plating efficiency of 40%–80%. The DNA content of these cells was determined by cytophotometry of the cell nucleus at two wavelengths. The expanding cells of the roots of haploid and diploid tobacco plants were used as controls, since the majority of these cells are in the G_2 phase of the mitotic cycle.

The data from these investigations are presented in Fig. 10 and 11. The histogram of the relative nuclear DNA content in newly isolated protoplasts from haploid plants resembles a statistically normal distribution with a single maximum, providing evidence of the homogeneity of the population with respect to this character and demonstrating the absence of significant numbers of cells synthesizing DNA in the preparations. The studies on the nuclei of diploid protoplasts reveal an analogous, normal distribution. Comparisons of the mean nuclear DNA content of mesophyll cells to that of root cells from the extension zone leads to the conclusion that the leaf mesophyll tobacco cells are diploid in diploid plants and haploid in the haploid plants. These results are in agreement with the data of other investigators obtained by the direct counting of chromosomes in young tobacco leaves (Nagata and Takebe 1971). Therefore, the leaf mesophyll cells of tobacco represent genetically regular homoploid tissue of the sporophyte. The histograms of the relative nuclear DNA content of cells from the colonies produced after 2 weeks of in vitro culture are presented in Figs. 10 and 11. The process of culturing is accompanied by a significant divergence in the cultures, resulting in an assymetrical, bimodal distribution of the nuclear DNA content. These alterations in the histograms may be interpreted as a doubling of the ploidy in most of the cells of both populations (evidenced by the bimodal distribution with a euploid maxima) and that DNA synthesis is occurring in a part of the population (causing the assymetry of the peak). The analogous histograms for the cells regenerated from protoplasts treated with the mutagen and fusogen are also presented in Fig. 10 and 11. The treatment with both agents results in an increase in the proportion of cells with a doubled DNA content, relative to the initial DNA content. Moreover, the treatment with the mutagen led to a shift of the maxima from the euploid values to higher ones, probably as a result of the arrest of a large number of the cells in the S phase of the mitotic cycle.

The chromosome study which was performed in addition to the cytophotometric analysis has yielded results that are in good agreement with the data

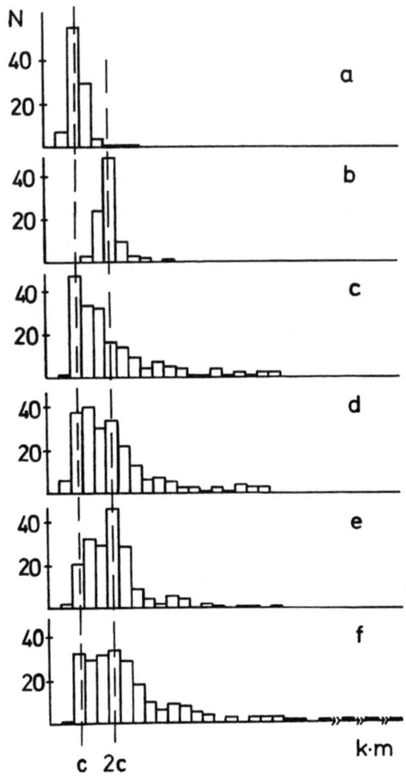

Fig. 10

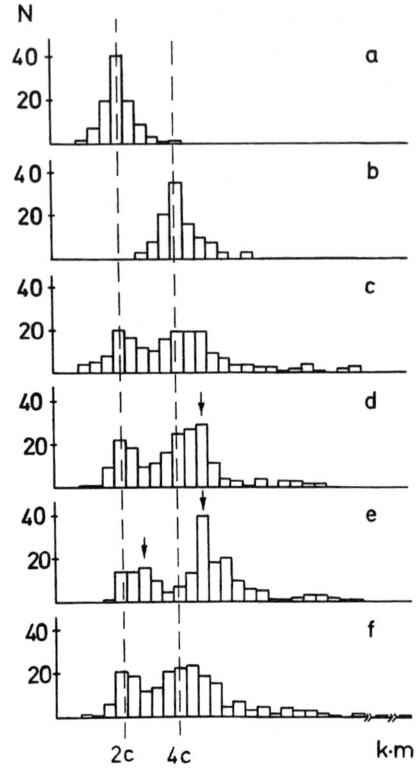

Fig. 11

Fig. 10. Cytophotometric study of nuclear DNA content of haploid mesophyll tobacco protoplasts, var. Dubeck 44, in culture as well as after treatment with a mutagen and fusion inducing agent. The histograms represent the relative nuclear DNA content (N = number observed; $k \cdot m$ = DNA content in arbitrary units) of: **a** freshly isolated haploid protoplasts; **b** haploid tobacco plant root cells from the expansion zone in G_2 phase; **c** cells of colonies formed after 14 days of in vitro culture of the protoplasts; protoplasts after 14 days of culture after treatment for 1 h with **d** 2 m*M* or **e** 4 m*M* N-nitroso-N-methylurea or **f** with polyethylene glycol. (Gleba and Gleba 1978a)

Fig. 11. Cytophotometric study of nuclear DNA content of diploid tobacco mesophyll protoplasts, var. Dubeck 44, in culture as well as after treatment with a mutagen and fusion inducing agent. The histograms represent the relative nuclear DNA content (N = number observed, $k \cdot m$ = DNA content in arbitrary units) of: **a** freshly isolated protoplasts; **b** diploid tobacco plant root cells from the expansion zone in G_2 phase; **c** cells of colonies formed after 14 days of in vitro culture of the protoplasts; protoplasts after 14 days of culture after treatment for 1 h with **(d)** 2 m*M* or **(e)** 4 m*M* N-nitroso-N-methylurea or **(f)** with polyethylene glycol. The deviations of the distribution maxima from euploid values are indicated by the *arrows*. (Gleba and Gleba 1978a)

on the doubling of the nuclear DNA in cultured protoplasts. The chromosomes of 5- to 6-day-old cultures of initially haploid protoplasts were counted. The exact number of polyploid plates was difficult to determine, since the divisions of filial cells regenerated from the same protoplast are usually synchronous and the superimposition of two or more haploid metaphases during the pressing of the preparations is possible, resulting in the scoring of artifact polyploid metaphases.

The analysis of 112 metaphase plates revealed 39 di- and polyploid metaphases. Since such a high number of artifacts is unlikely, we believe that we have unequivocally demonstrated that polyploidization occurs during the initial stages of protoplast culture.

Studies on the ultrastructure of cultured tobacco mesophyll protoplasts (Belitzer et al. 1977; Sidorov et al. 1977) have demonstrated that the first division in cells regenerated from protoplasts is accompanied by various abnormalities of the cytokinetic process, which are most probably caused by a lack of coordination of mitosis and cytokinesis. In particular, only a partial cell wall is formed between the filial cells in most cases (Fig. 3). If the filial cells are only partially separated, it is reasonable to postulate that reduplication of genetic material may take place via the formation of a common spindle during synchronous mitosis of the physically unseparated filial nuclei during subsequent cell divisions. It becomes clear that any nonspecific treatment, such as the conditions of culturing or application of a mutagen or fusogen, which interferes with the coordination of the first mitotic cycle may stimulate the formation of cells bearing a doubled nuclear DNA content relative to the initial cells.

3 Protoplast Fusion and Parasexual Hybridization of Higher Plants

3.1 Perspectives

3.1.1 Terminology

Several synonyms are currently in use to designate the phenomenon and process of hybridization by protoplast fusion, as well as to name the hybrid cells and hybrid plants obtained via this method. Unfortunately, each of these terms has its drawbacks in describing the production of hybrids without sexual crossing. The terms "parasexual hybridization" and "parasexual hybrids" were introduced by Carlson (Carlson et al. 1972). A serious argument against the use of these terms is that the natural process of genetic exchange in fungi has been previously designated "parasexual" (Pontecorvo 1958). We wish to describe a completely artificial method of hybridization that is applicable for all plant species. Nonetheless, at least in the case of higher plants, Carlson's terminology should not result in ambiguous interpretation and, for this reason, will be used throughout this book. The terms "somatic hybridization" and "somatic hybrids" probably have been adopted from animal somatic cell genetics (Ephrussi 1972; Harris 1974; Ringertz and Savage 1976) and were first proposed in the context of plant protoplast fusion by Melchers (Melchers and Labib 1974). In our opinion, the inadequacy of these terms is revealed by the following considerations. First, the hybridization by protoplast fusion is not limited to the use of somatic parental cells since, theoretically, protoplasts of gametes may also be utilized. Second, the protoplasts used for fusion in most experiments were isolated from callus and these cells are therefore not somatic since they are not the cells of soma in a physical sense nor in the sense of their epigenetic organization (in contradistinction to the cultured animal cells) (Ringertz and Savage 1976). Finally, the term somatic hybridization may be misinterpreted as the semantic antithesis to the term "cell hybridization" and inappropriate associations with the term "vegetative hybridization" may also arise. Since the use of the terms somatic hybridization and somatic hybrids is so widespread in the literature of protoplast fusion, they will be used in this book in instances where they will not result in added confusion. We have proposed the term "nonsexual hybridization" (Gleba 1979), which is semantically equivalent to parasexual hybridization, but has not been used previously in another specific context.

In a similar fashion, the designation of hybridizations and hybrids obtained via protoplast fusion may be confusing. In place of the symbol × (A × B) used to designate the sexual cross, hybrids produced by protoplast fusion have been described using either the symbol + (for example, *Nicotiana tabacum* + *Nicotiana sylvestris*) (Winkler 1908; Melchers and Labib 1974) or by the symbol ⊗ (Power et al. 1977). Since the + symbol has been previously proposed for grafts and the ⊗ is often typographically inconvenient, we shall follow the recommendation of Binding et al. (1984), who proposed the symbol (×) [A(×)B]. The term "cytoplasmic hybrid" ("cybrid") will be used to designate the hybrid carrying the nuclear genes of one of the parents and the cytoplasmic (extranuclear) genes of both parents (or of the second parent only) (Power et al. 1975; Cocking 1976a, b). Those hybrids possessing alternative cytoplasmic genes from both parents will be termed "cytoplasmic heterozygotes" ("cytogetes") (Sager 1975). It should be mentioned that the terminology used to describe the hybridization by protoplast fusion has not yet been standardized and does not conform with the generally strict system of terms traditionally employed by geneticists (Rieger et al. 1976).

3.1.2 History of Parasexual Hybridization of Higher Plants

The interspecific hybrid *Nicotiana glauca* (×) *N. langsdorffii* isolated by Carlson et al. (1972) in Brookhaven National Laboratory seems to be the first reported higher plant hybrid obtained using the techniques of parasexual hybridization. The extreme brevity of this communication as well as the author's subsequent estrangement from studies of hybridization by protoplast fusion have incited several investigators to express doubts about the orginal results obtained by Carlson (Melchers and Labib 1974; Melchers 1977). Since the critical remarks concerning this work have not provided unambiguous evidence that Carlson's work was the result of an artifact, we date the history of experimental parasexual hybridization from the appearance of this communication.

It is not out of place to report here a brief chronology of the principle events in the history of somatic cell hybridization in higher plants. The protoplasts of higher plants were first obtained using mechanical isolation by von Klercker (1982). Spontaneous fusion of isolated protoplasts was first observed by Küster (1909). However, isolated protoplasts remained queer things until 1960, when Cocking (1960) published work dealing with the enzymatic method of protoplast isolation. Further investigations have demonstrated that protoplast fusion may be controlled, i.e., may be induced (Power et al. 1970). The experiments of Takebe et al. (1971), Nagata and Takebe (1971), Nitsch and Ohyama (1971), Ohyama and Nitsch (1972), and Grambow et al. (1972), based on the widespread techniques of plant cell tissue culture (originally elaborated by Gautheret 1966 and White 1954), first demonstrated the potentials for the regeneration of a whole plant from naked protoplasts. Development and subsequent perfection of techniques for somatic cell hybridization (Keller and Melchers 1973; Eriksson et al. 1974; Kao and Michayluk 1974) made possible the first experiments with the production of hybrid plants by somatic cell fusion (Carlson et al. 1972; Melchers and Labib 1974; Schieder 1975a; Gleba et al. 1975; Kameya 1975a). Simultaneously, the first data concerning the genetic novelty of the plants regenerated from protoplast fusion products was obtained (Melchers and Labib 1974; Gleba et al. 1975). Today, three principal trends exist and are being further developed in investigations of parasexual hybridization: studies on and reconstruction of novel combinations of cytoplasmic genes (Gleba et al. 1975, 1978 a; Chen et al. 1977; Belliard et al. 1977 a, b, 1978, 1979; Zelcer et al. 1978; Izhar and Power 1979; Aviv et al. 1980; Aviv and Galun 1980; Medgyesy et al. 1980; Menczel et al. 1981, 1982; Nagy et al. 1981; Glimelius and Bonnett 1981; Glimelius et al. 1981; Sidorov et al. 1982 a); studies of cell hybridization between distantly related plant species (Kao 1977; Wetter 1977; Gleba and Hoffmann 1978, 1979, 1980; Melchers et al. 1978; Krumbiegel and Schieder 1979, 1981; Dudits et al. 1979, 1980; Wetter and Kao 1980; Gleba et al. 1982 a, b, c; Skarzhynskaya et al. 1982); and the use of protoplast fusion for production of hybrids of practical interest for plant breeding (Butenko and Kuchko 1979; Komarnitsky et al. 1980, 1981, 1983; Power et al. 1980; Schieder 1980 a; Douglas et al. 1981 a, b, c; Evans et al. 1981, 1982).

Two significant aspects of the history of plant cell hybridization are worth stressing. First, the methodology of this field has arisen from experiments with in vitro plant cell and tissue cultures and even today most of the investigators working with parasexual hybridization are plant physiologists or "culturists". The present achievements in plant cell and plant hybridization via protoplast fusion represent a logical extension of the methods of in vitro cell manipulation. Therefore, one can state that the most significant successes in the exploration of in vitro cultured cells (such as examination and elaboration of the conditions for maintainance of unorganized plant cell growth, experimental demonstration of cell totipotency, induction of differentiation and plant regeneration from cultured cells, analysis of genetic changes connected with the unorganized growth in vitro, etc.) and the eminent investigators who have worked with cultured plant cells and tissues also belong to the history of parasexual hybridization of higher plants. Second, the theoretical basis for plant cell hybridization is rooted to a great extent in animal cell hybridization, a field of endeavor which had already experienced an impetuous development (see the monographs of Ephrussi 1972; Harris 1974; Ringertz and Savage 1976). However, although the ideology and, to some extent, the methodology of handling animal and plant cells are similar, the investigations on plant cell hybrids retain not only their original aspects, but also possess the added potential for regeneration of entire hybrid plants from the cell hybrids, thereby opening new perspectives for the application of parasexual hybridization in plant breeding. Since the extranuclear genomes are inherited maternally in sexual crossing, the hybridization of somatic cells by protoplast fusion is a potent and novel tool for the genetic reconstruction and analysis of unique combinations of extranuclear and/or nuclear genomes.

During the last 10 years, more than 60 experiments on somatic hybridization of plant cells have been reported and a number of hybrid plants and hybrid cell lines have been isolated. The hybrid combinations obtained are listed in Table 2 (for the intraspecies combinations) and Table 3 (for the interspecies combinations). The mutant forms of the parental material(s) are also listed for the intraspecies combinations where appropriate. Two works (Power et al. 1975; Binding and Nehls 1978 a) have been omitted from Table 3, since, in our opinion and as discussed in 3.4, the results are unconvincing. As can be seen from the data presented in Table 2, 26 interspecific hybrids have been obtained. Fifteen of them are hybrids between different varieties of *Nicotiana tabacum*, while seven of the hybrids belong to other members of the Solanaceae and hybrids of the moss

Table 2. Intraspecific parasexual hybrids produced by plant protoplast fusion

Parental forms [a]		Ref.
Nicotiana tabacum chl$^-$ "sublethal"	(×) *N. tabacum* chl$^-$ "virescent"	Melchers and Labib (1974)
Spherocarpus donelli nic_2	(×) *S. donellii* gly$^-$ pol_2	Schieder (1975 a)
Nicotiana tabacum chl$^-$ (P)	(×) *N. tabacum* chl$^-$ Su/Su	Gleba et al. (1975)
Nicotiana tabacum chl$^-$ WS_1 WS_2	(×) *N. tabacum*	Kameya (1975 a)
Datura innoxia chl$^-$	(×) *D. innoxia* chl$^-$	Schieder (1977 a)
Nicotiana tabacum cms (*debneyi*)	(×) *N. tabacum*	Belliard et al. (1977 a, 1977 b)
Physcomitrella patents	(×) *P. patens*	Grimsley et al. (1977 a, 1977 b)
Nicotiana tabacum cms (*debneyi*)	(×) *N. tabacum* chl$^-$ (P)	Gleba et al (1978 a)
Nicotiana tabacum NR$^-$ cms	(×) *N. tabacum* NR$^-$ nia	Glimelius et al. (1978 a)
Nicotiana tabacum mtr	(×) *N. tabacum* chl "sublethal"	Gleba and Berlin (1979)
Petunia hybrida cms	(×) *P. hybrida*	Izhar and Power (1979)
Nicotiana tabacum NR$^-$ cnx	(×) *N. tabacum* NR$^-$ nia	Wallin et al. (1979)
Nicotiana sylvestris mt-r	(×) *N. sylvestrix* aec-r	White and Vasil (1979)
Nicotiana tabacum st-r	(×) *N. tabacum*	Wullems et al. (1980)
Nicotiana tabacum cms (*suaveolens*)	(×) *N. tabacum* NR$^-$ cnx	Glimelius et al. (1981)
Nicotiana tabacum chl$^-$ (p)	(×) *N. tabacum* NR$^-$ cnx	Glimelius and Bonnet (1981)
Daucus carota mt-r	(×) *D. carota* aec-r	Harms et al. (1981)
Nicotiana debneyi	(×) *N. debneyi*	Scowcroft and Larkin (1981)
Daucus carota ch-r	(×) *D. carota* chr-r	Lazar et al. (1981)
Nicotiana tabacum cnx	(×) *N. tabacum* cms (*undulata*)	Bonnett and Glimelius, pers. comm.
Nicotiana tabacum cnx	(×) *N. tabacum* cms (*suaveolens*)	Bonnett and Glimelius, pers. comm.
Nicotiana tabacum	(×) *N. tabacum* cms (*glauca*)	Bonnett and Glimelius, pers. comm.
Nicotiana plumbaginifolia NR$^-$	(×) *N. plumbaginifolia* NR$^-$	Marton et al (1982)
Nicotiana plumbaginifolia aux	(×) *N. plumbaginifolia* chl$^-$	Sidorov and Maliga (1982)
Nicotiana plumbaginifolia chl$^-$	(×) *N. plumbaginifolia* chl$^-$	Sidorov and Maliga (1982)
Nicotiana tabacum P$^-$	(×) *N. tabacum*	Nakata and Oshima (1982)

[a] The mutants utilized are listed after the species name where appropriate

Physcomitrella patens and of the liverwort *Sphaerocarpus donellii* were also obtained.

A total of 55 communications concerning the successful hybridization by fusion of protoplasts from different species have been reported. In particular, 41 interspecific, 1 intergeneric, 9 intertribal, and 3 interfamilial hybrids were produced. The majority of the hybrids are members of the Solanaceae (33 hybrids of

Table 3. Interspecific parasexual hybrids obtained by fusion of plant protoplasts

Parental forms	Ref.
Nicotiana glauca (×) *N. langsdorffii*	Carlson et al. (1972)
N. glauca (×) *N. langsdorffii*	Smith et al. (1976)
Petunia hybrida (×) *P. parodii*	Power et al. (1976a)
N. tabacum (×) *N. sylvestris*	Melchers (1977)
Daucus carota (×) *D. capillifolius*	Dudits et al. (1977)
N. sylvestris (×) *N. knightiana*	Maliga et al. (1977)
Glycine max (×) *N. glauca*	Kao (1977)
Lycopersicon esculentum (×) *Solanum tuberosum*	Melchers et al. (1978)
Arabidopsis thaliana (×) *Brassica campestris*	Gleba and Hoffmann (1978)
N. knightiana (×) *N. sylvestris*	Menczel et al. (1978)
N. knightiana (×) *N. tabacum*	Maliga et al. (1978)
Datura innoxia (×) *D. discolor*	Schieder (1978a)
Datura innoxia (×) *D. stramonium*	Schieder (1978a)
N. tabacum (×) *N. glauca*	Evans et al. (1978)
N. glauca (×) *N. langsdorffii*	Chupeau et al. (1978)
N. tabacum (×) *N. sylvestris*	Zelcer et al. (1978)
Solanum tuberosum (×) *S. chacoense*	Butenko and Kuchko (1979)
Datura innoxia (×) *Atropa belladonna*	Krumbiegel and Schieder (1979)
Daucus carota (×) *Aegopodium podagraria*	Dudits et al. (1979, 1980b)
N. tabacum (×) *N. glutinosa*	Nagao (1979)
N. tabacum (×) *N. alata*	Nagao (1979)
Petunia parodii (×) *P. inflata*	Power et al. (1979)
Petunia parodii (×) *P. parviflora*	Power et al. (1980)
N. tabacum (×) *N. sylvestris*	Aviv and Galun (1980)
N. tabacum (×) *N. glauca*	Evans et al. (1980)
N. tabacum (×) *N. sylvestris*	Medgyesy et al. (1980)
Datura innoxia (×) *D. candida*	Schieder (1980a)
Datura innoxia (×) *D. sanguinea*	Schieder (1980a)
N. glauca (×) (*Glycine max* (×) *N. glauca*)	Wetter and Kao (1980)
Daucus carota (×) *Petroselinum hortense*	Dudits et al. (1980a)
N. tabacum (×) *N. debneyi*	Piven (1980); Gleba (1980)
Daucus carota (×) *D. capillifolius*	Kameya et al. (1981)
N. tabacum (×) *N. rustica*	Douglas et al. (1981a)
N. knightiana (×) *N. tabacum*	Menczel et al. (1981); Nagy et al. (1982)
N. tabacum (×) *N. plumbaginifolia*	Sidorov et al. (1981a)
N. tabacum (×) *N. nesophila*	Evans et al. (1981)
N. tabacum (×) *N. stocknii*	Evans et al. (1981)
N. tabacum (×) *N. glutinosa*	Uchimiya (1982)
Atropa belladonna (×) *N. chinensis*	Gleba et al. (1982a, c)
Atropa belladonna (×) *N. tabacum*	Gleba et al. (1982b)
Solanum tuberosum (×) *N. tabacum*	Skarzhynskaya et al. (1982)
Solanum sucrense (×) *N. tabacum*	Skarzhynskaya et al. (1982)
N. tabacum (×) *N. plumbaginifolia*	Menczel et al. (1982)
N. tabacum (×) *N. plumbaginifolia*	Maliga et al. (1982)
N. tabacum (×) *N. plumbaginifolia*	Kanevsky and Gleba (unpublished)
N. tabacum (×) *Salpiglossis sinuata*	Nagao (1982)
N. tabacum (×) *N. repanda*	Nagao (1982)
N. tabacum (×) *Petunia hybrida*	Xianghui et al. (1982)
Solanum nigrum (×) *Solanum tuberosum*	Binding et al. (1982)
Brassica campestris (×) *Brassica oleracea*	Schenck (1982)
N. tabacum (×) *N. rustica*	Nakata and Oshima (1982)
N. tabacum (×) *Glycine max*	Chien et al. (1982)
N. chinensis (×) *Pisum sativum*	Gleba et al. (1983b)
N. tabacum (×) *N. plumbaginifolia*	Gleba et al. (1983c)

Nicotiana, 5 of *Datura*, 5 of *Solanum*, and 1 of *Lycopersicon* genera). A few positive results have also been obtained from species belonging to the Cruciferae, Umbelliferae, and Fabaceae families. In all the experiments cited in Table 3, either mesophyll protoplasts or the protoplasts derived from callus tissues have been utilized as the parental cell sources.

There are numerous reviews of the investigations in parasexual hybridization, including: Cocking 1965a, 1971, 1973, 1976b, 1981; Eriksson 1971; Power and Cocking 1971, 1977; Carlson 1973a, b; Chupeau et al. 1973; Melchers and Labib 1973; Chaleff and Carlson 1974; Heyn et al. 1974; Gamborg et al. 1974; Smith 1974, 1979; Chaleff and Gamborg 1975; Takebe 1975; Gamborg et al. 1975; Vasil 1976; Potrykus et al. 1976, 1977a; Bajaj 1977; Melchers 1977, 1980, 1981; Giles 1977; Gleba 1978c, 1979; Constable 1978; Butenko 1979; Vasil et al. 1979; Binding 1979; Thomas et al. 1979; Schieder and Vasil 1980; Cocking et al. 1981; Gleba and Evans 1982; Evans 1982; Binding et al. 1984.

3.2 Methods of Screening for Parasexual Hybrids

3.2.1 Introduction

Most experiments in plant cell hybridization use methods similar to those used in animal somatic cell and microbial genetics, where the rare genetic events produced by the different agents or treatments may be studied. In other words, large populations of cells of both parents are induced to fuse and since the fusion of the cells in a mixed population is a statistical phenomena, it is expected that all possible combinations of fusion products are formed. Therefore, it is necessary to separate the heteroplasmic fusion products from the unfused parental cells and the homoplasmic fusion products using various selection techniques.

Current methods enable or even require the use of populations as large as 10^6–10^8 cells. Since the frequency of genetic events induced by existing techniques is rather significant (Keller and Melchers 1973; Kao and Michayluk 1974; Kao 1977), the problem of selection can be simplified and may sometimes be solved by regeneration of a sufficiently large number of plants from the total preparation with subsequent visual selection, as has been performed by Belliard et al. (1977a, b) and Scowcroft and Larkin (1981). The selection employed in most experiments is applied at either the hybrid cell level or the regenerated plant stage. All of the selection methods impose some restrictions on the spectrum of the hybrids recovered. Moreover, some of these methods seem to induce additional genetic changes, particularly mutations, which complicate the analysis of the hybridization process. Some selection techniques are severely limited because of the necessity of using special mutants or conducting exhaustive preliminary studies to determine the physiological peculiarities of the parental cells (or of the sexually obtained hybrid cells, if possible). Often, the subsequent process of culturing the hybrid cells is complicated by the application of the selection procedure. In some cases, the efficiency of the selection technique is very low and therefore theoretically inappropriate; nonetheless, these same techniques have been successfully ap-

Table 4. Methods for selection of plant parasexual hybrids

Basis of the selection method	Ref.
I. Genetic complementation by cell fusion of:	
A) Chlorophyll deficient mutants	Melchers and Labib (1974)
	Gleba et al (1975)
	Schieder (1977 a)
	Gleba (1979)
	Douglas et al. (1981 a, b)
B) Chlorophyll deficient plastome mutants with wild type cells [a]	Gleba et al. (1978a)
	Gleba (1980)
C) Chlorophyll deficient semidominant mutants with wild type cells [a]	Nagao (1978, 1979)
	Evans et al. (1981)
D) Auxotrophic mutants	Schieder (1974)
	Grimsley et al. (1977 a)
	Glimelius et al. (1978 a)
	Wallin et al. (1979)
E) Mutants of resistance to antimetabolites	White and Vasil (1979)
	Harms et al. (1981)
F) Mutants of different classes	Glimelius and Bonnett (1981)
II. Physiological complementation	Carlson et al. (1972)
	Smith et al. (1976)
	Power et al. (1976 a)
	Power et al. (1977)
	Chupeau et al. (1978)
III. Mechanical isolation of heterokaryons and their cloning	Kao (1977)
	Gleba und Hoffmann (1978)
	Menczel et al. (1978)
	Gleba and Berlin (1979)
	Wetter und Kao (1980)
	Skarzhynskaya et al. (1982)
	Menczel et al. (1981)
	Gleba et al. (1982 a, b)
IV. Reactivation of hybrids after fusion of cells inactivated by chemical and physical factors	Sidorov et al. (1981 a)
V. Combined methods using:	
A) Mutants and application of selective culture media	Melchers (1977)
	Cocking et al. (1977)
	Schieder (1978 a, 1980 a)
	Dudits et al. (1979)
	Power et al. (1979, 1980)
	Evans et al. (1980 a)
	Glimelius et al. (1981)
	Dudits, in press
B) Mutants plus inactivation of cells	Dudits et al. (1980 a)
	Medgyesy et al. (1980)
	Sidorov et al. (1981)
	Menczel et al. (1982)
	Lazar et al. (1981)
C) Application of selective media plus cell inactivation	Zelcer et al. (1978)
VI. "Semiselective" methods	Kameya (1975 a)
	Dudits et al. (1977)
	Melchers et al. (1978)
	Izhar and Power (1979)
	Butenko und Kuchko (1979)
VII. Absence of specific selection technique	Belliard et al. (1977 a, b)
	Binding and Nehls (1978 a)
	Scowcroft and Larkin (1980)

[a] This type of gene interaction is reported here to make the picture complete. However, description in terms of complementation is incorrect

plied in actual experiments. These problems will be discussed in detail in the following sections.

The methods of selection for plant parasexual hybrids are highly diverse. Unfortunately, no generally accepted systematics exist to describe these methods (for example, see Power and Cocking 1977; Cocking 1979). The current methods of hybrid selection and their functional basis are given in Table 4.

3.2.2 Genetic Complementation

The utilization of genetic complementation for hybrid plant cell selection has been adopted from the experiments on animal somatic cell genetics. Since it was originally impossible to utilize auxotrophic or resistance mutants for this purpose, the original work in this field by Melchers and Labib (1974) used chlorophyll deficient mutant parental cells which subsequently resulted in complementation in the hybrid products. The authors performed their experiments on tobacco species which possess a number of well-studied chlorophyll mutants and were successful in using a combination of two nuclear recessive mutations that cause light-dependent chlorophyll deficiency. The plants homozygous for any of these genes are bleached and die when grown under intense illumination (greater than 10 klx) (Povilaitis and Cameron 1963; Melchers and Labib 1974). After protoplast fusion and cell colony regeneration, calli were transferred onto a medium that induces shoot organogenesis and they were then cultured under nonpermissive light conditions. The purported hybrids were identified as green colonies. In our experiments (Gleba 1974; Gleba et al. 1975), also performed with tobacco, a plastome chlorophyll deficient mutant obtained from tissue bearing only defective plastids and the nuclear genome mutant of aurea type ("sulphur") with an intermediate type of dominance (Su/Su plants homozygous for this character) were used. Both of these mutants have also been well-studied (see Chap. 4). The hybrid products capable of performing photosynthesis were selected after regeneration. In these last experiments, the choice of mutations (markers) that are distinguishable in the heterozygous state and belonging to different idiotypes has permitted a more detailed analysis of the various hybrids obtained (see 4.1).

The logical extension of these results has been realized in experiments where only one parent was marked by a photosynthetic mutant and the other parent was the wild type. The hybridization of mutants homozygous for a semidominant mutation with the wild type cells results in the production of nuclear heterozygotes that differ phenotypically from both parental types. This approach was the basis for interspecific hybrid selection in the works of Nagao (1978, 1979) utilizing the tobacco mutant "aurea" and of Evans et al. (1981) using heterozygosity for the Su gene. The employment of plastome chlorophyll deficient mutants as a marker for hybrid selection also permits the screening in the absence of a marker in the other parental form since a part of the hybrid plants will be variegated (Gleba et al. 1978a). However, this method of screening is applicable only for the detection of cytoplasmic heterozygotes, which are not necessarily also nuclear heterozygotes (see Chap. 4).

Several works have recently appeared in which genetic complementation of photosynthetic mutants has been used to isolate hybrids of *Nicotiana* (Melchers

Fig. 12. Selection of plant somatic hybrids via complementation of chlorophyll deficient mutants. **A** A green and therefore hybrid colony is clearly distinguishable from the other colonies regenerated after protoplast fusion. **B** A somatic hybrid plant of *Datura innoxia* (in the *middle*) regenerated from the fusion of protoplasts from two chlorophyll deficient mutants. The parental protoplasts regenerate albino plants (*left* and *right*)

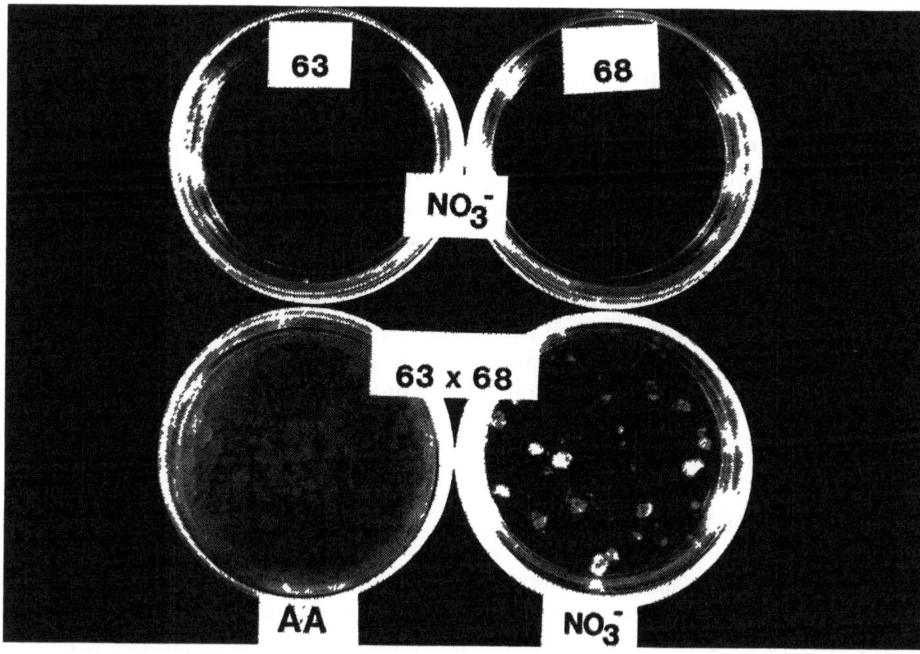

Fig. 13. Selection of somatic hybrids of *Nicotiana tabacum* via complementation of nitrate reductase deficient mutants. Polyethylene glycol treated mixtures of cnx-68 and nia-63 protoplasts were grown in selective *(NO_3^-)* and nonselective (*AA*) medium for 30 days. In the *upper row*, control cultures of polyethylene glycol treated protoplasts of both mutants were plated into the selective *(NO_3^-)* medium. (Glimelius et al. 1978)

1977; Douglas et al. 1981 a) and *Datura* (Schieder 1977 a). Uncharacterized chlorophyll deficient mutants, produced in some cases especially for hybridization purpose (Schieder 1976), have been utilized in these experiments (Fig. 12). Chlorophyll deficient mutants belong to the best known mutants in higher plants, so genetic complementation based on their use represents one of the most trustworthy and simple ways to select for parasexual hybrids.

Nonetheless, it should be mentioned that applying the selection at the level of regenerated plants is a serious drawback of this method, since it makes it difficult or even impossible to select individual hybrid cells.

Selection of parasexual hybrids based on the use of auxotrophic mutants was first successfully demonstrated with the mosses. Schieder (1975 a) isolated hybrids of the liverwort *Spherocarpus* by fusing protoplasts of a male strain auxotrophic for glucose with protoplasts of a female strain defective for nicotinic acid. Other investigators obtained a number of hybrid lines of *Physcomytrella patens* using protoplast fusion of different auxotrophic mutants (nicotinic acid, thiamine, and para-aminobenzoic acid dependence) (Grimsley et al. 1977 a, b). More detailed description of these works is presented in connection with the possibilities of genetic analysis using parasexual techniques (see Chap. 6).

Similar work performed on flowering plants (tobacco) appeared in 1978. Extant tobacco cell mutants with genetically different nitrate reductase deficiencies (methods for their selection and reversion are well-studied; Müller and Grafe 1978; Mendel and Müller 1978, 1979) were used as the initial material. A large number of wild type hybrid strains were isolated after protoplast fusion (Glimelius et al. 1978a; Wallin et al. 1979) (Fig. 13). Mutations of resistance to different antimetabolites have also been employed for the selection of hybrid products after protoplast fusion. White and Vasil (1979) used cell lines resistant to the amino acid analogs S-2-aminoethylcysteine and 5-methyltryptophan to select somatic hybrids of *Nicotiana sylvestris*. This same combination of mutants was also used for the screening of somatic hybrids of carrot (Harms et al. 1981). As long as the production and selection of new plant cell mutants continues to be successful, the selection of parasexual hybrids based on the genetic complementation of auxotrophic and/or other mutant parental cell lines may prove to be the optimum method for hybrid production and identification.

3.2.3 Physiological Complementation

The term "physiological complementation" is not sufficiently specific. We shall use this term in a somewhat broader sense in the description of the capacity of hybrid cells to survive and propagate, as well as to sustain organized growth, under culture conditions which are unacceptable for growth and/or differentiation of the parental cells. This inability of the parental cells to grow or differentiate is not associated with some definite mutation, but is rather a normal physiological response of the cells to the culture conditions employed.

A variation of physiological complementation was suggested by Carlson (Carlson et al. 1972; Smith et al. 1976; Chupeau et al. 1978). In these studies, the interspecific hybrids *Nicotiana glauca* (×) *N. langsdorffii* were obtained by parasexual hybridization. The sexual hybrids between these species had been previously produced. It was demonstrated that these hybrid plants form tumors and that the hybrid cells cultured in vitro are capable of propagation in media devoid of exogenous phytohormones, i.e., they are hormone independent. Therefore, the selection of the parasexual hybrids in these experiments was based on this peculiarity (hormone independence) of the cultured hybrid cells. Leaf mesophyll protoplasts of both species were fused and, after a certain period to permit the phenotype expression in the hybrid cells, the cells were transferred onto nutritive medium containing no exogenous phytohormone. Subsequently, the colonies capable of growing under these conditions were selected. It should be noted that this approach is applicable only for this particular combination of species and as a result this specific technique is now relegated to primarily theoretical rather than practical considerations.

A broader approach based on the same idea was proposed by Cocking (Cocking et al. 1974; Cocking 1976a, b; Power and Cocking 1977). After defining alternative culture conditions suitable for the growth of the cells of one of the parental types, but unsuitable for growing the second one and vice versa, the mixed cell suspension obtained after fusion was placed into the first culture medium to eliminate the cells of one parental type and was then subsequently transferred to conditions causing the similar elimination of the cells of the second parental type. These selective conditions may sometimes be combined in the same culture medium. For example, it was demonstrated that protoplasts of *Petunia parodii* were incapable of regenerating into callus tissue when cultured in Murashige and Skoog (1962) nutrient medium, whereas the protoplasts of *Petunia hybrida* divide normally under these conditions. Likewise, it has been shown that hybrid cells of both sexual and parasexual hybrids are capable of growing in this same medium in most cases. The cells of *P. hybrida* are much more sensitive to actinomycin D as compared to *P. parodii*. The culturing of a mixture of both parental and hybrid cells in Murashige and Skoog medium supplemented with actinomycin D results in selective and formation propagation of only hybrid colonies (Power et al. 1976).

Extensive work has been performed at Nottingham University to investigate the natural resistance of tobacco and petunia protoplasts to different substances in order to establish selective culturing conditions for the isolation of parasexual hybrids (Cocking et al. 1974). Three different selection procedures have already been proposed for the species combination *Petunia hybrida* (×) *P. parodii* (Cocking et al. 1977; Power et al. 1977). The development of such procedures is facilitated by the fact that a phenomena similar to heterosis takes place in many hybrid fusion products (i.e., they divide and grow much more actively than the parental cells) (Schieder 1977a; Gleba and Hoffmann 1978). In the work of Schieder, the first colonies to develop turned out to be hybrid ones, although the selection and identification was later conducted on the basis of genetic complementation between two chlorophyll deficient mutants of *Datura*. In spite of its successful application to selection of parasexual hybrids, we have some apprehensions concerning the use of the physiological complementation method. It is evident that the significant physiological heterogeneity of in vitro cultured cells, composed of the complete spectrum of cell responses to the selective conditions, must play an important role in the application and success of this method. As far as we know, this problem still has not been seriously studied.

3.2.4 Restoration of Growth Capacity in Hybrids Upon Fusion of Inactivated Cells

The principle of this method is derived from the work of Wright (Wright and Heyflick 1975; Wright 1978) in which animal cells from two parental forms were treated before fusion with different chemical poisons, such as ethidium bromide and sodium azide, causing the irreversible inactivation of various macromolecules and consequently the inactivation of the cells themselves. The hybridization may be considered to be a type of rescue process which enables the survival and subsequent division of heterokaryons only. The method was also found suitable for the selection of fusion products of cells and enucleated cytoplasts. Nehls (1978) attempted to fuse protoplasts of *Petunia* and *Datura* inactivated by monoiodoacetate and diethylpyrocarbonate, but the results (development of a single cell colony whose hybrid nature was confirmed only by visual microscopic observation) can hardly be discussed seriously. However, the use of iodoacetate for plant cell inactivation has recently been refined to the point of being a routine technique in the laboratory of Maliga (Medgyesy et al. 1980; Sidorov et al. 1981 a; Menczel et al. 1982). The irradiation with X- and gamma rays (50–300 J kg^{-1}) was successfully employed as an inactivating agent initially by Zelcer et al. (1978) and later by Hungarin researchers (Sidorov et al. 1981 a; Menczel et al. 1982). Although there are a number of investigations in which the inactivation of one of the parents is used, to date the combination of two inactivation agents for hybrid production and plastid transfer has been used only in the work of Sidorov et al. (1981 a). As might be expected, the high doses of irradiation result in the inactivation of the nuclei of the treated parental cells so that the resulting fusion products are cytoplasmic hybrids (cybrids). This approach seems to be quite promising, since it requires neither preliminary production of mutants nor the performance of special experiments to determine the selective physiological conditions for hybrid screening. However, the procedure of inactivation may have grave genetic consequences which are only beginning to be studied and appreciated (see Chap. 4).

3.2.5 Mechanical Isolation

The idea of mechanical isolation and culturing of single fusion products can be attributed to Kao (Kao 1977; Kao and Wetter 1977) and has been successful thanks to the elaboration of techniques for culturing single plant cells (Bergmann 1960; Jones et al. 1960; Reinert 1963; Reinert and Ardenne 1964; Vasil and Hildebrandt 1965; Backs-Hüsemann and Reinert 1970; Kao and Michayluk 1975; Gleba 1978 a) (see Sect. 2.1). Morphologically different types of parental protoplasts are fused, the heteroplasmic products of fused protoplasts are identified directly via microscopic observation, isolated by micropipette and then cultured individually in special enriched media (Kao 1977) or in microdroplets (Gleba 1978 a). The hybrids selected via this method are described in Chap. 3 and 4 (Kao 1977; Gleba and Hoffmann 1978; Gleba and Berlin 1979; Gleba et al. 1982 a, b). Although this method requires rather refined methodological training and the technique is relatively difficult to reproduce, the cloning of the individual heterokaryocytes that accompanies the selection nevertheless opens up new perspectives which can not be provided by any of the other reported screening techniques.

For example, hybrids are produced without the use of mutants or special conditions of physiological selection and there is an automatic production of clones by the selection of individual fusion products, as well as the potential for visual monitoring of the culture. As this method is based on the use of individual cells rather than entire cell populations, the rare and often disturbing genetic events, such as mutations, need not be taken into consideration for interpretation of the results. Thus, the mechanical isolation of single fusion products with the subsequent culturing of individual cells is presently the sole efficient method for isolation of parasexual hybrids between distantly related plant species.

A main shortcoming of mechanical isolation of parasexual hybrids is the necessity for utilization of genetically irregular in vitro cultured cells as one of the parental forms. Moreover, the use of morphologically and hence metabolically contrasting parental cells to enable visual identification of hybrids evidently affects the fate of parental genes upon hybridization (see Chap. 4).

In less stringent variations of this method, the isolated fusion products are transferred by micropipette to a feeding medium containing the cells from one of the parents which are incapable of morphogenesis and thus provide optimal conditioning of the medium for the hybrid cells. Afterwards, the hybrids are re-identified at the stage of organogenesis (Menczel et al. 1978). In another variation, the fusion products are cultured together in a common droplet of feeding medium (Menczel et al. 1982). However, these variations, while extending the application of the method of mechanical isolation, lose some of the advantages of the original techniques of Kao since one can not be sure that the hybrid colonies isolated in this way are really independent clones or that aggregation of hybrid and parental cells has been excluded, etc.

There are also other potential ways for screening and selecting somatic cell hybrids. For example, Galbraith and Galbraith (1979) have demonstrated that protoplasts can be easily marked via staining with vital fluorescent dyes (rhodamine B and fluorescein conjugated with octadecylamine and octadecanol were used) and therefore the hybrid cells (and unfortunately agglutinated cells) will possess fluorescence characteristic for both dyes which can be measured with a microspectrofluorimeter. This method of vital staining does not noticeably lower the survival rate of the cells. Of course, if mesophyll protoplasts are used as one of the parental forms, they and their fusion products may be recognized either without additional vital staining owing to the fluorescence of the chlorophyll or they may be stained with fluorescein isothiocyanate (Hess 1973). Another method for vital staining of plant protoplasts has been proposed by Meadows and Potrykus (1981) using the fluorescent dye Höchst 33258. In combination with one or more of these vital stains, it is feasible that plant parasexual hybrid cells may be selected by using flow-sorting instruments similar to those used for sorting and selecting animal hybrid cells (for example, the FACS system produced by Becton Dickinson, USA).

Redenbaugh et al. (1982) have characterized protoplasts of *Euphorbia lathyris*, *Nicotiana glauca*, *N. langsdorffii*, *Petunia parodii*, and *P. inflata* by flow cytometry on the basis of laser light scattering and chlorophyll fluorescence or via exogenously added stains such as fluorescein isothiocyanate and rhodamine isothiocyanate. Further analysis and mechnical sorting of differentially stained mesophyll protoplasts of these species was done with a Becton Dickinson Fluorescence-Activated Cell Sorter (FACS IV). The procedure permitted the sorting and recovery of up to 2000 intact protoplasts per hour under sterile conditions. It has yet to be demonstrated that this type of cell selection works accurately in the case of populations treated according to existing fusion protocols, where a large proportion of the cells are agglutinated. But even if this technique proves to be useful only for the enrichment rather than the rigid selection of heterokaryons, it greatly expands the horizons of hybrid cell manipulations by enabling the selection of hybrids after the fusion of morphologically and genetically indistinguishable cells.

3.2.6 Physical Enrichment

It has been demonstrated that plant protoplasts can be separated into fractions with different specific densities using isopycnic centrifugation in density gradients of iso-osmolal solutions (Harms and Potrykus 1978 a, b). This is especially useful since protoplasts from different tissue such as those from leaf mesophyll and from callus also have different buoyant densities. Since the fusion products of two parental cells differing in buoyant density must have intermediate values of buoyant density, isopycnic centrifugation might be used for the isolation or fractionation of fusion products. The use of this method is greatly complicated by the fact that the standard techniques of fusion result in significant cell aggregation and the aggregates naturally also have intermediate buoyant densities as compared to the parental cells. Nevertheless, it has been shown experimentally that the frequency of tobacco hybrid colony formation, counted on the basis of genetic complementation (using a combination of mutants adopted from Melchers and Labib 1974), is significantly higher in those fusion products selected via centrifugation as compared to the control (unselected) fusion products. Therefore, this method provides an efficient enrichment for hybrid fusion products, even if it is not a screening system in the strict sense of the word. Obviously in some cases, this approach in combination with subsequent visual selection may be entirely sufficient for the isolation of parasexual hybrids.

3.3 Frequency of Formation of Parasexual Hybrids

The frequency of hybrid formation is an important parameter characterizing the process of parasexual hybridization by protoplast fusion. As will be evident from the following discussion, as well as from Chaps. 4 and 5, the question is concerned not with a single frequency, but rather with several frequencies, each of which is a measure of the many genetically distinct hybrids that are formed. The process of parasexual hybridization may be studied by cytological methods (analysis of cells immediately after fusion or any time thereafter) as well as by genetic methods (statistical evaluation of hybridization events based on study of the progenies). It is evident that the hybridization process is influenced by the number of parental cells used, the genetic, epigenetic, and physiological characteristics of these cells and the methods employed for inducing hybridization. Out of all the possible cell types, only a very few combinations have been used for parasexual hybridization, including the fusion of mesophyll with mesophyll protoplasts (Carlson et al. 1972; Melchers and Labib 1974; Gleba et al. 1975; Power et al. 1976 a; Smith et al. 1976; Melchers 1977; Schieder 1977 a, 1978 a; Gleba et al. 1978 a; Zelcer et al. 1978; Krumbiegel and Schieder 1979), callus with mesophyll protoplasts (Power et al. 1975, 1977; Kao 1977; Maliga et al. 1977, 1978; Dudits et al. 1977, 1979; Gleba and Hoffmann 1978; Gleba and Berlin 1979), callus with callus protoplasts (Glimelius et al. 1978; White and Vasil 1979), as well as fusion between protenema protoplasts (Grimsley et al. 1977a). In these studies, the fusion was induced by sodium nitrate (Carlson et al. 1972; Power et al. 1975), gelatin (Kameya 1975a), a solution with both a high pH and a high concentration of calcium ions (Melchers and Labib 1974; Grimsley et al. 1977a), polyethylene glycol (Gleba et al. 1975; Smith et al. 1976; Maliga et al. 1977, 1978; Dudits et al. 1977; Schieder 1977a; Belliard et al. 1977a, b; Gleba et al. 1978a; Glimelius et al. 1978a; Zelcer et al. 1978), or by combined PEG-high pH-high Ca^{2+} techniques (Kao 1977; Schieder 1978a; Gleba and Hoffmann 1978; Menczel et al. 1978; Krumbiegel and Schieder 1979).

Most of the experiments on hybridization by protoplast fusion have been aimed at the production of hybrid plants, rather than at a special analysis of the hybridization process itself. Therefore, the procedures used for induction and selection of hybrids as well as the conditions of culturing and plant regeneration, while enabling the regeneration of hybrids, nonetheless result in significant shifts in the quantitative aspects of the experiments, making a statistical interpretation of the genetic events impossible. The protoplasts are subjected to a number of treatments (isolation and fusion) and pass through different media (protoplast isolation, culture and regeneration media). All of these influences result in specific and nonspecific elimination of a portion of the cell population and since the protoplasts are often isolated from material that is in a physiologically heterogenous state, the regenerated plants may not accurately reflect the initial cell population, especially in hybridization of different plant species. These obstacles may be illustrated by the complications encountered during the interpretation of results in the selection of hybrids when mutants of photosynthesis are used as the parental forms: first, the photosynthetic activity is higher in the hybrids; second, many of the parental forms are extremely sensitive to certain light regimes; and third, if the resulting hybrids differ from each other, they may also experience differing degrees of selection under the same physiological conditions. All of these facts must be taken into consideration for a thorough appreciation of the quantitative and qualitative data obtained in hybridization experiments. In most experiments, 10^5 to 10^7 protoplasts are treated simultaneously, the cells are cultured en masse in the same plate and the callus tissues obtained are then transferred into many petri dishes for organogenesis. In such a situation, one can not be certain that all of the hybrid plants derived from the same petri dish are not the progeny of a single hybrid cell. Therefore, it is sometimes difficult to properly correlate the number of genetic events resulting from cell hybridization with the number of hybrid plants regenerated.

As was previously mentioned (Sect. 2.2), the techniques available for the induction of protoplast fusion are efficient and produce fusion events with a frequency as high as 10^{-1}, as determined by cytological methods. Cytological studies on the percentage of heterokaryons after fusion has revealed 2% (Power et al. 1977), 1%–10% (Maliga et al. 1977), and 5%–12% (Glimelius et al. 1978a) of the total number of cells in the preparation. On the other hand, attempts to measure the frequency of fusion events using genetic methods indicate that such events are much more rare; they arise with a frequency of 10^{-5} to 10^{-6} of the total number of parental cells plated when mesophyll protoplasts or a mixture of mesophyll and callus protoplasts are used (Gleba et al. 1975, 1978a; Power et al.

1977; Dudits et al. 1977; Schieder et al. 1977a, 1978a) or 10^{-3} to 10^{-4} if protonema protoplasts (Grimsley et al. 1977a) or callus protoplasts are utilized (Glimelius et al. 1978; White and Vasil 1979). The induction of fusion by polyethylene glycol treatment used in almost all of these studies results in aggregation of the protoplasts, which hampers the quantitative analysis of the hybridization process. Another pitfall is the disruption of cell colonies that originate from a single fusion product that takes place at the later stages of culturing. As a consequence, the estimations of fusion frequency should be regarded as only preliminary.

The most rigorous assessment of the frequency of formation of nuclear hybrids has been performed by Melchers and Labib (1974) and Grimsley et al. (1977 a). In the first study, tobacco mesophyll protoplasts were fused via the high pH-high Ca^{2+} method (Keller and Melchers 1973), which prevented the aggregation of the protoplasts. The suspension was cultured in "liquid" agar to prevent the possible aggregation and disruption of the regenerating colonies. The volume of the culture was gradually increased during the propagation of the cells in order to ensure both the normal self-conditioning of the cells and sufficient dilution of the suspension to decrease the probability of contact between the colonies. The large colonies were then plated onto solid agar and the number of wild type colonies produced by genetic complementation in the hybridization of the parental cells was determined. A total of 2.2×10^6 callus colonies were regenerated in this experiment, of which at least 12 were found to be hybrids of an independent origin. This means that the frequency of hybrids displaying genetic complementation was 5.4×10^{-4}, which is in good agreement with the data of other investigators who have employed other methods of fusion induction. A later study in the same laboratory on a larger number of colonies obtained from fused protoplasts has demonstrated approximately the same frequency of hybridization (Melchers and Sacristan 1977). In the second work (Grimsley et al. 1977a), fusion was also induced by high pH and calcium ions and the protoplasts were highly diluted after treatment (this is possible because of the ability of protoplasts from moss protonema to grow at low plating densities). The frequency of hybrid formation measured in this experiment constituted 4.2×10^{-4} of the total number of cells of the parental type that was present in the lower concentration in the original mixture. However, this data can not automatically be extrapolated to experiments with flowering plants.

The use of extranuclear genes as markers for selection of parasexual hybrids has shown that most cytoplasmic hybrids (see Chap. 4) are not true hybrids for nuclear markers, probably as a result of nonfusion and segregation of nuclei of the parental types during the culturing of the heterokaryons that originate from protoplast fusion products (Gleba et al. 1975, 1978a; Belliard et al. 1977a, b; Gleba 1978c). Therefore, the hybridization frequency determined on the basis of nuclear gene complementation really represents the frequency of nuclear fusions in hybrid cells, whereas the true frequency of formation of viable heterokaryons may be four to eight times higher in some cases (Belliard et al. 1977a, b; Menczel et al. 1978; Gleba and Berlin 1979; Gleba 1980). As an exception to this rule, there has been a report of fusion of nuclei taking place in all of the hybrids (Sidorov et al. 1981a). These estimates of fusion frequency reflect the situation in heterokaryons obtained by the fusion of two mesophyll protoplasts or of mesophyll with callus protoplasts. It appears that the frequency of nuclear fusion in hybrids formed from the fusion of two callus protoplasts surpasses that of hybrids between mesophyll and callus protoplasts. This seems to be a good explanation for the higher frequency of production of true hybrids in the experiment of Glimelius et al. (1978) and of White and Vasil (1979). However, even after taking this into account, the maximum frequency of hybridization determined by genetic methods is approximately 10^{-3}, which is still much lower than the frequency of hybridization determined by cytological methods. This fact may probably be explained by the damage or destruction of a high percentage of the cells during protoplast fusion so that the number of hybrid cells that survive sustained culture or regeneration is much lower than the number of hybrids immediately after fusion. This process may be more pronounced in interspecific hybridization. For example, according to Schieder (1978 a), colonies regenerate from about 40% of the protoplasts from *Datura* mesophyll after treatment with polyethylene glycol, whereas colonies regenerate from only about 2% of the total number of cells after fusion of *Datura innoxia* and *D. straminium*. Moreover, the majority of these fusion products consist of more than two parental cells (see Chap. 3) and multicellular heterokaryons are usually less viable. To conclude the discussion of this question, it should be mentioned that the physiological variability of plant cells and the complications arising in exact evaluation of hybrid frequency in the parasexual hybridization process decrease the potential of this method for quantitative studies in the genetics of hybrid plant cells.

3.4 Protoplast Fusion and the Bypassing of Sexual Barriers

The analysis of the data available on plant hybridization by protoplast fusion provides good evidence that this method allows the crossing of plant forms which can not be crossed sexually because of their inability to undergo normal morphogenesis (gamatogenesis). This was first demonstrated by the experiments on parasexual hybridization which resulted in the direct hybridization of haploid tobacco plants achieved by fusion of haploid protoplasts derived from them (Melchers and Labib 1974). Sexual crosses between haploid plants are ruled out because of their complete sterility. Alternative, classical methods for hybridization require the production of diploid plants from such haploids before hybridization. Similar successful experiments have been performed by other investigators using haploid somatic cells as one of the parental types for parasexual hybridization (Gleba and Berlin 1979). It has also been possible to perform parasexual hybridization with the protoplasts of cultured cells defective for morphogenesis as one or both of the parental forms (Maliga et al. 1977; Grimsley et al. 1977 a; Kao 1977; Schieder 1978 a; Gleba and Hoffmann 1978; Glimelius et al. 1978 a; Gleba and Berlin 1979), as well as with protoplasts from conditional lethals (i.e., cells incapable of living outside the test tube) (Gleba et al. 1975; Schieder 1977 a, 1978 a; Dudits et al. 1977; Power et al. 1977).

3.5 Analysis of the Parasexual Hybridization Process and the Genetic Composition of Plant Forms Arising from Protoplast Fusion

3.5.1 General Guidelines

The proper evaluation and interpretation of the results obtained from plant hybridization by protoplast fusion requires the definition of criteria by which to judge them. From a geneticist's point of view, the analysis is complicated by the fact that the experimental system contains large populations (10^5 to 10^7) of parental cells. The desired genetic events take place with a relatively high frequency due to the induction of protoplast fusion and are subsequently revealed by differential screening methods. Nonetheless, at the same time other undesirable genetic changes, such as polyploidy, chromosome rearrangements, and other mutations are also taking place and give rise to cells or plant forms that are phenotypically indistinguishable from the hybrids. These changes may not be rejected by the screening methods employed and thus introduce an amount of uncertainty into the evaluation and interpretation of the events under investigation. An additional source of error in the interpretation of results is the production of chimaeric plant forms (see Chaleff and Carlson 1974) by aggregation of the parental cells upon morphogenesis. Therefore, it is extremely important to perform experiments in such a way as to avoid possible misinterpretations and to provide means to distinguish true hybrids from other cells or plants, which are phenotypically identical to the true hybrids, and/or from chimaeric callus tissues or plants.

The problem of identifying mutations is extremely important in experiments on intraspecific hybridization and is complicated by the fact that the frequency of reversion for most mutations is not known. This in turn makes the statistical evaluation of genetic events in the cultured protoplasts after fusion even more complicated (see Sect. 3.2). Moreover, it is both theoretically and practically possible to produce a broad spectrum of genetically different hybrids via parasexual hybridization (see Chap. 4). Thus, most plant forms selected by the various screening methods must be subjected to additional analysis to verify their hybrid origin and to study their genetic constitution in more depth.

The design of a parasexual hybridization experiment plays a deciding role in the subsequent analysis of the results. Therefore, the following list of rules or guidelines that we try to follow in our experiments may prove useful.

1. Use only well-studied, simple Mendelian or extranuclear characters and try to avoid the use or polygenic or morphological characters as much as possible. Use phenotypically distinct mutations and try to simultaneously involve as many marker characters as possible.
2. Try to show that the forms obtained after hybridization are really heterozygous for the marker genes or characters. This normally means that the study of sexual progeny is stipulated.
3. Evaluate the possible role of the mutational origin of phenotypically novel forms by measuring the mutation rate in control experiments.
4. Explore the possibility that the regenerated plants are chimaeras of the parental cells. True hybrid plants usually pass on both parental characters in the sex-

ual process, so that their progeny retain the heterozygosity. Alternatively, the purportedly hybrid cells may be simply cloned.

5. Keep in mind not only the genetic variability induced by selective agents, but also that caused by the isolation of the protoplasts and the processes of culturing and unorganized cell growth (i.e., polyploidy, aneuploidy, and mutations).
6. Take into consideration the potential genetic diversity of forms arising during hybridization and subsequent propagation of hybrid cells or heterokaryons. Heterozygosity for nuclear genes is not always accompanied by heterozygosity for extranuclear genes and vice versa, as a result of segregation of plasma genes during cytokinesis and because of nonfusion of nuclei and their segregation. These problems are discussed in Chap. 4.

Several studies have been published in which insufficient attention has been paid to these basic guidelines and therefore have yielded equivocal results. For example, Power et al. (1975) carried out the hybridization of mesophyll protoplasts from *Petunia* with a hormone independent strain of callus protoplasts from *Parthenocissus*. The protoplasts were fused with sodium nitrate and plated on B medium (Nagata and Takebe 1971), in which the cells of *Parthenocissus* are incapable of dividing. After several weeks of culture, the cells were transferred onto B medium without exogenous phytohormones, on which the colonies regenerated from *Petunia* protoplasts can not grow. Three different callus lines were selected in this way and were subjected to chromosome and isoenzyme analysis. The study of chromosome size and morphology, both of which differ significantly in these species, revealed only the *Parthenocissus* type of chromosomes. The analysis of peroxidase isoenzymes demonstrated the presence of both *Petunia* and *Parthenocissus* isoenzymes in the callus lines. The progressive loss of *Petunia* isoenzymes was observed during continued culturing. The authors discuss these results in terms of the hybridization and selection of cell hybrids between the two species and explain the absence of *Petunia* chromosomes and the subsequent disappearance of the *Petunia* isoenzymes in the various lines by a selective elimination of *Petunia* genetic material, similar to the chromosome elimination of one of the parental types observed in animal somatic cell hybrids. However, another explanation is equally plausible: chimaeric calli are formed in which the *Parthenocissus* cells survive the first selection stage without dividing and then divide in the medium without hormones. At the same time, the colonies of *Petunia* cells stop growing in the medium without hormones (or grow very slowly as a result of the conditioning of the medium provided by the other cells). The result would be the absence of *Petunia* chromosomes in the metaphase plates of the chimaeric rather than hybrid callus, in spite of the temporary presence of the *Petunia* isoenzymes. The authors obviously ought to clone the cells under investigation to exclude this alternative explanation.

3.5.2 Genetic Methods of Analysis

3.5.2.1 Hybrid Crosses

If the screening method employed for the selection of parasexual hybrids is based on the use of recessive mutants, the subsequent self-pollination (sexual crossing) of the hybrids will result in the segregation and the expression of the mutant phenotype in the F_1 progeny. This sort of analysis elegantly confirms the hybrid nature of the plant forms regenerated from fusion products. The first example of this type of investigation on parasexual hybrids may be attributed to Melchers (Melchers and Labib 1974). Hybrid analysis also contributes to the study of the fate of extranuclear genes in parasexual hybrids (see Sect. 4.2) and permits the demonstration of cytoplasmic heterozygosity as well. Hybrid analysis is sometimes complicated by the genetic irregularity of the plants regenerated after fusion as compared to the regularity of seedlings produced via self-pollination. Schieder (1977, 1978 b) has shown the sorting out of chlorophyll deficiency among the androhaploid plants obtained after parasexual hybridization via protoplast fusion and has used this result as proof of the hybrid origin of these plants. This sort of analysis is correct only if the parental albino mutants used are phenotypically distinguishable, since the sorting out of chlorophyll deficient androhaploids is also possible in revertants heterozygous for one of the parental genes for chlorophyll deficiency and is not perforce associated with the hybrid origin of these plants. The segregation in androgenesis can not be used for quantitative analysis of segregation frequencies of definite phenotype classes, in contrast to the segregation in F_1 progeny (Melchers 1972).

3.5.2.2 Cloning

The cloning of individual plant cells may now be practically performed in many cases (see Sect. 2.1). The term "clone" is used here to designate the totality of cells or organism(s) originating from a single cell. The cloning of single cells from an organism of supposed parasexual hybrid origin is a simple and elegant method to exclude the formation of a chimaera of the parental cells as an alternative to true hybridization. Surprisingly, this sort of confirmatory study on parasexual hybrids has not yet been performed. However, some authors have cloned single protoplast fusion products as a method of hybrid selection (Kao 1977; Gleba and Hoffmann 1978; Menczel et al. 1978, 1981; Gleba and Berlin 1979; Gleba et al. 1982 a, b). This use of cloning also permits the exclusion of the formation of chimaeras as an explanation of the experimental results; moreover, as was mentioned in Sect. 2.2, the techniques of single cell culturing automatically reduce the complications associated with the influence of rare genetic events (mutations) in the experimental system. However, as will be discussed in more depth in Chaps. 4 and 5, the cloning of single heterokaryons results in many cases not only in the segregation of the extranuclear genes, but also of the nuclear genes, because of the nonfusion of the nuclei in the heterokaryons. So, in a genetic sense, the cloning of heterokaryons is not equivalent to the cloning of uninuclear cells and the use of this method for screening of hybrids is not a guarantee for the isolation of nuclear hybrids (i.e., hybrids in the common sense of the word). Therefore, a clone of cells derived from a single heterokaryon may be a mixture of distinct cells of both parental types, at least with respect to the nuclear genes. Thus, the cloning of heterokaryons must often be repeated or supplemented by another method of analysis.

3.5.2.3 Cytogenetic Studies

These studies involve the analysis of the chromosome number and chromosome morphology in the hybrids, assuming, of course, that the parental chromosomes are distinguishable. The data on chromosome numbers in plants obtained after intraspecific hybridization may either support or disprove the hybrid nature of their nuclei, but these data are often relatively vague because of the significant chromosome variability that arises as a result of in vitro culturing and regeneration of cultured cells (Sunderland 1973; D'Amato 1978; see Sect. 2.3). However, such an analysis can unequivocally exclude the formation of a hybrid for nuclear genes, as, for example, in the case of cytoplasmic heterozygotes (Belliard et al. 1977a, b; Gleba et al. 1978). The study of chromosome size and morphology is extremely efficient, particularly for hybrids between distantly related species (Kao et al. 1974; Kao 1977; Wetter and Kao 1980; Gosch and Reinert 1976; Constabel et al. 1977; Gleba and Hoffmann 1978, 1979; Krumbiegel and Schieder 1979; Gleba et al. 1982a, b). No other method of analysis gives comparably reliable and convincing proof of the hybrid nature of these cells. During cell division, the chromosomes of certain hybrids undergo extensive rearrangements which complicates the cytogenetic analysis (see Chap. 5). The chromosomes of long-term in vitro cell cultures are sometimes highly modified as compared to the chromosomes of the initial material. These facts must also be taken into consideration in the cytogenetic analysis of hybrids obtained by protoplast fusion of cultured cells.

Attempts to use the differential staining of *nuclei* for studies of the hybridization process by fusion of protoplasts of different species have been described (Constabel et al. 1975a; Kao 1977), but methods of differential *chromosome* staining for analysis of hybrid cells and plants obtained by parasexual techniques have still not been reported. Analogous approaches have proved to be very informative in studies of animal somatic cell hybrids (Bobrow et al. 1972; Bobrow and Cross 1974; Friend et al. 1976).

3.5.2.4 Cytophotometric Measurement of Nuclear DNA Content

These investigations have been performed for callus lines of clones and regenerated plants from the intergeneric hybrids *Solanum tuberosum* (×) *Nicotiana tabacum* (Skarzhynskaya et al. 1982), *Atropa belladonna* (×) *Nicotiana tabacum* (Gleba et al. 1982b), and *Atropa belladonna* (×) *Nicotiana chinensis* (Gleba et al. 1982c). The adsorption of Feulgen stained nuclei was measured at two different wavelengths and thus the DNA content determined in these studies; however, other methods for the determination of nuclear DNA content via cytofluorimetry have been described. Moreover, during the last few years, flow cytofluorimetry has become more exact, rapid, and automated and, as a result, is finding widespread application, especially in studies with animal cells. It even appears that flow fluorimetry of suspensions of isolated plant nuclei is possible. Cytophotometry is relatively time-consuming and imprecise and hence its routine application for the analysis of the hybrid nature of parasexual hybrids is rather limited. However, when properly automated, these methods become valuable for the quantitative characterization of the relative DNA content in large cell populations and permit the detection of increases in the heterogeneity of the cell population, etc. This, in turn, may also enable to detect the selection of cells of some definite ploidy as a result of a certain experimental manipulation or event (for example, upon transition to differentiated growth).

3.5.3 Biochemical Analysis

A number of investigations have employed biochemical studies of the cells and plants obtained by parasexual hybridization to demonstrate their hybrid nature. Such analyses are especially appropriate in the case of interspecific hybridization, where major biochemical differences in the parental forms might be expected. The molecules, primarily macromolecules, that differ in their physical and/or biochemical characteristics in the parental forms are used as markers for the components of the parental cells. The analysis consists of the demonstration of the molecular markers characteristic for both parents in the hybrids under investigation. This sort of analysis permits the exclusion of the mutational origin of phenotypically novel forms observed after hybridization, but can not exclude the possible existence of chimaeras rather than true hybrids (see, however, Sect. 3.5.3.2). The biochemical methods that have been used for the study of the nature of parasexual hybrids are given in the following parts of this section.

3.5.3.1 Multiple Molecular Forms of Enzymes

The separation of isoenzymes by electrophoresis in polyacrylamide gels and subsequent staining or localization of proteins via specific enzymatic activity is the simplest and, in many cases, the most efficient method for biochemical analysis of parasexual hybrids. The comparison of the spectra of isoenzymes in the parental species and in the hybrids enables general conclusions to be derived concerning the hybrid nature of the plants obtained from fused protoplasts. Unfortunately, this information is sometimes too general, since the genetic control of isoenzyme forms of most plant enzymes is still far from being sufficiently elucidated. However, some enzymes, such as lactate dehydrogenase, malate dehydrogenase, and some others are the exceptions in this respect (Scandalios 1974). Another complication arising during this analysis results from the pronounced physiological variability of different isoenzymes and from the dependence of their expression on the differentiation of the cells involved (see Chap. 5). In most cases, this variability can not be efficiently excluded by standardization of the physiological parameters of culturing and therefore one must be aware of it while interpreting the data. A number of different enzymes have been used in isoenzyme analysis of parasexual hybrids, including peroxidase (Carlson et al. 1972; Power et al. 1975, 1976a; Dudits et al. 1977; Gleba and Hoffmann 1978), malate dehydrogenase (Wetter and Kao 1976), lactate dehydrogenase (Wetter and Kao 1976; Medgyesy et al. 1980), alcohol dehydrogenase (Kao and Wetter 1977; Wetter 1977; Maliga et al. 1977; Gleba and Hoffmann 1978), esterase (Wetter and Kao 1976; Maliga et al. 1977, 1978; Menczel et al. 1978; Gleba and Hoffmann 1978, 1979; Medgyesy et al. 1980;

Sidorov et al. 1981 a; Evans et al. 1981; Douglas et al. 1981 c; Maliga et al. 1982; Gleba et al. 1982 b, c), aminopeptidase (Wetter and Kao 1978; Evans et al. 1980 a), phosphodiesterase (Maliga et al. 1978), aminotransferase (Wetter 1977; Wetter and Kao 1980; Evans et al. 1980 a; Douglas et al. 1981 c), and amylase (Lönnendanker and Schieder 1980; Gleba et al. 1982 a). The isoenzymes of peroxidase have been often employed in these studies even though it is not the enzyme of choice due to the presence of a large number of physiologically labile isoenzyme forms (Scandalios 1974). Some parental isoenzyme forms are absent in the hybrids as a result of the repression of certain isoenzymes. Additionally, isoenzymes with new mobilities are also observed in hybrid cells, probably as a result of the formation of hybrid enzymes (Maliga et al. 1977; Gleba and Hoffmann 1978). Still other new isoenzymes must result from the derepression of genes which are not expressed in the parental cells or from posttranslational modification of the enzyme in the hybrid cell (Wetter 1977; Gleba and Hoffmann 1978). The existence of hybrid molecules of isoenzymes may be used as proof of the hybrid rather than chimaeric origin of the plants analyzed, at least in the case of well-studied enzymes like lactate dehydrogenase (Scandalios 1974).

3.5.3.2 Analysis of Fraction I Protein

Wildman and Bonner (1947) found a high molecular weight protein in very high concentration in the soluble proteins of spinach leaves and named this the Fraction I protein. Due to the investigations of Wildman's research group at UCLA and of many other laboratories, the Fraction I protein has become one of the best studied of all plant proteins (see the reviews of Akayana 1970; Kawashima and Wildman 1970; Kung 1976; Bottomley 1980). It constitutues about 40% of the total leaf soluble protein, is localized in the stroma of the chloroplasts and is therefore relatively easily prepared in large amounts. The Fraction I protein from tobacco may be isolated by crystallization (Chan et al. 1972), although this technique has not proven successful for the protein from other species. The Fraction I protein has been shown to have both ribulose-1,5-bisphosphate carboxylase and oxygenase activity and is responsible for catalyzing the first step in photosynthesis as well as in photorespiration. The Fraction I proteins from a variety of species show many similarities: the holoenzyme has a sedimentation coefficient of about 18S which corresponds to a molecular weight of approximately 560 000 and is composed of 16 subunits, 8 large subunits each with a molecular weight of approximately 54 000–56 000, and 8 small subunits each with a molecular weight of approximately 12 000–14 000.

The large and small subunits of this protein may each be separated into two or more reproducible bands or variants that display different isoelectric points after isoelectric focusing in polyacrylamide gels. Different plant species show different patterns for both types of subunits and may thus be distinguished in mixing or hybridization experiments. The analysis of inheritance of this protein after interspecific hybridization in *Nicotiana* has revealed that the isoelectric variants of the small subunit are inherited biparentally and in accordance with Mendelian laws and, therefore, are coded for by nuclear genes. On the other hand, the isoelectric variants of the large subunit are inherited uniparentally and maternally and, therefore, are coded for by plastome genes (Kawashima and Wildman 1970; Chan and Wildman 1972; Kung 1976, 1977; Jensen and Bahr 1977; von Wettstein et al. 1978). Thus, the Fraction I protein is an unique marker for the analysis of interspecific parasexual hybridization, since the data on its isoelectric variants provide information concerning the fate of both nuclear and extranuclear (plastome) genes in the hybridization process. However, it should also be mentioned that additional bands or variants (believed to be artifacts; Gianazza and Righetti 1980; O'Connell and Brady 1981; Leible et al. 1982) often arise during isoelectric focusing (the usual method of analysis of the subunits of the Fraction I protein) so that the analysis is greatly complicated, especially for plant forms that are heterozygous for the genes coding for the large subunit of Fraction I (Chan et al. 1976). Nonetheless, although some experiments have clearly shown that certain aspects of the pattern obtained via isoelectric focusing of the Fraction I protein are methodological artifacts, they have at the same time also provided support for the existence of unequivocal and reproducible differences in the isoelectric variants of this protein from different species.

Since the analysis of the Fraction I protein is performed on a highly purified protein and since the techniques of its analysis are thoroughly established, this method represents a profound, trustworthy, and feasible approach to the biochemical analysis of parasexual hybridization. The Fraction I protein from a number of interspecific parasexual hybrids has been analyzed in order to confirm their hybrid origin (Kung et al. 1975; Chan et al. 1977; Gleba et al. 1978a; Melchers et al. 1978; Zelcer et al. 1978; Evans et al. 1980a; Poulsen et al. 1980; Iwai et al. 1980, 1981; Komarnitsky and Gleba 1981; Douglas et al. 1981c; Kumar et al. 1981a; Glimelius et al. 1981). The analysis of Fraction I protein often enables the alternative explanation of the formation of chimaeras rather than true hybrids to be excluded, since the segregation of extranuclear genes usually results in the elimination of one or the other of the parental plastomes prior to the analysis and, consequently, of one of the large subunit types, while at the same time, the hybrid nature of the nuclear genes as shown by the small subunits is retained. Such a result argues against the chimaeric origin of such a plant. The analysis of the Fraction I protein has also been successfully applied to the study of extranuclear genes in parasexual hybridization (see Sect. 4.2).

3.5.3.3 Restriction Endonuclease Analysis of Chloroplast and Mitochondrial DNA

Intact circular DNA molecules may be isolated from plant chloroplasts and mitochondria (Kolodner and Tewari 1972; Herrmann et al. 1975) and can be analyzed by digesting them with purified restriction endonucleases that cleave the DNA molecules at specific base sequences, thereby producing a few or many discrete fragments of lower molecular weight that may be separated and characterized via agarose or polyacrylamide gel electrophoresis (Amerkhanova et al. 1976; Bedbrook and Bogorad 1976; Vedel et al. 1976; Atchison et al. 1976; Levings and Pring 1976). Several hundred different restriction endonucleases have been described (Roberts 1982) and over 50 are now commercially available. Each restriction endonuclease exhibits a preference for a particular, usually relatively short, base sequence and it cleaves the DNA within or adjacent to that sequence. Since the extranuclear genomes of different species are quite different in size and sequence, the selection of and digestion by the proper restriction endonucleases enables the DNA fragments obtained to be used as species-specific markers for the organelle genomes during the parasexual hybridization process. This, in combination with markers for nuclear genes, provides conclusive evidence for the existence of hybrids.

Due to the tremendously rapid development of the field of molecular biology, this method of hybrid analysis is now beginning to become quite popular and threatens to take over the role formerly played by the Fraction I protein for several excellent reasons. First, the use of isoelectric focusing is avoided. Second, the restriction endonuclease analysis of plastid DNA permits the detection of less than 0.1% of the parental types of DNA, whereas the analysis of the Fraction I protein would not detect less than 10% contamination by one of the parental types (Schiller et al. 1982). A potential drawback to this method is that most preparations of chloroplast DNA are contaminated with mitochondrial DNA and vice versa, due to problems in the fractionation of plant cell organelles (see Sect. 2.2.3). However, this problem can be overcome by the careful selection of the restriction endonucleases employed, so that with the proper nuclease one of the DNA types is completely digested while the other is not or only partially digested. It is interesting to note that the restriction endonuclease analysis of parasexual hybrids has supported the conclusions derived from the analysis of Fraction I protein (Schiller et al. 1982).

3.5.3.4 Molecular Hybridization of Nucleic Acids

This method has proven suitable for studies on evolution and taxonomic relations between different organisms and, in particular, for higher plants (Bendich and McCarthy 1970; Flavell et al. 1977). It has been applied to the analysis of parasexual hybrids only once (Dudits et al. 1979). In this method, the extent of molecular reassociation upon hybridization is taken as a measure of the systematic and phylogenetic relationship of the species from which the nucleic acid was isolated. DNA with DNA as well as DNA with RNA hybridizations are possible. The analysis of the degree of hybridization of specific DNA fractions or fragments, obtained by preliminary fractionation of total nucleic acid preparations, has significantly improved this technique. Nucleic acid hybridization may prove to be extremely promising for studies of the nature and origin of parasexual hybrids, especially in hybrids of distantly related species. There is good reason to believe that measurements of the stringency of nucleic acid hybridization can be used for a more definitive analysis of asymmetric hybrids (see Chap. 5), especially in those cases where analysis of chromosome morphology fails to reveal the presence of chromosomal material of one of the parental species in the hybrid cells.

3.5.3.5 Analysis of Low Molecular Weight Substances

The work of Ninnemann and Jüttner (1981) is the only study utilizing low molecular weight compounds for the analysis of somatic cell hybrids. They separated volatile, low molecular weight substances via gas chromatography from the parental forms and the somatic hybrids potato (×) tomato (Melchers et al. 1978). Since the compounds studied are widely spread in higher plants, the observed "species-specificity" of the different compounds must be regarded as their occurrence in quantities sufficient for detection, i.e., the analysis was quantitative rather than qualitative, as pointed out by the authors themselves. The analysis of the hybrids has demonstrated the presence of species-specific molecules of both parents, as well as the appearance of a novel terpenoid, which has not been found in either of the parental cells. The analysis of low molecular weight substances may be of greater theoretical interest for the study of secondary compounds which are strictly species-specific. Of obvious interest would be the study of the result of the "hybridization" of unique parental biosynthetic pathways which might potentially result in the synthesis of novel types of "hybrid" molecules. Further development of this basic method, along with a critical selection of the compounds to be analyzed, may enable this method to become of practical interest in the analysis of parasexual hybrids.

4 Transmission Genetics of Parasexual Hybridization in Closely Related Crosses

4.1 General Considerations

4.1.1 Perspectives and Philosophy of Analysis of Transmission Genetics

The results of parasexual hybridization by protoplast fusion were initially interpreted and discussed from a rather unexpected point of view. Most investigators have used parasexual hybridization for the purpose of crossing the noncrossable forms and thereby obtaining hybrids with novel gene sets. In this respect, the more significant results were expected to come from experiments on hybridization of distantly related species. Astonishingly, the earliest experiments of this type (Carlson et al. 1972; Melchers and Labib 1974; Power et al. 1976a) were performed on closely related species and were used to demonstrate the identity of the parasexual hybrids with the sexual hybrids obtained by crossing the same parental forms, rather than to investigate the genetic novelty or true hybrid nature of these products. Thus, the most interesting hybridizations between distantly related species were not initially performed and the most interesting aspects of the results of parasexual hybridization of closely related species, their genetic novelty, were also not investigated. This trend (Vasil 1976; Bajaj 1977) has continued and is responsible for the fact that there are no reviews on the subject of transmission genetics in parasexual hybridization, in spite of the existence of a number of experiments that provide more than enough data to discuss this problem. Several dozen general reviews that often repeat each other have been published an protoplast fusion and parasexual hybridization.

Data was presented in Chap. 3 to illustrate that it is really possible to produce cell and plant hybrids using the parasexual techniques of protoplast fusion with subsequent regeneration and selection of cell colonies or entire plants. The preceding discussion has been concentrated on the question of how valid the conclusions are concerning the hybrid nature of these products and to what extent the explanation of the results in terms of genetic events other than hybridization may be excluded. The possibility that this method allows the production of hybrids that are genetically different from those obtained sexually has been only partially discussed. Therefore, this chapter is devoted to the most important aspects and implications of parasexual hybridization, namely, the genetic novelty of parasexual hybrids. As was noted in the introduction, the usual sexual crossing is a strictly regulated process whose mechanisms have evolved to allow not only sufficient genetic recombination to ensure adaptability and survival of the species, but also to control or maintain this recombination within necessary limits to provide a certain regular shuffle of the genetic information. Parasexual hybridization by protoplast fusion is a completely artificial method and bypasses the natural restrictions inherent in sexual crosses. The genetic makeup of progeny of parasexual hybridization products is determined by several factors, including: (1) the choice of parental cells; (2) the method of hybridization; (3) the general biological potentials of the hybrid cells; (4) the influence of the in vitro culturing and other manipulations; and, perhaps most importantly, (5) the human decision of how to combine these factors. Thus, it is readily apparent that parasexual hybridization is not a strictly determined process, but is rather a function of the totality of several relatively independent and asynchronous events, such as cell fusion, fusion of nuclei, segregation or fusion of other cell organelles, selective destruction of some cell orgenelles. Since these events all affect the fate of genetic material, our main objective should be the elucidation of the regulation of these processes. This chapter presents an analysis of the fate of genetic information in the parasexual hybridization systems that are currently in use.

The term "transmission genetics" is included in the title of this chapter to underscore the fact that this inheritance is only partially a function of the internal characteristics of these biological systems and that it may be greatly affected by the particular experimental design. Likewise, Mendelian laws reflect not only the organization of genetic information within the cell, but also the mechanisms of the sexual process used to perform hybrid analysis.

The following events apparently play an important role in determining the fate of genetic information in the parasexual hybridization process:

1. fusion of the protoplasts;
2. fusion of the cell nuclei;
3. segregation of unfused nuclei during subsequent divisions of fusion products;
4. recombination of genetic information within the fused cells;
5. elimination of chromosomes of one or the other parent;
6. segregation or sorting out of organelles during divisions of fused cells;
7. selective destruction of organelles from one or the other parent;
8. selective propagation of organelles from one or the other parent;
9. genetic instability and mutation processes induced by hybridization;
10. genetic variability induced by experimental conditions;
11. changes in epigenetic determinants upon hybridization.

4.1.2 Initial Experiments on Transmission Genetics of Parasexual Hybrids

The first experiments providing insight into and discussion of the genetic novelty of parasexual hybrids were independently performed by the research groups of Melchers and Gleba. Although most of the conclusions and inferences in these studies were not rigorously proven, they nonetheless served to establish the main trends of subsequent investigations on parasexual hybridization.

Melchers' research group investigated the hybridization of haploid tobacco protoplasts isolated from the mesophyll of two chlorophyll deficient, light sensitive mutants of the chlorina type (Melchers and Labib 1974; Melchers and Sacristan 1977). The hybrids were selected as regenerants of wild type since the mutations used were recessive and nonallelic. At least 20 parasexual hybrids of independent origin have been found in 4 independent experiments of this type (the significance of the hybridization frequency is discussed in Sect. 3.2). A total of 108 hybrid plants have been produced from these hybrids. Chromosome analysis demonstrated that the majority of the plants regenerated were diploid, as would be expected from the fusion of haploid parental cells. However, a total of 10 polyploid, predominately triploid as well as tetraploid, plants were also found among those regenerated. Since the mesophyll cells in haploid tobacco plants are strictly homoploid and haploid (Gleba et al. 1978 b) and since the period of unorganized cell growth which often results in chromosome changes was minimal, it seems logical to explain, as Melchers did, the triploidy of the hybrid plants by the fusion of three (not two) parental cells followed by fusion of their nuclei. Thus, this work was the first demonstration of an important aspect in the transmission genetics of parasexual hybridization by protoplast fusion: the hybrid may originate from the fusion of more than two parental cells and, therefore, may represent the genetic product of at least three parental cells. Analogous experiments are described in detail in Sect. 4.1.3.

The fate of nuclear and extranuclear parental determinants in parasexual hybridization was observed in our experiments (Gleba 1974; Gleba et al. 1975; Burgutin et al. 1977). A mutation of plastome chlorophyll deficiency and a semidominant genome mutation of aurea type chlorophyll deficiency were employed in these studies, since the homozygous and heterozygous states of both of these mutations are phenotypically different.

The plastome variegated mutant of *Nicotiana tabacum* var. Samsun, used as one of the parental forms, arose spontaneously and has been shown to be inherited strictly uniparentally and maternally. Electron microscopy of the variegated tissues of this mutant reveals the presence of heteroplastidic cells containing both normal plastids as well as chlorophyll deficient plastids. This mutation is expressed phenotypically by chlorophyll deficiency. This character is extremely stable and no reverse mutations have been observed (Wildman et al. 1973). Studies on lines of white, aseptically grown plants of this mutant obtained in 1974 and cultured for 5 years have shown that the rate of reverse mutation is practically zero.

Biochemical analysis of the tissues of this mutant revealed that the function of the chlorophyll-protein complex I was affected. Moreover, it has been shown that this mutation is correlated with changes in the physical and chemical properties of the chloroplast DNA. A region of homology comprising about 500–1 000 base pairs was found to be absent in the chloroplast DNA of the mutant (Wong-Staal and Wildman 1973; Wildman et al. 1973). These data suggest that a deletion is responsible for this mutation and thus provide a good explanation for the absence of reverse mutations. It would be of interest to verify these conclusions using a more sensitive method of analysis, such as restriction endonuclease analysis of the chloroplast DNA (see Sect. 3.5.3.3).

For experimental purposes, callus tissues were produced from the variegated areas of the leaves. Plants were then regenerated from these callus tissues (Gleba 1974). Two different types of plants were produced: (1) periclinal chimaeras with a stable distribution of mutant and normal plastids in different embryonic layers and (2) plants possessing only defective plastids in all of their cells. Populations of protoplasts containing only defective plastids were obtained from the marginal areas of leaves of periclinal chimaeras with the formula of embryonic layers L_1:L_2:L_3 = defective : defective : normal (for the first three experiments) as well as from leaves of aseptically grown plants of albino type (for the fourth experiment).

Protoplasts of *Nicotiana tabacum* John Williams Broadleaf var., homozygous for the Su (sulphur) gene, were used as the second parental form in these hybridization experiments. This genome mutant of aurea type also arose spontaneously. This mutation is semidominant: the heterozygotes Su/su are yellow-green and viable, whereas seedlings carrying two mutant alleles (Su/Su) are yellow and die within 14 days of germination in the absence of exogenous sugar. Direct as well as reverse mutations are frequent and somatic crossing over also seems to take place in heterozygotes. The leaves of aseptically grown homozygous mutant plants were used as the source of protoplasts for these experiments.

In four independent hybridization experiments, a mixed suspension containing 1 to 3×10^6 plastome mutant protoplasts and 2 to 5×10^5 genome mutant protoplasts was treated with polyethylene glycol to induce fusion (after Kao and Michayluk 1974) and then cultured for plant regeneration. A similar untreated suspension was cultured as a control.

Although the mutations belong to different genetic systems, they are both expressed as defects in photosynthesis. Consequently, a genetic complementation must occur in hybrid cells produced via protoplast fusion. Therefore, plants regenerated from hybrid cells must be capable of photosynthesis. Use of the plastome mutation in the one parental form and the genome semidominant mutation in the other has permitted the identification of at least two classes of parasexual hybridization. In the first class, the plastids are inherited either uniparentally, resulting in the production of nonvariegated forms capable of photosynthesis, or biparentally, in which case at least some of the progeny should be variegated. In the second class of hybridization, the fusion of the protoplasts is followed by fusion of their nuclei. In this case, the regenerated plants should be true hybrids containing the complete set of parental genes and should therefore be yellow-green and variegated. If for any reason the nuclei do not fuse in the hybrid cells, the nuclei may segregate to different progeny cells. In this case, the integration of the cytoplasm and the mixing of the organelles will take place, resulting in the formation of cybrids, so that any regenerants selected will be green due to the absence of the Su gene and probably variegated as well. A mixture of parental protoplasts which were not treated with polyethylene glycol was also studied as a control to measure the rate of reverse mutations. The complication of the evaluation and interpretation of the data due to the aneuploidy of some of the plants obtained was also taken into consideration.

Plants capable of photosynthesis were produced in all four experiments where the mixture of protoplasts was treated with polyethylene glycol. Yellow-green as well as green plants were found and the majority of them were variegated (Table 5). Cytological analysis has demonstrated that all of the yellow-green plants were polyploid, either nearly tetraploid or, in one case, hexaploid. The majority of the green plants were triploid or aneuploid. Some of the green, variegated plants were grown to the flowering stage and displayed a reduced fertility. Hybrid analysis of these plants demonstrated that the variegation is inherited maternally (Table 6).

Table 5. Plant types regenerated from parasexual hybridization of *Nicotiana tabacum* protoplasts[a]

Experiment	Genotype of the hybrid plants	Phenotype of the hybrid plants	No. of plants
1	P^+/P^-, Su/su	Variegated, yellow-green	2
	P^+/P^-, su/su	Variegated, green	3
	P^+/P^+, Su/su	Yellow-green	9
2	P^+/P^-, Su/su	Variegated, yellow-green	1
	P^+/P^-, su/su	Variegated, green	3
	P^+/P^+, Su/su	Yellow-green	0
3	P^+/P^-, Su/su	Variegated, yellow-green	4
	P^+/P^-, su/su	Variegated, green	7
	P^+/P^+, Su/su	Yellow-green	1
4[b]	P^+/P^-, Su/su	Variegated, yellow-green	2
	P^+/P^-, su/su	Variegated, green	2
	P^+/P^+, Su/su	Yellow-green	0

[a] A suspension containing 1 to 3×10^6 protoplasts of a plastome mutant (P^-) and 2 to 5×10^5 protoplasts of a nuclear, semidominant mutant (Su) for chlorophyll deficiency was treated with PEG and the resulting hybrids regenerated. No plants capable of photosynthesis were regenerated from the control, nonfused mixture of parental protoplasts

[b] In this experiment, albino plants (a homoplastid line, L_1:L_2:L_3 = defective:defective:defective) were used as the source of protoplasts with the plastome mutant

Table 6. Analysis of F_1 progeny of variegated tobacco plants obtained after protoplast fusion of two chlorophyll deficient mutants (see Table 5)

Plant designation and crossing type	No. of seedlings with phenotypes			Total no. of seedlings
	Green	Variegated	White	
H1b, self-pollination	4	8	6	18
H2a, self-pollination	4	0	2	6
♀H1c × ♂ *Nicotiana tabacum*, var. Samsun	30	12	5	47
♀H2a × ♂ *Nicotiana tabacum*, var. Samsun	28	0	2	30
♀H2b × ♂ *Nicotiana tabacum*, var. Samsun	23	0	1	27
♀H4b × ♂ *Nicotiana tabacum*, var. Samsun	49	19	27	95

No plants capable of photosynthesis were found among the regenerants obtained from the control, nonfused mixture of parental protoplasts.

We concluded that these plants capable of photosynthesis are parasexual hybrids originating as a consequence of protoplast fusion for the following reasons. (1) Both alleles of two different genes, located initially in the different parental protoplasts, are present in the yellow-green, variegated regenerants. Two independent mutations would be required to explain this phenomena if parasexual hy-

Fig. 14. A Segregation of the Su gene following self-fertilization of a *Nicotiana nesophila* (×) *N. tabacum* Su/Su somatic hybrid (Evans et al. 1981). **B** Leaves of *Nicotiana tabacum*, *N. tabacum* (×) *N. glauca* somatic hybrid, and *N. glauca* (from left to right). (Evans et al. 1980)

Fig. 15. A A somatic hybrid plant of *Nicotiana nesophila* (×) *N. tabacum* Su/Su (*top*) and plants of the parental species, (*left*) *N. nesophila* (2n=48) and (*right*) *N. tabacum* (2n=48). **B** Flowers of *N. nesophila*, *N. nesophila* (×) *N. tabacum* Su/Su somatic hybrid and *N. tabacum* (*from left to right*). (Evans et al. 1981)

bridization did not occur. (2) The presence of parental cells heterozygous for plastids with the plastome mutation can be excluded in all four experiments. (3) Plants capable of photosynthesis were also found in an experiment where a pure line of plastome mutant plants was used as one of the parental forms (experiment four). (4) The variegation character was transferred to progeny obtained by sexual crosses of the hybrid plants. This is impossible if the plants in question were derived from chimaeras. (5) In all cases, no plants capable of photosynthesis were found among the regenerants in the control experiments.

These results indicate that: (1) both parental plastomes are transferred to the progeny via protoplast fusion; (2) in some cases, protoplast fusion is accompanied by nuclear fusion, producing a true hybrid cell; and (3) in some cases, nuclear fusion does not accompany fusion of the protoplasts and/or genomal or chromosomal segregation takes place during mitosis of the fusion products.

Thus, the genetic variability of the hybrid somatic plant cells produced by the hybridization process and the culture conditions permits the synthesis of totally new genetic combinations that were not previously obtainable. The data support the hypothesis that more than two parental cells may be fused in parasexual hybridization (this is confirmed by the production of the hexaploid hybrid plant). This conclusion is also in good agreement with that of Melchers (Melchers and Labib 1974).

A number of theoretically possible alternative explanations of the results of our experiments may be ruled out by careful consideration of the data. The sole alternative to the biparental inheritance hypothesis is that the mutation frequency of extranuclear genes in the parental cells was very high, resulting in de novo generation of defective plastids. This was not observed in the control experiments. The fact that the green plants obtained were aneuploid rules out the possibility of nuclear segregation in these heterokaryons. Likewise, if plants capable of photosynthesis were regenerated from cells with initially hybrid nuclei that later underwent chromosomal rather than genomal segregation, at least some of these plants should contain yellow-green areas carrying the Su gene. This was not observed. We later learned that the analysis of the hybrid plants could also include leaf shape as a polygenic nuclear encoded character (van der Veen 1957). Tobacco plants of Samsun var. have petiolated leaves and plants of John Williams Broadleaf var. have sessile leaves. Subsequent analysis of the green variegated plants revealed that all of these plants have petiolated leaves.

In other experiments, the segregation of the Su gene and leaf shape in various *Nicotiana* somatic hybrids has been studied (Evans et al. 1980, 1981) (Fig. 14 and 15).

A more detailed discussion of studies on the fate of nuclear and extranuclear genes in parasexual hybridization in closely related crosses is presented in the following sections.

4.1.3 Hybridization of More Than Two Parental Cells

The standard techniques of protoplast fusion induction, due to their statistical approach, result in the formation of fusion products consisting of different parental cell numbers and combinations. In contrast to sexual hybridization, which involves the fusion of only two cells from two parents, parasexual hybridization may theoretically involve the fusion of many cells from many parents. The problem of evaluating the number of parental cells that contributed genetically or, more broadly, have given rise to the progeny might be solved by hybridizing three genetically different cell types and producing "tribrids", etc. However, there have been no reports of such experiments.

In addition to the use of genetic markers, there are other methods to determine the number of parental cells involved in fusion and hybrid origination. One of the most common ways is to determine the chromosome number of parasexual hybrids to detect those products with a ploidy level that could only arise after fusion of more than two cells. For example, it is clear that the production of triploids by the fusion of initially haploid protoplasts must involve the fusion of more than two cells. This reasoning was used in the previously discussed experiments to support the conclusion that some parasexual hybrids are the products of the fusion of at least three parental cells (see Sect. 4.1.2; Melchers and Labib 1974; Gleba 1974; Gleba et al. 1975; Melchers and Sacristan 1977). In the experiments of Melchers' group, 18 of 63 green colonies gave rise to triploid or nearly triploid plants, even when the stage of unorganized growth was reduced to the relatively short period of not more than 2 months. In our work, one hexaploid plant was found among the total of nine plants capable of photosynthesis after hybridization of diploid tobacco cells.

Schieder (1977a) studied the ploidy of intraspecific somatic hybrids of *Datura innoxia* obtained by fusion of mesophyll protoplasts of diploid chlorophyll deficient mutants. Four tetraploids, eight hexaploids, three octaploids, and four aneuploids were found among the 20 regenerants from independent hybrid cell colonies capable of photosynthesis. Schieder has proposed that the proportion of aneuploids is really much higher, since the aneuploid cells of *D. innoxia* cultured in vitro grow relatively slowly and are therefore easily overlooked. A number of slow growing clones were observed, but not evaluated in this experiment. In this case, the existence of hexaploid regenerants may also be interpreted as evidence of the origin of these forms from multiple fusion events, although, unfortunately, it is not known whether the leaf mesophyll cells of *D. innoxia* are diploid or mixoploid.

Similar results were obtained in investigations of interspecific combinations produced by parasexual hybridization. Smith et al. (1976) studied 23 hybrid plants of *Nicotiana glauca* (×) *Nicotiana langsdorffii*. Practically all of these hybrids contained approximately 60 (2n) chromosomes. Diploid mesophyll protoplasts of *N. glauca* ($2n=24$) and *N. langsdorffii* ($2n=18$) had been utilized for this hybridization. Therefore, the significant deviation of the chromosome numbers from the theoretically expected amphidiploid chromosome number ($2n=42$) was interpreted as evidence for the origination of the hybrids by fusion of three rather than two parental cells. If this conclusion was true, the hybrids should certain 60 chromosomes (two sets of *N. langsdorffii* and one set of *N. glauca* chromosomes) or 66 chromosomes (one set of *N. langsdorffii* and two sets of *N. glauca* chromosomes). The authors concluded that the selection for this type of genetic variant took place as a result of the culture conditions employed. However, other explanations are equally plausible: (1) the initial mesophyll cells of one (or both) parental types might be mixoploid (as far as we know, the ploidy of leaf mesophyll cells in these species has not been studied); or (2) the hybrid cells are genetically unstable and their in vitro growth is accompanied by changes in the chromosome numbers (this factor can not be ruled out since the majority of hybrids studied possessed aneuploid chromosome numbers). Additionally, the study of chromosome numbers in hybrids of *N. glauca* (×) *N. langsdorffii* performed by Chupeau et al. (1978), has shown that these cells display an exceptionally high rate of chromosome variability. For a similar reason, the interpretation of the results and the conclusions from the experiments of Schieder (1978 a, 1980 a) on chromosome numbers of *Datura* interspecific hybrids still remain equivocal. In these experiments, regular hexaploid plants were observed among the other regenerants in all cases.

Another way to determine the number of parental cells that fused to form a hybrid is to clone single fusion products and to search for clones with a complex type of segregation. In some cases, the progeny of a single heterokaryon would consist of both true hybrid cells as well as segregants with the nucleus of one of the parents. In intraspecific hybridization, where the species specific elimination of chromosomes is less probable, these results might be interpreted with a high degree of certainity as being indicative of clone origination from a heterokaryon that initially contained several nuclei, two (or more) of which have fused and formed a hybrid nucleus, while at least one nucleus did not fuse and segregated upon subsequent cell division. Of course, in this case, the cells of this clone are the progeny of at least three parental cells in a physical rather than in a (nuclear) genetic sense.

A specific example of the application of this method is found in the work of Gleba and Berlin (1979) and Gleba (1980) (discussed in detail in Sect. 4.3). Tobacco protoplasts from a long-cultured, aneuploid cell line with resistance to 5-methyltryptophan and no capacity for morphogenesis were fused with haploid or diploid mesophyll protoplasts of a light sensitive, chlorophyll deficient mutant. The hybrid clones were obtained by mechanical isolation and individual cloning. From a total of 25 hybrid cell clones, 2 clones were found to give rise to not only hybrid plants, but also plants of parental type, presumably by segregation of unfused parental nuclei from a "complex" heterokaryon.

In the work of Sidorov et al. (1981 a), the mechanical isolation and cloning of the fusion products of a plastome chlorophyll deficient mutant of tobacco and mesophyll protoplasts of *Nicotiana plumbaginifolia* were used as a control for experiments on chloroplast transfer after cell inactivation (see Sect. 4.4). From ten hybrid clones, one clone was detected that regenerated plants hybrid for nuclei and tobacco plants, as a result of segregation (the analysis was based on the study of esterase isoenzymes). The remaining clones gave rise to only nuclear hybrids. Thus, in several experiments, the segregation of parental variants and nuclear hybrids has been observed among the progeny of individual fusion products, even though the vast majority of protoplast fusions resulted in the formation of true hybrids. The only satisfactory explanation for the complex segregation observed in the progeny of these clones seems to be the origination of the clone from multiple cell fusion.

In these same experiments (Sidorov et al. 1981 a) using irradiation to inactivate the parental cells for the screening of recombinants, 2 colonies out of 40 regenerated colonies were found to produce both hybrid plants and plants of one of the parental types (*N. plumbaginifolia*). However, in this case, the interpretation of these results is difficult because of the complications due to the interaction of the inactivation and selection processes after the fusion of the inactivated cells, particularly the possible reactivation of some of the parental cells which were physically close to the growing hybrid cells. This limitation is also inherent in the analysis of the similar experiments of Menczel et al. (1982), in which chloroplast transfer was carried out by fusion of mesophyll protoplasts of a streptomycin resistant mutant of tobacco, inactivated by using different doses of irradiation (or monoiodoacetate treatment in the control experiment) and mesophyll protoplasts of *N. plumpaginifolia*. In this experiment, the screening for chloroplast transfer was based on the resistance to the antibiotic. After the fusion of cells inactivated with monoiodoacetate or with low doses of radiation (50 J kg^{-1}), 4 out of 69 and 6 out of 38 clones, respectively, were found which displayed sorting out of parental types, in addition to hybrids for nuclei, upon regeneration.

To summarize, these results indicate that more than two cells may fuse during the parasexual hybridization process and various types of unique hybrid plants may be generated from these complex fusion products.

4.2 Fate of Extranuclear Genetic Determinants

4.2.1 Perspectives

It should be stressed that the present information concerning the behavior of plasmagenes in the parasexual hybridization process do not permit an unambiguous interpretation of the genetic changes that take place in the cytoplasm of fused cells and their progeny. The complexity of this problem is associated not only with the lack of experimental data involving the use of somatic cell fusion for producing hybrids heterozygous for plasmagenes, but with other factors as well. It is also caused by insufficient general knowledge of the extranuclear genes of higher plants, by the complex structure and interaction of the higher plant plasmon, as well as by the complexity and diversity of the theoretically possible genetic events which might result from cell fusion. It is extremely difficult to devise a sufficiently elegant scheme for an experiment on transmission genetics, since a number of simultaneous genetic events may occur, including: species specific as well as intraspecific selective elimination of genophores, mutation processes, mitotic segregation, and genetic recombination. Indeed, most of these processes actually take place, as will be dicussed in the following sections.

Although the experiments in the following sections are grouped and therefore separated under specific categories, such as cytoplasmic heterozygotes or recombination of plasmagenes, it may very well be that other interpretations of the results are equally possible. It is not our intention to stifle or hamper the appreciation of the relationship between the experimental results and their interpretation by the use of our classification scheme for the various categories in the behavior of plasmagenes. Rather, we hope to facilitate this discussion and analysis by the simultaneous categorization and presentation of the most logical explanations of these results.

Extranuclear (cytoplasmic or extrachromosomal) genetic determinants were described in higher plants in 1909 (Baur 1909; Correns 1909). In these experiments, it was shown that a character of green-white variegation was transmitted during sexual crossing either maternally (in *Mirabilis*) or biparentally, but not in accordance with Mendelian laws (in *Pelargonium*). It was originally assumed that the genetic determinants coding for the variegated character are located in the plastids themselves, due to the discovery in variegated plants of "mixed" cells that contain normal green and chlorophyll deficient plastids. Other extranuclear encoded characters were later found in higher plants, including: cytoplasmic male sterility and flower malformation (Edwardson 1970), resistance to streptomycin (Maliga et al. 1973, 1975; Umiel and Goldner 1976), resistance to tentoxin (Durbin and Uchitil 1977), and the polypeptide composition of Fraction I protein large subunit (Chan and Wildman 1972). The extranuclear genes in most higher plants are inherited strictly uniparentally and maternally (Hagemann 1959; Sager 1975; Kirk and Tilney-Bassett 1978).

From studies mostly with lower plants such as *Chlamydomonas* and *Saccharomyces*, it is now widely recognized that the plant cell contains at least two types of genophores in addition to the nuclear genome: the plastome in the plastids and the chondriome in the mitochondria. Together, they are responsible for the inheritance or transmission of extranuclear characters. Both genophores are circular DNA molecules with molecular weights of about 10^7–10^8 and have coding capacities for about 200 proteins. Both are present in several copies per organelle, although there may be more copies of chloroplast DNA (chlDNA) per chloroplast than copies of mitochondrial DNA (mtDNA) per mitochondrion. The specific DNAs, RNAs, and total protein machinery for the replication, transcription, and translation of genetic information are unique and distinct in chloroplasts and mitochondria. These features of the organelles are also dramatically different from those utilized in the expression of nuclear encoded characters. Good evidence exists that some of the genetic processes in organelles are semi-autonomous, so that to a certain extent, an organelle may be regarded as a cell within a cell. There are even theories that plant organelles are transformed symbiotic organisms that have lost their independence during evolution (for example, see Taylor 1970). Recent studies on the genetic control of the biosynthesis of the proteins of the 70 S ribosomes and the Fraction I protein (both of which are major constituents of organelles) show, however, that the genetic autonomy of the cytoplasm is rather limited and is subject to a strong nuclear control.

Despite extensive work on the biochemistry and genetics of organelles in lower plants, it still remains to be proven in higher plants that the genes coding for extranuclear characters are actually located in chlDNA and mtDNA. As far as we know, it has been demonstrated in only a few cases that a change in an extranuclearly encoded character is associated with a change in chloroplast DNA. In a chlorophyll deficient plastome mutant of tobacco, the chlDNA has a deletion of about 500–1 000 base pairs and, in other studies, a plastome mutation of *Oenothera* was correlated with small deletions in the chlDNA (Kirk and Tilney-Bassett 1978; Wong-Staal and Wildman 1973; Birky et al. 1974; Sager 1972, 1975; Bogorad 1977, 1981; Kung 1976, 1977).

The maternal inheritance of plasmagenes in sexual process is well studied and has been performed mostly on the agronomically important higher plant species. The restrictions of sexual crossing are the main obstacles to a more detailed genetic investigation of extranuclear genes, as well as to the production of cytoplasmic heterozygotes for the purpose of cytoplasmon engineering. Thus, parasexual hybridization should open new avenues in the study of extranuclear genes.

4.2.2 Protoplast Fusion and the Formation of Cytoplasmic Heterozygotes

The major portion of this section is devoted to a detailed description and discussion of two of our experiments that illustrate some of the possibilities and problems associated with the production of cytoplasmic heterozygotes by parasexual hybridization.

In the first experiment (Gleba et al. 1978a; Gleba 1978c, 1979), protoplasts were obtained from albino tobacco plants regenerated from tissue cultures of the white leaf areas of the variegated plastome mutant of *Nicotiana tabacum,* var. Samsun (Gleba et al. 1975). This mutant is described in detail in Sect. 4.1.2. The protoplasts isolated from aseptically grown, chlorophyll deficient plants containing only defective plastids were used as one of the parental forms (Fig. 16). Protoplasts of a cytoplasmic male sterile analog of *Nicotiana tabacum* L bearing the wild type plastome were used as the second parental form. This analog was produced by 13 successive backcrosses of a *Nicotiana debneyi* × *N. tabacum* hybrid to *N. tabacum*, var. Dubeck 44. Therefore, the genetic makeup of this analog should be a combination of Dubeck 44 nucleus with the cytoplasm of the Australian species, *N. debneyi* (2n = 48). The phenomenon of cytoplasmic male sterility (CMS) is usually regarded as a result of the noncoordination of nuclear and extranuclear genes and is well-known for tobacco hybrids synthesized using the cytoplasm of *N. debneyi* and some other *Nicotiana* species (Clayton 1950; Burk 1960; Ternovsky and Grebenkin 1967; Sand 1968; Burk and Mann 1970; Sand and Christoff 1973). No genes capable of restoring this lost fertility have been found in the different varieties of tobacco. In this particular case, the CMS character is expressed as a total absence of stamens, resulting in a unisexual flower (Fig. 17). It should also be mentioned that several investigators have successfully synthesized other analogs with phenotypically distinct forms of CMS using the cytoplasm of *Nicotiana debneyi* (Clayton 1950; Ternovsky and Grebenkin 1967; Sand and Christoff 1973).

The analysis of the parasexual hybrids of these two parental forms was performed in several ways. The biochemical analysis was based on the polypeptide composition of the large subunit of Fraction I protein. Leaf shape in these species is known to be a character encoded by several nuclear genes (van der Veen 1957). The Dubeck 44 var. has a sessile leaf shape, whereas the Samsun var. has a petiolated leaf shape. The distinctions of the initial parental forms that served as markers for the selection and analysis of the parasexual hybrids are listed below:

Parental form 1: *Nicotiana tabacum*, var. Samsun (mutant):

1. Plastome encoded chlorophyll deficiency.
2. Wild type flower.
3. *Nicotiana tabacum* type polypeptides of the large subunit of Fraction I protein.
4. Petiolated leaf shape.

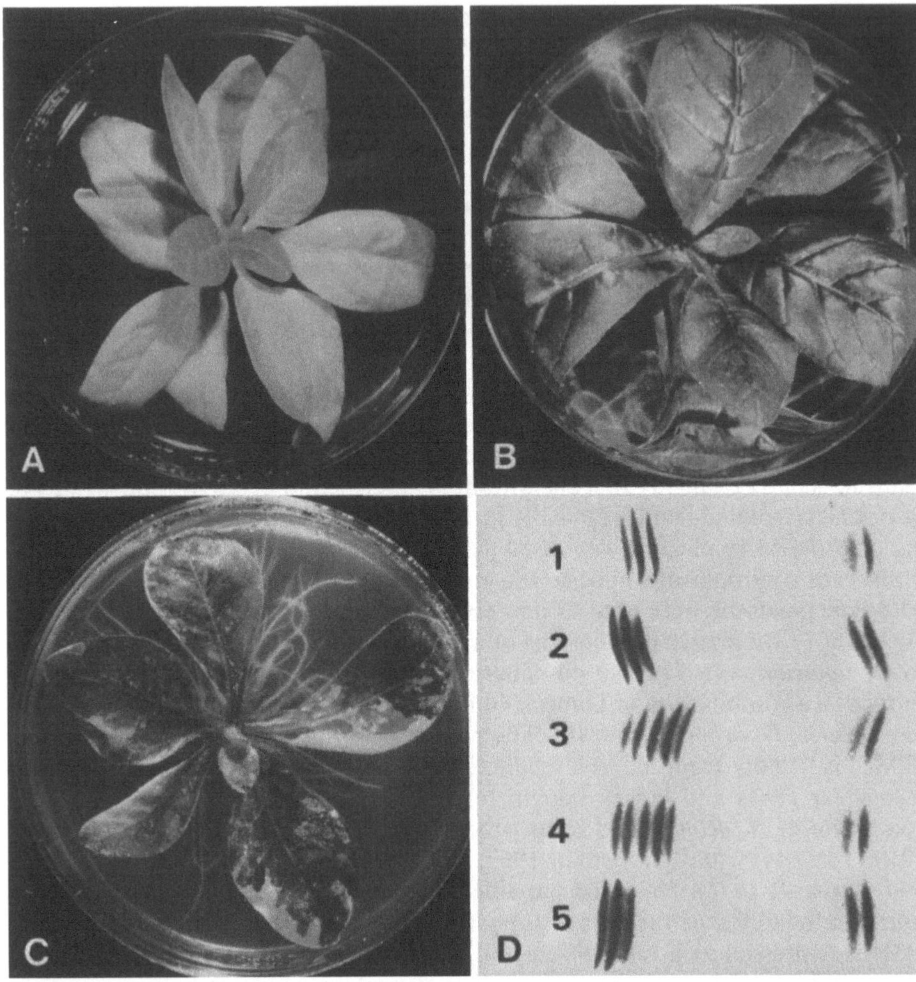

Fig. 16. Cytoplasmic hybrids (cybrids) of *Nicotiana tabacum* (×) *N. debneyi* produced by protoplast fusion. **A** The parental form of the tobacco plastome chlorophyll deficient mutant. **B** The other parental form, a cytoplasmic male sterile analog of tobacco, var. Dubeck 44, bearing the cytoplasm of *N. debneyi*. **C** One of the variegated plants regenerated from the fused protoplasts. **D** Isoelectric focusing patterns of the large (*left*) and small (*right*) subunits of Fraction I protein from: *1* the parental plastome chlorophyll deficient mutant of tobacco; *2* the other parental type with the cytoplasm of *N. debneyi*; *3* a variegated plant regenerated from fused protoplasts; *4* a variegated plant obtained via tissue culture techniques from a leaf fragment of a cytoplasmic hybrid; and *5* a green plant obtained via tissue culture techniques from a leaf fragment of a cytoplasmic hybrid. (Gleba et al. 1978a)

Parental form 2: *Nicotiana tabacum* L (analog of Dubeck 44 var.):

1. Wild type plastome.
2. Cytoplasmic male sterility.
3. *Nicotiana debneyi* type polypeptides of the large subunit of Fraction I protein.
4. Sessile leaf shape.

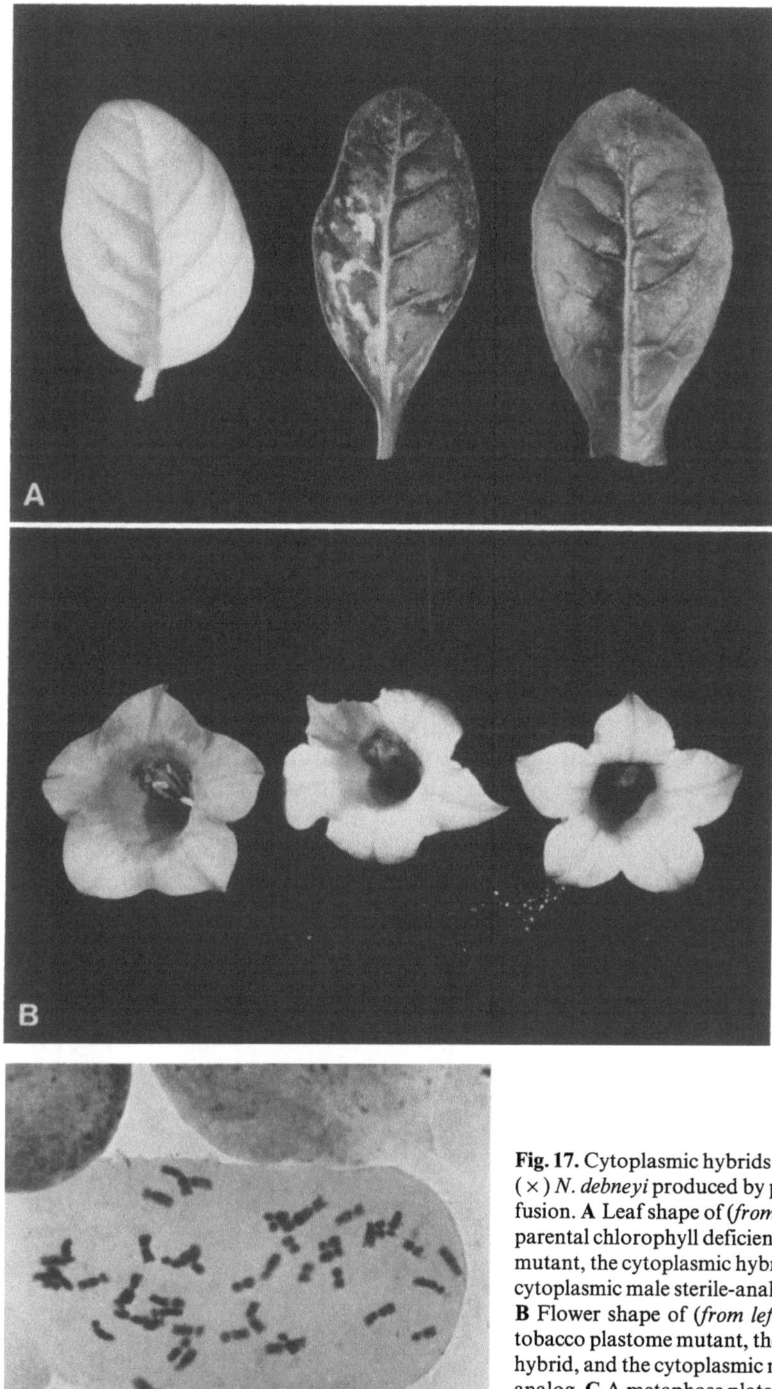

Fig. 17. Cytoplasmic hybrids of *N. tabacum* (×) *N. debneyi* produced by protoplast fusion. **A** Leaf shape of (*from left to right*): the parental chlorophyll deficient plastome mutant, the cytoplasmic hybrid, and the cytoplasmic male sterile-analog parent. **B** Flower shape of (*from left to right*): the tobacco plastome mutant, the cytoplasmic hybrid, and the cytoplasmic male sterile-analog. **C** A metaphase plate in a root cell of a cytoplasmic hybrid plant with 2n = 48 chromosomes, × 1 000. (Gleba 1978c)

The parasexual hybrids were selected at the stage of regenerated plants, as in our previous studies (Gleba et al. 1975). The selection was aimed at the recovery of variegated plants, since the protoplasts of the one parent gave rise to white plants and the protoplasts of the other parent gave rise to green plants.

It might be expected a priori that plastids from different plant species might successfully out-compete other plastids during propagation in a hybrid cell. This would lead to the segregation of the plastids and might result in the formation of nonvariegated plants from hybrid cells. This type of hybrid would not be detected in this experiment. On the other hand, the variegated plants might arise not only as a result of hybridization, but theoretically also because of other events, such as mutations and/or chimaera formation.

Four independent experiments were performed. In each of them, 10^6 to 10^7 protoplasts of both parental types were fused. A total of six variegated plants were found among the regenerants. Five of them were morphologically normal. They displayed no differences compared to the initial parental form Dubeck 44 var. and they possessed a sessile leaf shape. The sixth plant was morphologically abnormal and displayed an intermediate leaf shape. All of these plants produced only one type of flower. They were unisexual, male sterile, and identical to the flowers of the CMS analog, Dubeck 44 var. (Fig. 17).

The six variegated plants regenerated from the fused protoplasts were back-crossed to plants of the initial green parental form, Samsun var. (Table 7). Variegated plants, as well as green and white plants, were revealed in the F_1 progeny in all cases. This is the expected result if the hybrids are intracellular (but not cellular) chimaeras, i.e., cytoplasmic heterozygotes. Counting of the chromosomes in the root cells of these six hybrids has revealed that four of the morphologically regular plants were diploid (N1, N2, N3, and N4 in Table 7), one was tetraploid (N5 in Table 7), and one was nearly triploid (N6 in Table 7). It may be concluded on the basis of all these data that these plants are true cytoplasmic heterozygotes, at least for certain extranuclear genes.

The subsequent biochemical analysis of the polypeptides of Fraction I protein of these variegated hybrid plants has shown that these plants are also heterozygous for the extranuclear genes coding for the polypeptide composition of the large subunit of Fraction I protein (Fig. 16). Isoelectric focusing of the polypep-

Table 7. Analysis of F_1 progeny obtained by crossing variegated plants, produced by protoplast fusion, to one of the parental forms, *Nicotiana tabacum*, var. Samsun (wild type)

Plant designation and crossing type	No. of seedlings with phenotypes			Total
	Green	Variegated	White	
N1, ♀ H2 × ♂ Samsun	760	117	47	1073
N2, ♀ HP × ♂ Samsun	178	134	65	390
N3, ♀ HY × ♂ Samsun	385	96	303	784
N4, ♀ H8 × ♂ Samsun	128	168	72	368
N5, ♀ HX × ♂ Samsun	0	0	0	1968
N6, ♀ HZ × ♂ Samsun	No seeds were obtained			–

tides of the large subunit of Fraction I protein has demonstrated the presence of polypeptides of both *N. tabacum* and *N. debneyi* types not only in the primary variegated forms obtained via hybridization, but also in the secondary forms derived from them by somatic segregation and regenerated from the callus tissue produced from the primary leaves. This means that the sum of the parental polypeptides has been found in these hybrid plants. For this analysis, crystalline Fraction I protein was either isolated separately from the white or green leaf areas or from the white and green secondary forms derived from the same plant. The isolated protein was then mixed in a mass ratio of 1:1 and the analysis performed. This approach was necessary, since the chlorophyll deficient tissues usually have a protein content that is three to five times lower than that of green tissues (Wong-Staal and Wildman 1973). The genes coding for the polypeptide composition of the large subunit of Fraction I protein cosegregate in this case with the genes for the plastome encoded chlorophyll deficiency. (This problem will be discussed further in Chap. 6.) Thus, the biochemical study of these variegated forms rules out the possibility that they originated via mutation.

Analogous studies of Fraction I protein in the variegated plants of the F_1 progeny from sexual crosses of these cytoplasmic heterozygotes with the parental green form, Samsun var., have also demonstrated the presence of large subunit polypeptides of both *N. tabacum* and *N. debneyi* type in these plants.

The growth of lateral shoots was induced and periclinal chimaeras were produced from the variegated hybrid plants obtained by protoplast fusion, in order to more carefully investigate the cytoplasmic male sterility. These chimaeras displayed a distribution of normal and defective plastids in the embryonic layers L_1:L_2:L_3 of defective: defective: normal. In these chimaeras, all of the flowers were also male sterile, despite the fact that all the reproductive organs, which originate from the embryonic meristem layer 2 (Stewart and Burk 1970), were homozygous for plastome and carried the plastids of *N. tabacum* type.

The significance of these results may be summarized as follows. The variegated plants produced by protoplast fusion are heterozygous for the cytoplasmic (plastome) genes for chlorophyll deficiency, as well as for the plastome genes for the large subunit of Fraction I protein. This heterozygosity is a consequence of the "hybridization" of the cytoplasm of the two parental forms, as well as of the biparental inheritance of the extranuclear genes in parasexual hybridization. It is not a result of mutations. At the same time, these hybrid plants have inherited the extranuclear encoded character for cytoplasmic male sterility from one parent only. This means that uniparental inheritance was observed for these extranuclear genes, in contrast to that observed for chlorophyll deficiency and Fraction I. Only one of the six variegated plants seems to be a true nuclear hybrid, as judged by its tetraploidy and intermediate leaf shape (which is typical for heterozygotes for the nuclear genes coding for this character). The remaining hybrids are diploids and probably originated from heterokaryons in which the nuclei of Samsun var. were eliminated for some reason.

The purpose of the second experiment was to repeat the hybridization of the cytoplasms of *N. tabacum* and *N. debneyi* in the presence of the nuclei of both species (Piven 1980; Gleba 1980). The tobacco plastome chlorophyll deficient mutant, var. Samsun (albino plants) was again used as one of the parental forms. The

Table 8. Characteristics of plants regenerated from the parasexual hybrid *N. tabacum* (×) *N. debneyi*

Plant designation	Plant morphology[a]	Chromosome no.	Leaf type	Fertility of the pollen (%)
D1	Intermediate	96	Variegated	0.4
D2	deb	48	Variegated	0.1
D3	deb	48	Variegated	1.2
D4	deb	48	Variegated	–
D5	deb	48	Variegated	0
D6	deb	48	Variegated	1.6
D20	deb	48	Green	–
D21	deb	48	Green	32.0
D22	deb	48	Green	–
D23	deb	–	Green	17.5
D24	deb	–	Green	2.28
D25	deb	–	Green	–
D27	deb	48	Green	3.2
D28	deb	48	Green	–
D30	deb	48	Green	–
D31	deb	48	Green	0.3
D35	deb	48	Green	–
D40	deb	48	Green	9.7
D41	deb	48	Green	1.8
D42	Intermediate	96	Green	–
D43	deb	48	Green	0
D44	deb	48	Green	0.7
D45	deb	48	Green	–
D46	deb	70	Green	–
D47	Intermediate	48	Green	–
Parental form, *N. debneyi*	*N. debneyi*	48	Green	70–80
Parental form, *N. tabacum*	*N. tabacum*	48	White	80–90

[a] deb = *N. debneyi* type.

Australian species *Nicotiana debneyi*, var. Domin. was utilized as the second parental form. These experiments were designed (1) to provide data on the fate of different extranuclear genes in parasexual hybridization; (2) to study the possible differences in the segregation of cytoplasmic heterozygotes in the presence of different nuclei; and (3) to carry out a more rigorous study of the fate of nuclear genes in these cytoplasmic heterozygotes.

In each of four independent experiments, 10^6 to 10^7 cells of both parental types were used. The protoplast fusion and subsequent plant regeneration were performed as described for the previous experiment. The selection and screening process was based partly on the variegation character and partly, at the stage of flowering plants, on the changes in flower morphology. In the 4 experiments, 6 variegated plants and 25 plants with different abnormalities in the morphology of the reproductive organs were found among the regenerants (Table 8). All of the variegated plants possessed morphologically abnormal flowers. The vast majority of the hybrid plants were morphologically similar to *N. debneyi*. Three

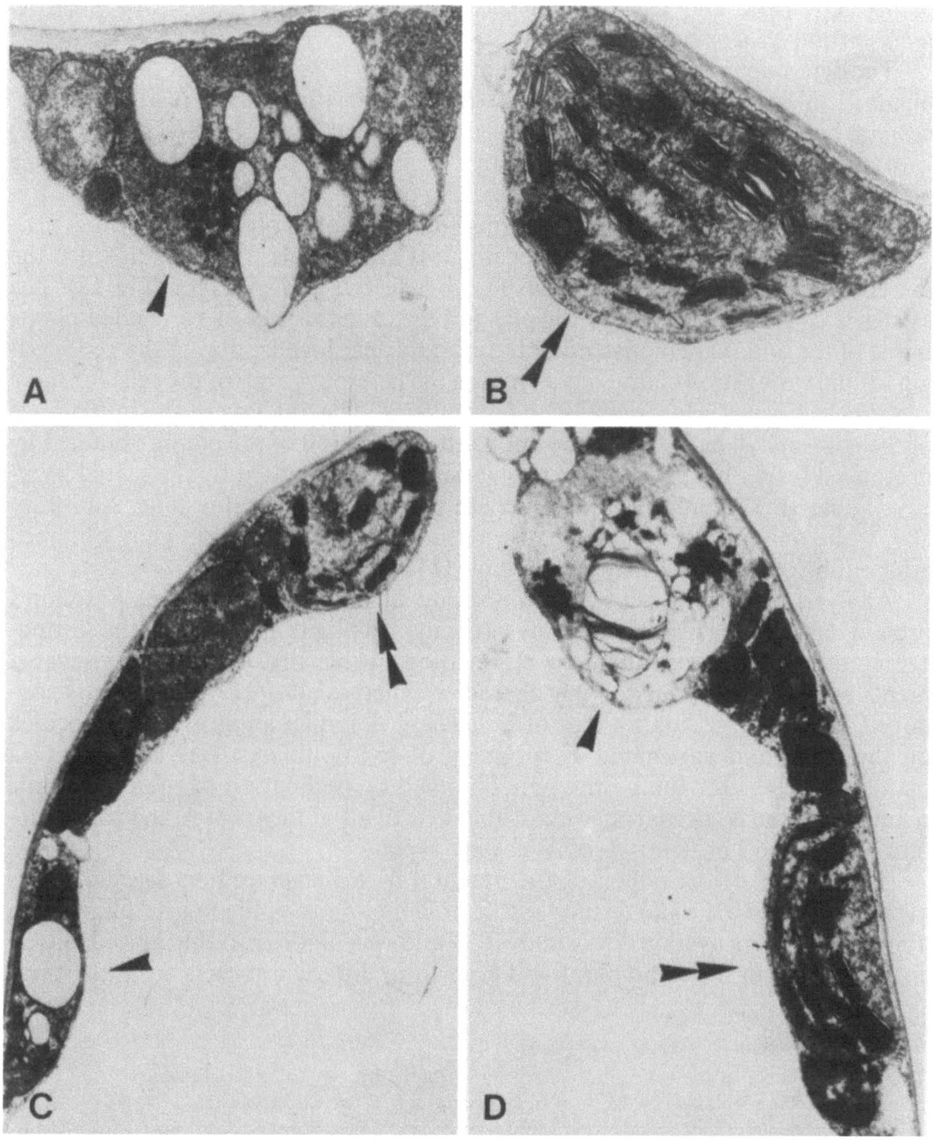

Fig. 18. Ultrastructural analysis of the various cell types in cytoplasmic hybrids of *N. tabacum* (×) *N. debneyi.* The plastids of the tobacco plastome mutant (**A**) (*single arrow*), ×23000, are clearly different from those of the other parent with the *N. debneyi* plastids (**B**) (*double arrow*), ×22000. **C–D** The mixed cell types in the leaf tissue of the variegated cytoplasmic hybrids with the presence of both tobacco mutant plastids (single arrow) and *N. debneyi* plastids (double arrow), ×9000. (Piven 1980; Gleba 1980)

plants (D1, D42, and D47) displayed a morphology intermediate between *N. tabacum* and *N. debneyi*.

The determination of chromosome number was possible in 19 of the 25 hybrid plants (Table 8). Sixteen of them were strict diploids (2n = 48; *N. tabacum* and *N. debneyi* have the same basic chromosome number). Two plants (D1 and D42) with a novel, intermedidate morphology were tetraploid (or allodiploid). One plant (D46) possessed 70 chromosomes. The overwhelming majority of these hybrid plants were genetically regular forms and were not hybrid for nuclei.

Electron microscopy of the variegated leaf areas in plant D2 has revealed the presence of cells containing plastids of both defective and wild type (Fig. 18). The existence of cells heterozygous for plastid type in the tissues of variegated plants is one of the criteria demonstrating the extranuclear localization of the gene coding for the variegation character, as well as the heterozygosity of the cells for these extranuclear genes (Hageman 1959; Sager 1972). The analysis of the inheritance of the chlorophyll deficiency gene upon self-pollination of the plants obtained by protoplast fusion has shown that this variegation is inherited by the sexual progeny (Table 9). Therefore, these plants are heterozygous for this gene. Additionally, white and green plants have also been found among the progeny obtained after selfing of the apparently green plants D41 and D42 (Table 9).

Analysis of the polypeptide composition of the large subunit of Fraction I protein has been performed on some of the hybrid plants and some of the secondary plants derived from them by segregation (Table 10). All of the variegated plants were heterozygous for this character, whereas all of the green plants contained only the large subunit type of *N. debneyi*. A similar analysis was performed on the green and variegated F_1 progeny of self-pollinated D1, D3, and D42 plants, as well as for the F_1 progeny obtained by pollination of the male sterile plant D5 by the *N. debneyi* parent (Table 11). In all of these cases, the small subunit of Fraction I protein was of *N. debneyi* type.

The fertility of the pollen of the parental *N. tabacum* and *N. debneyi* plants averages about 80%. The fertility of the pollen of most of the hybrid plants with abnormal flowers averaged less than 10% and was highly variable among different flowers from the same plant and even from different anthers from the same

Table 9. Inheritance of the variegation character in the F_1 progeny of self-pollinated plants that had been obtained by fusion of *N. tabacum* and *N. debneyi* protoplasts

Plant designation and phenotype	No. of seedlings produced with phenotypes			Total
	Green	Variegated	White	
D2, variegated	69	5	4	78
D3, variegated	34	9	3	46
D4, variegated	53	0	20	80
D6, variegated	58	0	2	60
D41, green	45	7	9	61
D42, green	46	0	1	47

Table 10. Analysis of the polypeptide composition of the large subunit of Fraction I protein in cytoplasmic heterozygotes *N. tabacum* (×) *N. debneyi* obtained after somatic segregation

Plant designation	Phenotype	Type of large subunit [a]
D1–7	Green, self-sterile	deb
D1–2	Green, self-fertile	deb
D1–18	White	tab
D2–1	Green	deb
D2–4	White	tab
D3–2	Green	deb
D3–4	White	tab
D4–1	Green	deb
D4–2	White	tab
D23	Green	deb
D27	Green	deb
D30	Green	deb
D31	Green	deb
D35	Green	deb
D40	Green	deb
D41	Green	deb
D42	Green	deb
D43	Green	deb
D44	Green	deb
D45	Green	deb
D46	Green	deb
D47	Green	deb

[a] deb = *N. debneyi* type; tab = *N. tabacum* type

Table 11. Analysis of the polypeptide composition of the large subunit of Fraction I protein in the F_1 progeny obtained from *N. tabacum* (×) *N. debneyi* cytoplasmic heterozygotes

Parental form and type of cross	Phenotype	Type of large subunit [a]
♀ D5 × ♂ N. debneyi	Green, male sterile	deb
D3, self-pollination [b]	Green, fertile	deb
D3, self-pollination [b]	Green, male sterile	deb
D3, self-pollination [b]	White	tab
D42, self-pollination	Green, fertile	deb
D42, self-pollination	White	tab

[a] deb = *N. debneyi* type; tab = *N. tabacum* type. The small subunit of Fraction I protein was of *N. debneyi* type in all cases

[b] These forms were obtained by somatic segregation from one F_1 variegated seedling

Table 12. Fertility of the pollen from different anthers of male sterile (plant D1–7) and partly male fertile (plant D1–2) plants obtained by somatic segregation of *N. tabacum* (×) *N. debneyi* cytoplasmic heterozygotes

Flower number	No. of anthers with phenotypes		Total
	Sterile	Partly fertile	
From plant D1–7:			
1	5	0	5
2	5	0	5
3	7	0	7
4	5	0	5
5	5	0	5
6	5	0	5
7	5	0	5
From plant D1–2:			
1	3	1	4
2	5	1	6
3	2	3	5
4	5	0	5
5	1	4	5
6	5	0	5

Table 13. Analysis of male sterility and flower abnormalities in the F_1 progeny of *N. tabacum* (×) *N. debneyi* cytoplasmic heterozygotes

Parental plant and type of cross	Phenotype of progeny	
	Partial or complete male sterility	Flower abnormalities
D3, self pollination	+	+
♀D3 × ♂ N. debneyi	+	+
♀*N. debneyi* × ♂ D3	–	–
D42, self-pollination	+	+
♀D42 × ♂ N. debneyi	+	+
♀*N. debneyi* × ♂ D42	–	–

flower (Table 8). In most of the plants, completely sterile and partly fertile lines were sorted out via somatic segregation upon vegetative propagation (Table 12).

The cytological study of the pollen has revealed only degenerate, flattened pollen grains in the completely sterile forms. The self-pollination of the partly fertile forms and the backcrossing of the sterile forms to the parental *N. debneyi* resulted in the production of viable seeds. The F_1 plants produced upon self-pollination of partly fertile forms were also heterogeneous for pollen fertility and both sterile and fertile F_1 plants were observed. The vegetative propagation of one variegated, partly fertile F_1 progeny, obtained from the self-pollinated D3 plant, has also permitted the production of completely sterile plants. Heterozygosity for the male sterility (fertility) character in analytic crosses of D3 und D42 plants with *N. debneyi* was inherited uniparentally and maternally (Table 13).

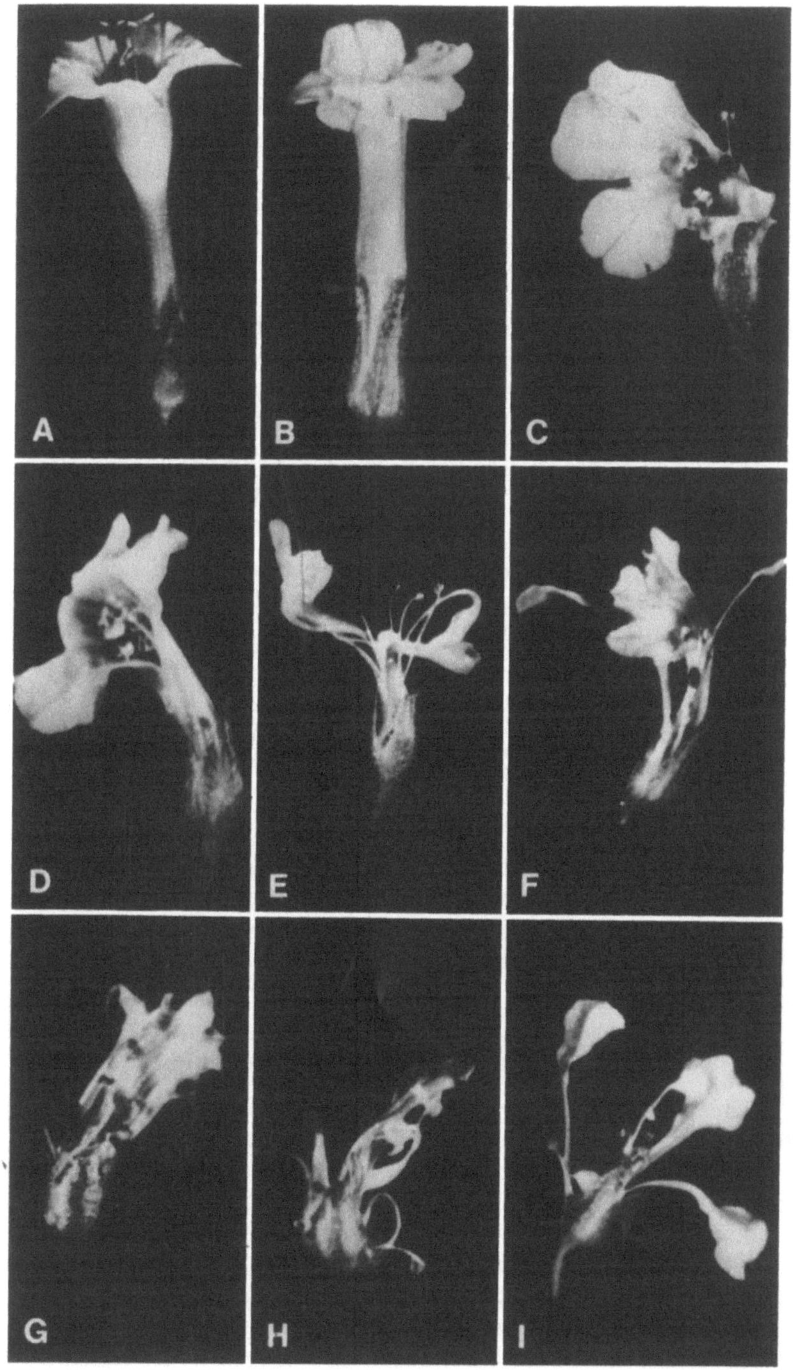

Fig. 19. Somatic segregation of morphologically different flower types in the cytoplasmic hybrids of *N. tabacum* (×) *N. debneyi* produced by protoplast fusion. The flowers of the parental forms of *N. tabacum* (**A**) and *N. debneyi* (**B**) are shown with a variety of diverse hybrid flowers (**C–I**). (Piven 1980; Gleba 1980)

The majority of plants produced via protoplast fusion possessed morphologically abnormal flowers. The abnormalities included the appearance of additional pistils or the feminization of the flowers. Additional defects included either nonviable pollen, underdeveloped anthers, irregularity in the number of anthers, complete reduction of the anthers producing unisexual flowers, and assymmetry or reduction of the corrolla. The shape of the flowers varied even within the same plant (Fig. 19). This variability was also observed in the F_1 progeny obtained either by self-pollination of the D1, D3, and D42 plants or by their pollination with *N. debneyi* pollen. All of the flower abnormalities were maternally inherited (Table 13). All of these plants seem to be heterozygous for the extranuclear genes coding for the characters of anther fertility (sterility) and flower abnormalities (for the sake of simplicity, both of the phenomena of sterility and abnormal flower morphology are regarded here as two aspects of one and the same character).

The following conclusions may be drawn from these two sets of experiments.

1. The hybrid plants obtained by parasexual hybridization possessed phenotypic characters which were absent in the parental forms (such as variegation, partial male sterility, and abnormal flower morphology).
2. These new characters are coded for by extranuclear genes [on the basis of the uniparental inheritance in sexual crossing, the transfer of maternal heterozygosity to the sexual progeny, the segregation of these characters upon vegetative propagation (mitotic segregation), and the presence of cells heterozygous for plastids].
3. The novel hybrid plant forms are heterozygotes for extranuclear genes (shown by simultaneous transfer of alternate characters to the sexual progeny and by mitotic segregation).
4. The heterozygosity of the novel plant forms is a consequence of the hybridization (fusion) of the parental cells, rather than a result of mutation or the formation of chimaeras (shown by the persence of large subunits of Fraction I protein of both parental types and the heterozygosity of the plants reflects the heterozygosity of the cells).
5. Some of the hybrid plant forms isolated after fusion are heterozygous for certain extranuclear genes and homozygous for others (i.e., variegated plants with CMS and green plants that show segregation of different types of CMS).

Thus, *protoplast fusion enables the production of cytoplasmic heterozygotes of higher plants bearing extranuclear genes from both parents.*

Good evidence for the biparental inheritance of plastome genes in hybridization by protoplast fusion and for the resulting plastome heterozygosity has been obtained in the laboratory of Wildman (UCLA). In these studies, the biochemical analysis of Fraction I protein was performed on *Nicotiana glauca* (×) *N. langsdorffii* hybrids produced by Carlson and Smith (Carlson et al. 1972; Smith et al. 1976). The principle of the selection of these hybrids was discussed in Sect. 3.2. All but one of the hybrids was aneuploid. One form with polypeptides of the large subunit of both parental types was found among the 17 nuclear hybrid plants investigated. Chimaera formation could be excluded, since the F_1 progeny obtained by self-pollination of this plant possessed a hybrid pattern of small subunits of

Fraction I protein. Additionally, another plant was found that had a large subunit of *N. glauca* type in the tissues of one shoot and that of *N. langsdorffii* type in the tissues of another shoot. Thus, these investigations confirm the biparental inheritance of extranuclear genes in hybridization by protoplast fusion. Of the remaining hybrids, eight possessed a large subunit of *N. langsdorffii* type and six of *N. glauca* type. These results indicate the occurrence of a sorting out of plastids among the hybrid products, in spite of the absence of obvious selection pressure at the genophore level.

Hybrid plant heterogeneity for plasmagenes coding for the polypeptide composition of the large subunit of Fraction I protein has also been demonstrated by Iwai et al. (1980, 1981). They studied the somatic hybrids *Nicotiana tabacum* (×) *N. rustica* obtained by Nagao (1978). The analysis of four hybrid plants revealed the large subunit from only one parental type in each plant (two plants possessed the large subunit of *N. rustica* type and two of *N. tabacum* type). However, the study of nine androgenic haploid plants, derived from one hybrid plant with polypeptides of tobacco type, revealed two plants with polypeptides of *N. rustica* type. In this case, as in the results of Chen et al. (1977), the small subunit polypeptides of Fraction I protein consisted of both parental types, thereby excluding the possibility of a chimaeric rather than a hybrid origin of these plants.

Extensive analysis of Fraction I protein in another species combination [*N. tabacum* (×) CMS analog of tobacco bearing the cytoplasm of *N. suaveolens*] enabled Glimelius et al. (1981) to demonstrate the production of plants heterozygous for plasmagenes. Protoplasts from tobacco callus tissues defective for nitrate reductase and incapable of shoot morphogenesis were fused with mesophyll protoplasts of the CMS analog. The fusion conditions were selected so that the nonfused parental mesophyll protoplasts were destroyed. The hybrids were selected on the basis of their morphogenetic capacity and their ability to grow on media free of reduced nitrogen. A total of 25 hybrid plants were selected in 7 independent experiments. Sixteen were found to be segregants bearing the plastids of *N. suaveolens* type, five were segregants with *N. tabacum* plastids, and four plants possessed the large subunits of Fraction I protein of both parental types.

Other workers have also produced cytoplasmic heterozygotes of tobacco by protoplast fusion (Belliard et al. 1977a, b, 1978). The parental protoplasts for these studies were derived from a cytoplasmic male sterile analog with *N. debneyi* cytoplasm (Tsikov and Tsikova 1971) and sessile leaf shape (a polygenic Mendelian character; van der Veen 1957) and from normal fertile Samsun or Xanthi varieties with petiolated leaf shape. In 2 experiments, 10 nuclear hybrids, as judged by leaf shape, and nearly 100 plants with abnormalities in flower morphology and different expression of the male sterility character were observed. These results may be easily explained in terms of the biparental inheritance of the cytoplasmic genes coding for the male sterility character, even though the analysis of the progeny indicating the maternal inheritance of novel morphological types can hardly be interpreted unambiguously (see Sect. 4.2.4).

Nakata and Oshima (1982) have found 2 variegated plants among 11 somatic hybrids (screened on the basis of leaf shape) from the progeny of hybrids between plastome chlorophyll deficient mutant and normal tobacco. The variegation was also transmitted to the next generation. They also observed some mottled plants among regenerants of parasexual hybrids of a plastome mutant of tobacco and *Nicotiana rustica*, but these plants were weak and abnormal.

Izhar and Power (1979) and Izhar and Tabib (1980) studied the fate of CMS genes after parasexual hybridization in *Petunia*. The parental forms used were a CMS line of *P. hybrida* with pigmented flowers and a male sterile line of *P. auxilliaris* white flowers. Among nearly 140 regenerants produced in two experiments, one fertile form that was morphologically intermediate between the parents was found (this plant was described as morphologically identical to *P. hybrida* by Izhar and Tabib, 1980). Moreover, after self-pollination of the regenerants, three plants showed segregation of male sterile forms. Heterozygosity of these three plants for male sterility resulted from the presence of CMS genes (which demonstrated maternal inheritance in analytical crosses) and from the heterozygosity of the *P. auxilliaris* parental cells for nuclear male fertility restorer genes (mfr genes; revealed by analytical crosses with fertile lines without mfr genes). The interpretation of these experiments is complicated and ambiguous because: (1) *P. hybrida* possesses numerous fertility restoring genes; (2) neither chromosome numbers nor the nuclear coded characters were studied in these hybrids; and (3) judging from the results of the interspecific hybridization, the *P. auxilliaris* parental line contains mfr genes in a heterozygous state.

A number of conclusions about the biparental inheritance of cytoplasmic genes in the parasexual hybridization of higher plants have been based only on the study of the segregation of plasmagenes rather than on the unequivocal demonstration of the production of cytoplasmic heterozygotes (see Sect. 4.2.3 and 4.2.4). Nonetheless, the studies discussed in this section provide not only direct evidence for the production of cytoplasmic heterozygotes in parasexual hybridization, but also show that this heterozygosity may persist for a rather long time.

The problem of screening for cytoplasmic heterozygotes is especially difficult since organelle segregation takes place in the hybrid products obtained by protoplast fusion (see Sect. 4.2.3). The investigations of Chen et al. (1977) provide good evidence that the screening for nuclear heterozygotes is a "poor sieve", in that segregants for cytoplasmic genes are immediately found. One of the most suitable methods for screening cytoplasmic heterozygotes entails the use of plastome chlorophyll deficiency as one of the markers of parental cytoplasm (Gleba et al. 1975), since it enables the selection of plastome heterozygotes as variegated forms at the stage of regenerated plants. Therefore, the entire selection procedure is relatively long and requires a considerable amount of technical work. The method of cytogete selection using the characters of cytoplasmic male sterility share this drawback, since flowering plants must be produced before screening. Moreover, the male sterility morphological defect may be induced de novo by a variety of genetic defects, such as aneuploidy or genome mutations, that may be a function or artifact of the parasexual hybridization process. The attempts to produce cytoplasmic heterozygotes using extranuclear characters such as resistance to streptomycin (Maliga et al. 1973, 1975) or kanamycin (Dix et al. 1977) are not generally applicable due to the selective pressure exerted on the plastome. At present, the combination of the techniques of individual protoplast fusion product cloning (Kao 1977; Gleba and Hoffmann 1978) with the use of the plastome chlorophyll deficient mutation and/or the cytoplasmic male sterility mutations seem to be the most elegant approach for the study of cytoplasmic heterozygotes.

4.2.3 Mitotic Segregation of Plasmagenes and the Sorting Out of Parental Genotypes

The segregation of plasmagenes is characteristic for the process of mitotic cell division. It leads to the sorting out and, hence, production of novel forms homozygous for certain plasmagenes (providing that genetic recombination does not take place). However, this sorting out may occur at different rates for different genes and the ability to detect heterozygosity for cytoplasmic genes depends on the rate of this segregation process. Therefore, it is no small wonder that most investigations have described cytoplasmic homozygotes rather than heterozygotes in the analysis of parasexual hybrids (Belliard et al. 1978; Poulsen et al. 1980; Douglas et al. 1981 d; Glimelius and Bonnett 1981; Scowcroft and Larkin 1981).

Some investigators have rashly concluded that cytoplasmic heterozygotes obtained by fusion of somatic cells are unstable and therefore experience a rapid sorting out. However, such an interference seems to be incorrect for several reasons. First in most hybridization experiments, the screening of parasexual hybrids has been designed to select for nuclear heterozygotes, whereas the fate of the plasmagenes was studied at a time long after the initial hybridization. Second, the methods of analysis of plasmagenes have generally been of low sensitivity and therefore of little use in demonstrating heterozygosity in those cases where one of the parental genophores prevails. This is the case when the polypeptide composition of Fraction I protein is analyzed. The isoelectric focusing and staining commonly employed are capable of detecting large subunit polypeptides at a level of about 10% of the total peptides (Chen et al. 1977). A much more sensitive, and therefore more appropriate, system of analysis is the electrophoretic separation of fragments of organellar DNA produced by cleavage with specific restriction endonucleases. This method permits the detection of specific DNA at a concentration of less than 1% of the total in admixtures of DNA (Schiller et al. 1982; Maliga, personal communication).

However, such detailed analysis has been performed on only about two dozen parasexual hybrid plants to date. Unfortunately, most authors did not specify the time elapsed between hybridization, in vitro cultivation, and the analysis of the cytoplasmic characters. Thus, the majority of somatic hybridization experiments are far from being perfect model systems for the analysis of plasmagene genetic processes.

It is inappropriate to state that the cytogetes obtained by the fusion of somatic cells undergo *rapid* segregation, because the term "rapid" may have the sole, relative meaning of "more rapid than in cytogetes obtained by means other than protoplast fusion." However, there are no facts to support such an assertion. The process of plasmagene segregation for cytogetes obtained by mutagenesis or sexual hybridization (in plant species with biparental inheritance of plasmagenes) has been studied insufficiently. The scanty data available concern only the segregation in variegated plants that are heterozygotes for plastome chlorophyll deficiency. For example, after self-pollination of variegated tobacco plants, the F_1 progeny consist mainly of segregates, with variegated forms constituting only a small percentage of the plants obtained. Hence, it follows that the majority of the gametes produced in the parental variegated plants are segregates for plasmagenes, even though these plants are heterozygous for these plasmagenes. In this case of intraspecific hybridization, the competition of plastids would be expected to be minimal. The influence of plastid competition on the rate of the segregation may be more pronounced in interspecific crosses (see the review of Kirk and Tilney-Bassett, 1978).

The plastome chlorophyll deficiency has been utilized in several experiments as a marker to select for the heterozygous state of plasmagenes after parasexual hybridization (Gleba et al. 1975, 1978a; Gleba 1978c, 1980; Piven 1980; Glimelius and Bonnett 1981; Sidorov et al. 1981). In all cases, some of the regenerated plants were variegated and were therefore heterozygous for plasmagenes. In some experiments, the frequency of such plasmagene heterozygotes was quite low (4.4% in the work of Glimelius and Bonnett, 1981).

As should be readily apparent by now, the process of plasmagene segregation is far from being a occasion I event. The complete analysis of plasmagene segregation results should evaluate the potential contribution of: (1) selective pressure in the process of mitotic segregation of plasmagenes; (2) linked segregation or cosegregation of plasmagenes; (3) production of new characters during the process of sorting out; and (4) modification or alteration of the segregation process itself.

Only the first of these points will be considered in detail here. The cosegregation of plasmagenes is discussed in Sect. 6.1 in connection with possibilities of genetic analysis of cytoplasmic genes using somatic cell fusion. However, it is appropriate to mention here that the plasmagenes sort out as two distinct cosegregation groups. The first group includes those characters that segregate with the chloroplast DNA (plastome) such as plastome chlorophyll deficiency, resistance to streptomycin, resistance to tentoxin, and the polypeptide composition of the large subunit of Fraction I protein. The second group includes the characters of cytoplasmic male sterility and abnormal flower morphology and cosegregates with mitochondrial DNA. The techniques of somatic cell hybridization have thus provided the first genetic evidence for the linkage of plasmagenes and the existence of at least two linkage groups (genophores) outside the nucleus of higher plant cells. The generation of new plasmagene encoded characters during the parasexual hybridization process and the influence of various factors upon the segregation of plasmagenes will be discussed in more detail in Sect. 4.2.4 and 4.4, respectively.

The apparent rapid sorting out of plasmagenes in the process of mitotic segregation might be explained by selective pressure at the level of cytoplasmic genophores, especially in the case of interspecific hybrids. In this respect, the analysis of species specific selection of genophores in somatic hybrids is of particular interest. The data tend to refute the existence of a species specific selection pressure. For example, in the work of Chen et al. (1977), the analysis of 16 somatic hybrid plants *Nicotiana glauca* (×) *N. langsdorffii*, obtained by Smith et al. (1976), has revealed that 8 plants possessed the large subunit of ribulose-1,5-bisphosphate carboxylase (Fraction I protein) and hence the plastome of *N. langsdorffii*. Likewise, six plants were segregates bearing the plastids of *N. glauca* type. Fraction I protein analysis of nine somatic hybrids potato (×) tomato demonstrated potato plastids in five and tomato plastids in four of the forms obtained (Poulsen et al. 1980). A similar analysis of four somatic hybrids *N. tabacum* (×)

N. rustica found two segregants with *N. tabacum* and two with *N. rustica* plastids (Iwai et al. 1980), whereas the studies of 14 somatic hybrids in the same species combination have revealed 11 plants with *N. rustica* and three plants with *N. tabacum* plastids (Douglas et al. 1981 a–d). Similarily, analysis of Fraction I protein from 25 plants, obtained by fusion of tobacco protoplasts bearing normal cytoplasm with protoplasts possessing the cytoplasmic male sterility of *N. suaveolens*, demonstrated 16 plants with *N. suaveolens* plastids and five plants with tobacco plastids (Glimelius et al. 1981). Analysis of restriction patterns of chloroplast DNA in nine tobacco hybrids, obtained by fusion of wild type cells with those of a cytoplasmic male sterile analog bearing the cytoplasm of *N. debneyi*, has revealed chloroplast DNA of tobacco in two of them and chloroplast DNA of *N. debneyi* in the other seven plants. Recent data suggest that some selective pressure against certain genophores may exist. Bonnett and Glimelius (personal communication) have estimated organelle segregation in somatic hybrids formed by fusing protoplasts of a tobacco nitrate reductase deficient mutant to those of three cytoplasmic male sterile analogs of tobacco with the plasmons of *N. suaveolens, N. undulata*, and *N. glauca*. The hybrid plants were analyzed for two chloroplast encoded characters (ribulose-1,5-bisphosphate carboxylase and ATPase) and cytoplasmic male sterility/fertility. The ratios of parental chloroplast types among the regenerated hybrid plants were 11 *N. tabacum*: 10 *N. glauca,* 5 *N. tabacum*: 15 *N. suaveolens*, and 0 *N. tabacum*: 22 *N. undulata*. The ratios of fertility of sterility characters were 3 *N. tabacum*: 15 *N. glauca,* 1 *N. tabacum*: 20 *N. suaveolens*, and 6 *N. tabacum*: 15 *N. undulata*. Thus, a limited selective pressure appears in the case of more divergent chloroplast DNAs (*N. tabacum–N. undulata*), operating against *N. tabacum* plastids, whereas fertility genes seem dominant over sterility genes (at least in this situation where the nuclear background is also hybrid). Likewise, Flick and Evans (1982) evaluated chloroplast content in nine somatic hybrids representing four species combinations: *Nicotiana tabacum* (×) *N. nesophila*, *N. tabacum* (×) *N. glauca*, *N. tabacum* (×) *N. otophora*, and *N. tabacum* (×) *N. sylvestris*. Most of the plants (seven out of nine) turned out to be sensitive to tentoxin, indicating some selective pressure against tobacco plastids in the limited number of somatic hybrids studied. Restriction pattern analysis of chloroplast DNA of three different interspecies *Petunia* somatic hybrids [*P. parodii* (×) *P. hybrida*, *P. parodii* (×) *P. inflata* and *P. parodii* (×) *P. parviflora*] has shown that all of the hybrids analyzed contained the chloroplast DNA of *P. parodii* (Kumar et al. 1982).

Selective pressure operating against a specific genophore should result in the sorting out of segregants bearing the plastids of only one of the parental types. In the studies cited and where a sufficient number of hybrid plants arising from independent cell fusion events was studied, the sorting out of not one, but rather of both parental types of plastids has generally been observed. Thus, it may be concluded that the process of segregation of plastids is random and that there is no compelling evidence that an overt selective pressure exists at the organelle level in cytoplasmic heterozygotes obtained via somatic cell fusion.

It is reasonably certain that segregation rather than another genetic process, such as recombination, is responsible for the results observed in the case of plastid

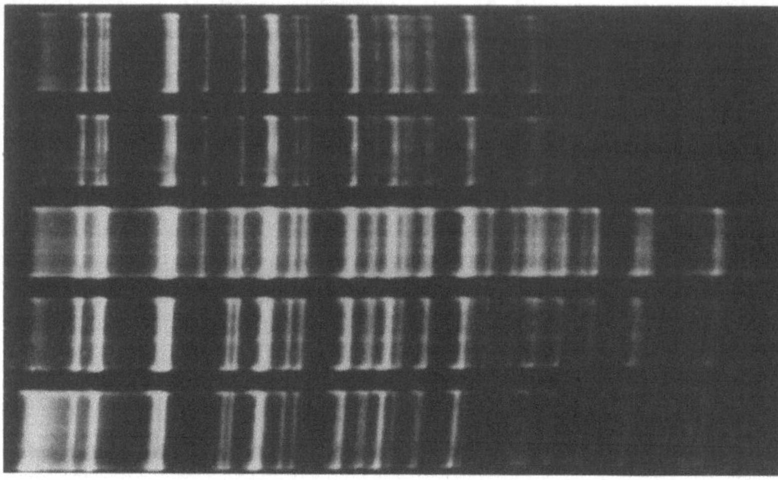

Fig. 20. Electrophoretic separation of restriction fragments produced by digestion with the restriction endonuclease Eco RI of plastid DNA from *Nicotiana tabacum* (*SR1*), *N. knightiana* (*N.kn*), a *mixture* of these two parental DNAs, and from plants *tk 17R* and *tk 9* of the somatic hybrid *N. tabacum* (×) *N. knightiana*. (Menczel et al. 1981 a)

genes. Studies of chloroplast DNA in somatic hybrids have revealed restriction patterns completely identical to one of the parental forms (Belliard et al. 1978; Aviv et al. 1980; Menczel et al. 1981 a, b; Sidorov et al. 1981 a; Scowcroft and Larkin 1981) (Fig. 20). Additionally, all of the characters associated with the plastome (such as plastome chlorophyll deficiency, resistance to streptomycin and tentoxin, polypeptide composition of the large subunit of Fraction I protein, and restriction pattern of chloroplast DNA) invariably cosegregate. On the other hand, quite another picture is observed with mitochondrial DNA restriction patterns and cytoplasmic male sterility characters of somatic hybrids. In all probability, these phenomena are associated with a heterogeneity of mitochondrial DNAs in higher plant cells that complicate the process of mitotic segregation of these characters and/or with other genetic interactions (recombination). These latter points are addressed in Sect. 4.2.4.

4.2.4 Recombination of Plasmagenes

Complex patterns of segregation of the cytoplasmic determinants for flower morphology (Belliard et al. 1978; Piven 1980; Gleba 1980) and of mitochondrial DNAs (Belliard et al. 1979; Nagy et al. 1981; Galun et al. 1982) have been found in cytogetes formed by parasexual hybridization.

Cytoplasmic male sterility and the flower abnormalities found in plants obtained by protoplast fusion of *N. tabacum* and *N. debneyi* (Piven 1980; Gleba 1980), as well as of tobacco with cytoplasmic male sterile tobacco analog with the cytoplasm of *N. debneyi* (Belliard et al. 1977 a, b, 1979), evidently can not be coded for by nuclear genes because (1) some of the hybrids are regular diploids in that they contain only the nucleus of one of the parents and (2) these characters are inherited maternally and show mitotic segregation. The plastome genes are not involved nor responsible for these modifications, since the sterility and flower abnormalities segregate independently of chloroplast DNA (Belliard et al. 1978) and independently of the characters coded for by plastome (Gleba 1978 c; Gleba et al. 1978 a). The nature of the complex segregation of abnormal flower morphology character (i.e., segregation with sorting out of more than the two relatively stable parental forms) may be explained in one or more of several ways.

1. The heterogeneity may be caused by the interaction or the combination of at least two different cytoplasmic genophores which segregate independently during mitotic divisions, thus giving rise to different plasma types with stable and morphologically distinct types of male sterility. Such an explanation requires that the character in question is coded for by at least two distinct cytoplasmic genophores. The mitochondrion could be one of them and the second must be undescribed as of yet, since as was mentioned, the plastome does not participate in the determination of cytoplasmic male sterility or flower malformation. Unfortunately, there is no experimental evidence to support the existence of a third extranuclear genophore.

2. The heterogeneity may be caused by the presence of genetically distinct copies of the same genophore coding for the cytoplasmic male sterility and thus is a result of the heterogeneity of its molecules in the parental cells. According to this hypothesis, the cytoplasmic male sterility is coded for by genes located on mitochondrial DNA and thus the number and morphology of every type of cytoplasmic male sterility are determined not only by the number of genetically distinct mitochondrial DNAs in the parental cells, but also by their mitotic segregation after parasexual hybridization.

3. The heterogeneity may be caused by intraspecific inter- or intramolecular rearrangement of mitochondrial DNAs that is, via as yet undescribed pathways, connected with or induced by somatic cell hybridization.

4. The heterogeneity may be a result of genetic interaction (such as fusion of organelles or molecular recombination of genophores) between the parental genophores, leading to the formation of organelles bearing different heterozygous sets of genophores or with novel DNAs. This hypothesis derives some support from the observation that recombinant forms of mitochondrial DNA have been found in animal somatic cell hybrids (Dawid et al. 1974). This fourth hypothesis does not necessarily exclude the second or third ones.

Several detailed studies on the recombination of the mitochondrial genome in parasexual hybrids have been performed via an analysis of restriction endonuclease fragments of mitochondrial DNA. Differences in the number and size of the DNA fragments generated by restriction endonuclease digestion of various DNA molecules is an elegant and sensitive measure of the degree of homology of these molecules, in some cases permitting the detection of single nucleotide changes.

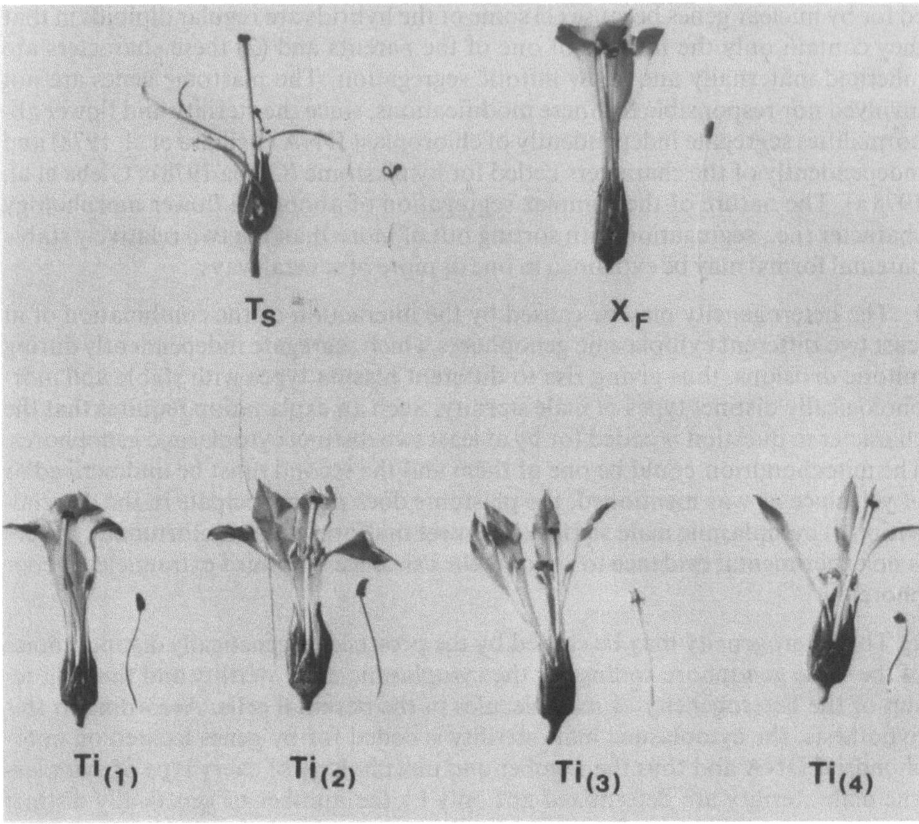

Fig. 21. Flower morphology classes of somatic hybrids (*Ti1–Ti4*) of *Nicotiana tabacum* var. Xanthi (×) *N. tabacum* var. Techne with *N. debneyi* cytoplasm. The Xanthi *(XF)* and Techne *(Ts)* flower types are also shown. (Belliard et al. 1978)

In the first of these studies (Belliard et al. 1979), hybrid plants produced by protoplast fusion of wild type tobacco with tobacco bearing the cytoplasm of *N. debneyi*, and thus possessing cytoplasmic male sterility, were used. The parasexual hybridization resulted in production of hybrid and cybrid plants with morphologically novel and stable types of cytoplasmic male sterility (Fig. 21). Earlier studies of the restriction endonuclease Eco RI fragments of chloroplast DNA of these plants revealed only pure types of chloroplast DNA of one or the other parents to be present in each particular hybrid, indicating no correlation between the chloroplast DNA and the novel forms of cytoplasmic male sterility (Belliard et al. 1977a, b, 1978) (Fig. 22; Table 14). The digestion of the parental mitochondrial DNA with the restriction endonuclease Sal I generated 31 fragments in wild type tobacco and 38 fragments in the cytoplasmic male sterile analog. Of these fragments, 20 were common to both cell types, 11 were specific for wild type tobacco, and 18 were specific for the cytoplasmic male sterile analog. Thus, 49 fragments can be resolved in admixtures of the parental mitochondrial DNAs. Anal-

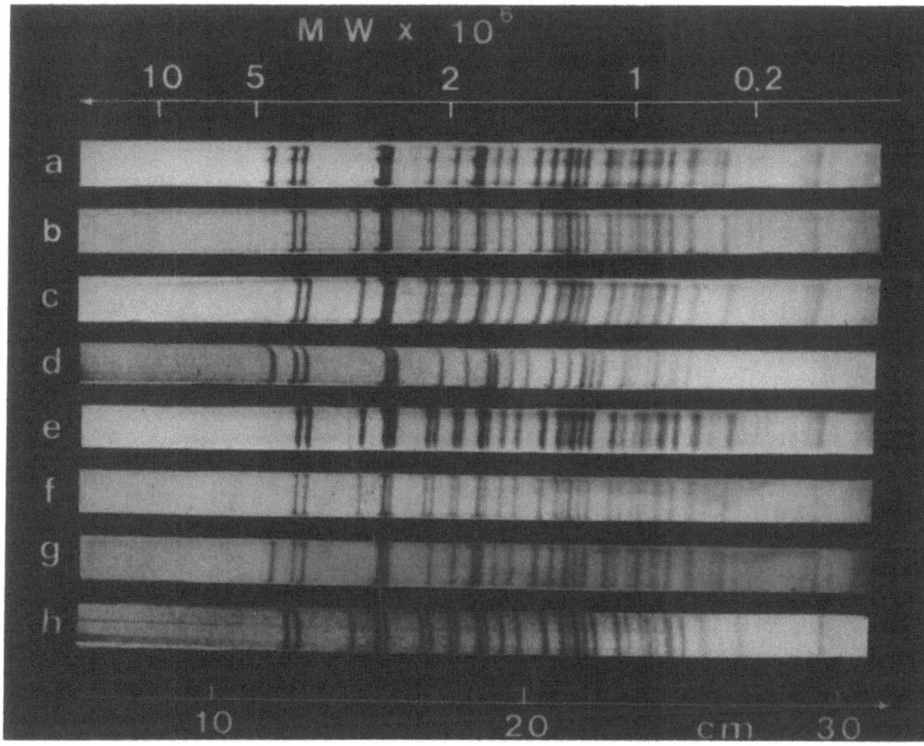

Fig. 22. Agarose gel electrophoresis of restriction endonuclease Eco RI digests of chloroplast DNA from (*a*) *Nicotiana tabacum* var. Xanthi; (*b*) *N. tabacum* var Techne, with *N. debneyi* cytoplasm; (*c*) *N. debneyi*; and (*d–h*) several somatic hybrids (cybrids). (Belliard et al. 1978)

ysis of mitochondrial DNA from nine cybrids belonging to four different morphological types of cytoplasmic male sterility demonstrated that each of them possessed an unique set of fragments, differing not only from the other hybrids, but also from both parents. The number of fragments in the different hybrids ranged from 35 to 41, among which fragments characteristic for *both* parents were always found. Moreover, in all but one of the hybrids, one to four novel fragments that were absent in the parental species were found. In spite of the complicated "biparental" inheritance of the types of fragments, a quantitative correlation between the degree of flower malformation and the number of mitochondrial DNA fragments species specific for *N. debenyi* was evident. These results are summarized in Table 15.

Similar results were reported by Nagy et al. (1981) on studies of mitochondrial DNAs of ten somatic hybrid plants *Nicotiana tabacum* (×) *N. knightiana*, which had been produced previously by joint culturing of three heteroplasmic fusion products (Menczel et al. 1981). The results of digestion of the mitochondrial DNAs with the restriction endonucleases Sal I and Bam HI are presented in Fig. 23. The treatment of parental mitochondrial DNAs with Sal I produced 31 frag-

Table 14. Relationship between floral phenotype and type of chloroplast DNA in the parents and various somatic hybrids of *Nicotiana tabacum* var. Xanthi (×) *N. tabacum* var. Techne (with *N. debneyi* cytoplasm)[a]

Plant	Flower class	Male organ	Type of chloroplast DNA
Parents			
var. Techne	–	Sterile	debneyi
var. Xanthi	–	Fertile	tabacum
Hybrids			
T_{161}	I	Fertile	tabacum[b]
T_{27}	I	Fertile	debneyi
T_{47}	I	Fertile	debneyi
T_1	II	Sterile	debneyi
T_{13}	III	Sterile	debneyi
T_{75}	IV	Sterile	debneyi
X_1	IV	Sterile	debneyi[c]
X_7	IV	Sterile	debneyi[c]
X_6	II	Sterile	tabacum

[a] Belliard et al. (1978). The restriction endonuclease analysis is presented in Fig. 22

[b] This plant presents an change of sterile Techne into fertile Techne

[c] These plants represent a change of fertile Xanthi into sterile Xanthi

Table 15. Summary of the analysis of the restriction fragments generated by digestion with Sal I of mitochondrial DNA from the parents and various somatic hybrids of *Nicotiana tabacum* var. Xanthi (×) *N. tabacum* var. Techne (with *N. debneyi* cytoplasm)[a]

Plants	Flower phenotype and class[b]	No. of restriction fragments					Molecular weight[c]
		Common	debneyi	tabacum	New	Total	
Xanthi[d]	Fertile	20	–	11	–	31	206
Techne[d]	Sterile	20	18	–	–	38	256
T_{161}	Fertile, I	20	8	7	3	38	293
T_{47}	Fertile, I	18	9	7	1	35	269
T_{27}	Fertile, I	19	12	7	1	39	264
T_1	Sterile, II	17	12	5	4	38	260
T_{75}	Sterile, III	17	15	3	2	37	267
T_{13}	Sterile, IV	20	18	1	–	39	259
X_6	Sterile, II	20	16	3	1	40	248
X_1	Sterile, IV	19	17	3	2	41	296
X_7	Sterile, IV	20	17	1	1	39	290

[a] Belliard et al. (1979). See also Fig. 22 and Table 14

[b] The flower class is as in Table 14

[c] $\times 10^{-6}$

[d] The restriction maps of *fertile* Techne and Xanthi are identical with wild type tobacco. The cytoplasmic male sterile Techne used in these experiments maps identical to *N. debneyi*

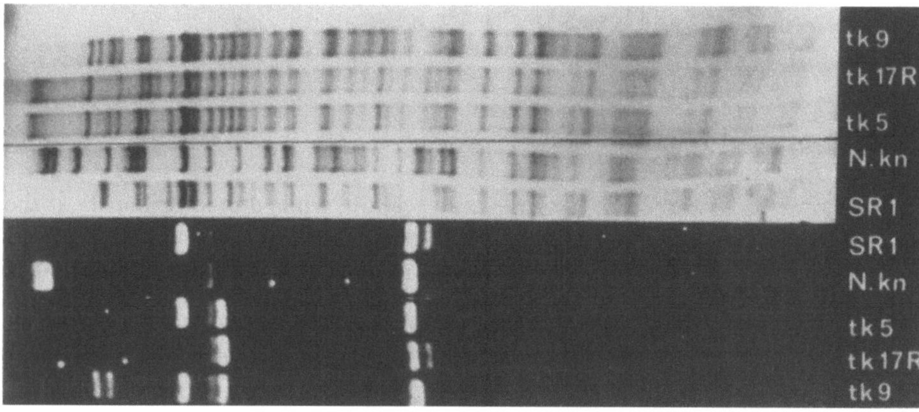

Fig. 23. Electrophoretic separation of Bam HI digests of mitochondrial DNA from *Nicotiana tabacum* (SR1), *N. knightiana* (N · kn), and three lines (tk 5, tk 9, and tk 17R) of the somatic hybrids *N. tabacum* (×) *N. knightiana.* The stained patterns are on the top and on the bottom are autoradiograms of these fragments after hybridization with a specific radioactive probe (^{32}P-labelled 2/12 plasmid DNA). Clear differences in the parental and hybrid bands are evident. (Nagy et al. 1981)

ments from *N. tabacum* and 26 fragments from *N. knightiana.* Altogether, 39 different fragments were found, of which 21 were species specific. Sal I restriction maps of the mitochondrial DNA from eight hybrid plants revealed an unique pattern for each of them, as well as a difference compared to the parental forms. The number of restriction fragments ranged from 29 to 35 for the different hybrids. Seven of the eight plants possessed one three new fragments which were absent in both parents. A total of five novel restriction fragments were observed. The parental and hybrid mitochondrial DNAs also differed from each other in restriction fragment analysis using Bam HI. In this case, the hybrid plants also revealed one to four new fragments. It is also important to note that these analyses were repeated four times over the span of one year and the results were reproducible and completely identical.

Galun et al. (1982) analyzed the mitochondrial DNAs of somatic hybrids produced by fusion of tobacco mesophyll protoplasts with mesophyll protoplasts of *N. sylvestris* bearing the cytoplasm of *N. suaveolens* (Aviv and Galun 1980). Seven plants bearing the *N. sylvestris* nucleus and tobacco plastids (four plants) or *N. suaveolens* plastids (three plants) were studied. Five of these plants were fertile and two displayed cytoplasmic male sterility. Restriction endonuclease analysis was performed using Sal I and Xho I. The cybrids possessed mitochondrial DNA that, on the whole, was similar to one of the parent; however, in all but one of the cases, the pattern of Sal I fragments contained not only one to four bands of the other parent, but also one or two new fragments. Likewise, the Xho I patterns contained one to four novel fragments. The restriction patterns remained stable over three generations for the two cybrids studied that long.

The problems associated with genetic analysis using segregation of plasmagenes from cytogetes obtained via protoplast fusion are discussed in Sect. 6.1.4.

4.3 Fate of Nuclear Genes in the Process of Somatic Hybridization

4.3.1 Segregation of Nuclei

The experiments described in the foregoing section, while aimed primarily at the elucidation of the fate of extranuclear genetic determinants, have demonstrated at the same time that the cytoplasmic heterozygotes are not always heterozygotes for nuclear encoded characters.

For example, in our works (Gleba et al. 1978a; Gleba 1980), described previously in detail (4.2.2), only 1 of 6 and 2 of 25 cytoplasmic heterozygotes, respectively, or their derivatives were polyploid and could be referred to as nuclear hybrids.

In the work of Belliard et al. (1977a, b, 1978, 1979) (Sect. 4.2), cytoplasmic male sterile analog of tobacco, var. Techne, possessing sessile type leaves, was hybridized with tobacco var. Samsun and Xanthi, both with petiolated type leaves. The leaf shape in tobacco is controlled polygenically and coded for by nuclear genes (van der Veen 1957). Thus, this character may be used as a marker for hybrid formation and may be compared with other markers for cybrid formation. Amphidiploid mesophyll protoplasts were utilized as parental material for fusion and the screening of supposed hybrids was carried out at the stage of flowering plants so that those forms with new combinations of parental traits (morphology of flower abnormalities associated with male sterility plus leaf shape) were selected. In 2 experiments on 936 regenerants, a total of 57 nuclear hybrids and 225 cybrid plants were selected. Subsequent biochemical studies of the cybrid forms with new flower morphology and modified characters of cytoplasmic male sterility/fertility completely confirmed the original supposition about the hybrid nature of the cytoplasm of these plants, whereas the analyses of chromosome numbers have demonstrated that most of them are amphidiploid and thus can not contain the chromosomes of both parents. The results obtained in these experiments unequivocally demonstrate that fusion of tobacco mesophyll protoplasts results in the formation of not only true hybrids, but also other recombinant forms with a cybrid nature that possess a novel cytoplasm and the nuclear genes of only one parent. Quantitative data can not be drawn from these experiments since the number of independent fusion events that gave rise to the plants studied was not monitored or controlled. Theoretically, all of the cybrid and hybrid forms obtained in these two experiments may have been derived from as few as four independent fusion events.

Glimelius and Bonnett (1981) studied the fate of nuclear genes after parasexual hybridization of diploid mesophyll protoplasts of a plastome chorophyll deficient tobacco mutant, var. Samsun, with petiolated leaves and a line of aneuploid tobacco cells, var. Gatersleben, defective for nitrate reductase activity, incapable of morphogenesis, and with a sessile type of leaves and augmented flowers. Hybrids were selected on the basis of their ability to grow on media free of reduced nitrogen and to regenerate green, photosynthesizing plants. A total of 87 green cell colonies were recovered. An analysis of 23 plants representing independently selected colonies revealed that 5 of them were diploid and identical in flower and leaf morphology to the Samsun parent; thus, they can not be nuclear hybrids. The other 18 regenerants were aneuploid, nearly triploid and possessed an intermediate leaf shape and augmented flowers, as well as reduced fertility, and thus displayed all of the characters expected for true nuclear hybrids.

In another experiment (Glimelius et al. 1981), diploid mesophyll protoplasts of tobacco cytoplasmic male sterile analog (plasma of *N. suaveolens*) were fused with callus cells of a tobacco line defective for nitrate reductase activity. The screening was based on growth in the absence of reduced nitrogen, capability for plant regeneration, and morphological and/or functional restoration of male fertility. Five out of the 25 plants selected were diploid, i.e., not hybrids for nuclear encoded characters, even though they possessed morphologically novel flowers and one plant was fertile. In addition, two plants also contained the plastids from the other (callus cell) partner, as demonstrated by analysis of Fraction I protein.

Unfortunately, these studies provide no definitive estimation of the frequency of nuclear fusion in protoplast fusion products.

To this end, we performed a special study of nuclear gene behavior upon hybridization by protoplast fusion (Gleba and Berlin 1979; Gleba 1980). The choice of our experimental system was determined by the following considerations. First of all, the selection of hybrids should not be based on use of nuclear characters via genetic or physiological complementations, since it is highly probable that a significant percentage of possible genetic combinations would be automatically rejected by the screening procedure. Second, a selection based on use of extranuclear encoded characters is also unlikely to be efficient, because the cytoplasmic heterozygotes described so far seem to represent a minor portion of the hybrids, while the majority are cytoplasmic segregants and are thus not selected. Therefore, we chose to mechanically isolate and culture the individual fusion products. This allows the cloning of individual genetic events (fusions) and enables the study of a great number of descendants from each line, whereas the selection of fusion products is not perforce connected with any genetic events inside a cloned product. Although it is impossible to theoretically exclude the selection of some genetic combinations out of the total spectrum of diverse forms arising within a clone, we tried nevertheless to minimize such a process by utilizing cells of the same species (tobacco) as parental forms.

Cells of an old, tobacco callus cell line, var. Xanthi (Fig. 24), initiated by Filner in 1965 (Heimer and Filner 1970) were used as one of the parental types. This line is defective in morphogenesis and has the character of resistance to 5-methyltryptophan (Fig. 25). The resistance was selected by growing cells in suspension culture on medium containing 50 μM of 5-methyltryptophan (Berlin 1976) and some characteristics of the resistant line have been described (TX6 line; Berlin and Widholm 1978). A cytological study of chromosome number showed that most of the 5-methyltryptophan resistant cells are aneuploid hyperdiploid.

Amphihaploid ($n = 24$) and amphidiploid tobacco plants, var. Rosa, homozygous for recessive gene of light sensitivity (sl mutant utilized by Melchers and Labib, 1974) were used as the second parent (Fig. 24). Thirteen individual products of the fusion of callus protoplasts with haploid mesophyll protoplasts, as well as 12 products of the fusion of callus protoplasts with diploid mesophyll protoplasts

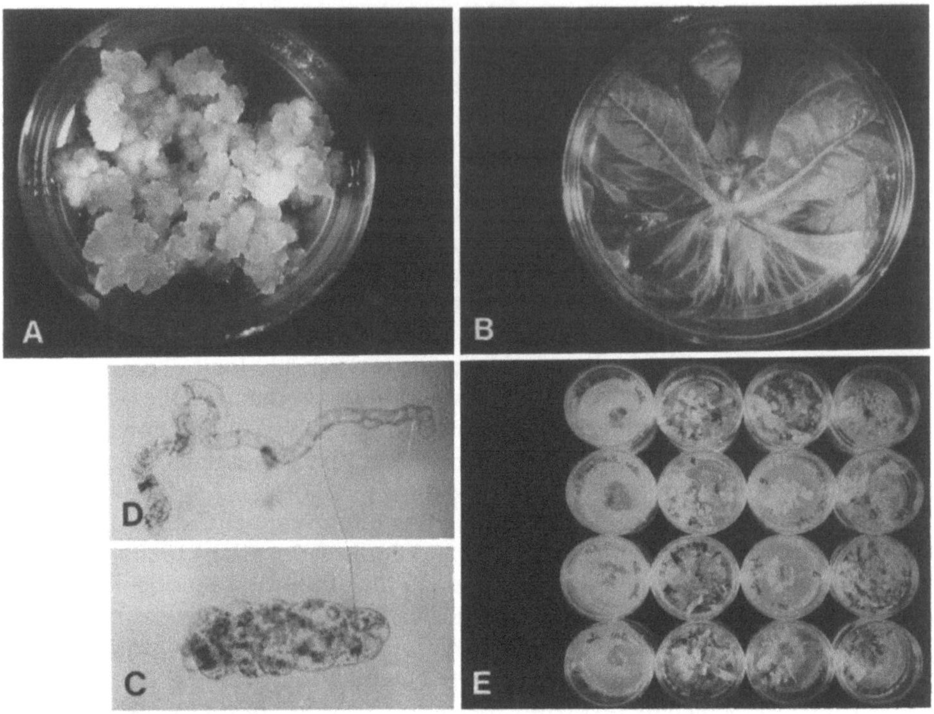

Fig. 24. Protoplast fusion and the fate of nuclear genetic determinants. **A** Tissue of the tobacco cell line, var. Xanthi, bearing the character of resistance to 5-methyltryptophan and defective for morphogenesis is on the left side, while wild type tissue is on the right. **B** A diploid plant of tobacco, var. Rosa, homozygous for the recessive gene of light sensitivity. **C** Morphology of a primary colony obtained from a single tobacco mesophyll protoplast. **D** Morphology of a primary colony regenerated from a fusion product of the somatic hybrids of the mutant tobacco cell line, var. Xanthi (×) mutant tobacco, var. Rosa. **E** The induction of morphogenesis in the culture of cells obtained from (from left to right): single protoplasts of the tobacco cell line, var. Xanthi, defective for morphogenesis and three individual lines of heteroplasmic fusion products. (Gleba 1980)

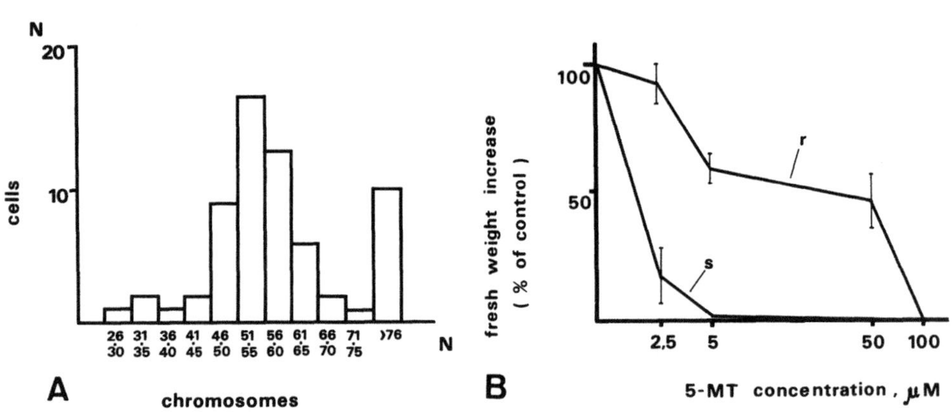

Fig. 25. Protoplast fusion and the fate of nuclear genetic determinants. **A** A histogram of the distribution of chromosome numbers in cells of the aneuploid cell line of tobacco, var. Xanthi, resistant to 5-methyltryptophan. **B** Ability of wild type (*s*) and resistant (*r*) cell lines of tobacco to survive and grow in the presence of 5-methyltryptophan (*5-MT*). (Gleba 1980)

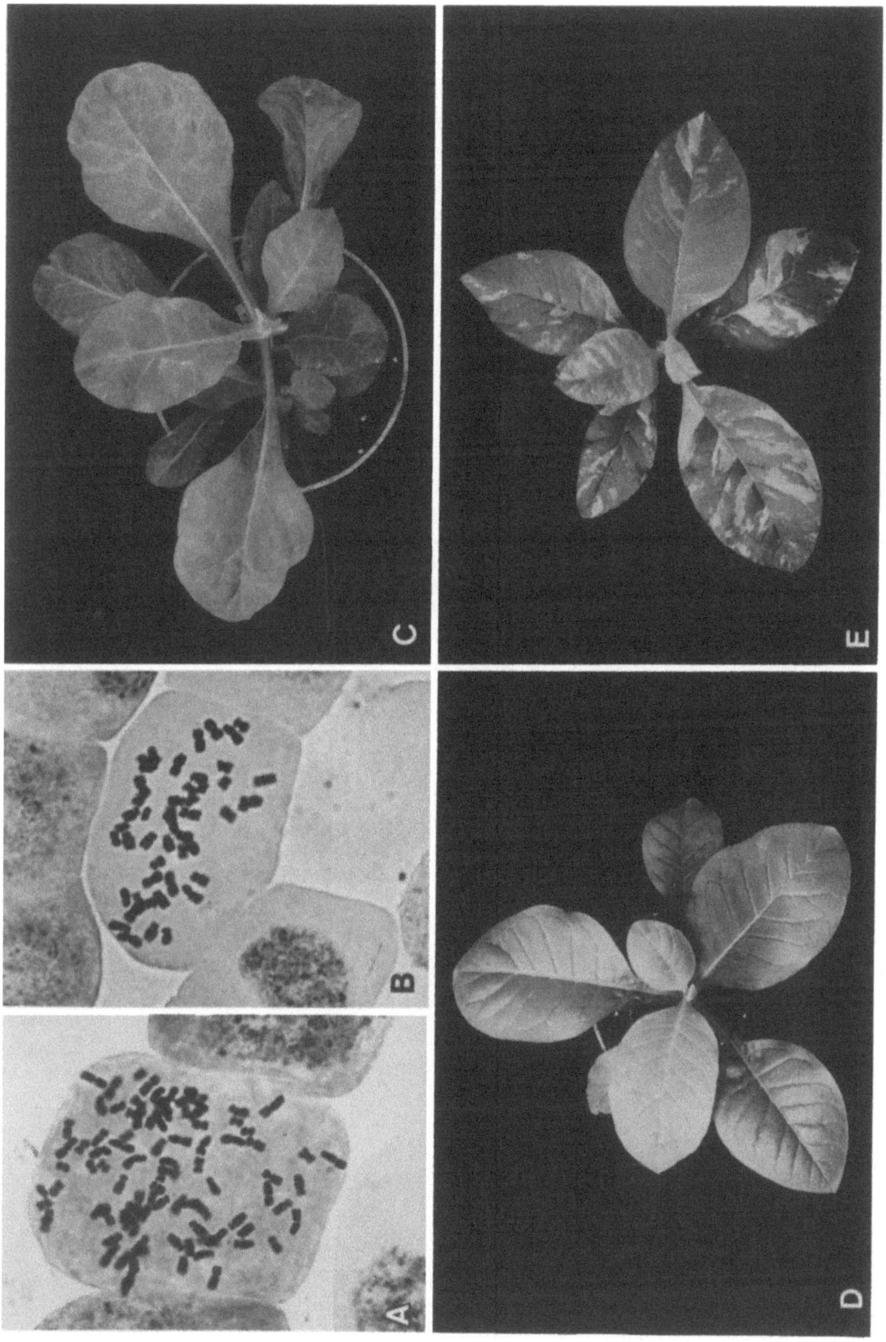

Fig. 26. Cytological and morphological analysis of somatic hybrids of *Nicotiana* (see Fig. 24 and 27). **A** A metaphase plate in a root cell of hybrid D 6-1 (Group A), ×1000. **B** A metaphase plate in a root cell of hybrid D 4-4 (Group B), ×1000. **C** Hybrid plant H 5-1 (Group A). **D** Hybrid plant H 6-5 (Group B). E Hybrid plant H 4-11 (Group B). (Gleba 1980)

were mechanically isolated in two independent experiments. In addition, 12 individually dividing parental callus cells and 6 mesophyll haploid protoplasts were isolated as controls (Fig. 24). Plants were regenerated from all the colonies obtained from the parental mesophyll protoplasts as well as from the fusion products (Fig. 24). Cells of one line obtained from a fusion product gave only severely malformed teratoma that failed to form roots and were not involved in the subsequent analysis. All of the colonies from the parental callus protoplasts were defective for morphogenesis, as was the initial parental callus line. A total of 350 mature plants were regenerated (5–25 from each clone). Study of these regenerants enabled the fusion products and thus the fusion events to be separated into three groups of clones with similar or identical characteristics.

Table 16. Chromosome numbers of tobacco plants regenerated from individual protoplast fusion products

Plant[a]	Chromosome no.	Plant	Chromosome no.
H1–1	24	D1–1	48
H1–2	24	D2–2	48
H1–3	24	D2–4	48
H1–4	24	D3–1	48
H1–6	24	D4–4	48
H1–11	24	D4–5	48
H2–2	24	D4–7	48
H3–1	48	D5–2	48
H6–1	48	D5–3	48
H6–4	48	D5–13	48
H6–6	48	D7–1	48
H7–2	48	D7–3	48
H7–3	48	D7–6	48
H8–5	48	D6–1	88–95
H8–11	24	D6–2	86–106
H8–12	24	D6–4	93–99
H8–13	24	D10–1	84–100
H13–3	24	D10–3	86–99
H5–2	86–94	D10–4	89–93
H5–5	75–92	D9–1	43
H5–6	82–98	D9–4	80–113
H9–1	68–73	D9–5	48
H9–2	71–72	D9–6	48
H9–3	66–72	D9–8	80–97
H9–4	72	D9–10	48
H9–5	71–75	D12–1	48
H9–9	68–74	D12–2	48
		D12–3	79–92

[a] Plants obtained from fusion products of callus and haploid mesophyll protoplasts are designated by H and from callus and diploid mesophyll protoplasts by D. The first number identifies a cloned fusion product and the second number designates one of the plants regenerated from that fusion product. Thus, for example, plants H1–1 through H1–11 represent six different regenerants from the same single fusion product H1

Group A included plants regenerated from cell clones H5 and H9 (obtained from fusion of callus with haploid mesophyll protoplasts) and cell clones D6 and D10 (from fusion of callus with diploid mesophyll protoplasts). All of the plants from these clones were morphologically aberrant with irregular leaf shape, leaf mosaicism, multicaulis plants, vine stem, and even in some cases variegation and albinism. A typical example of such a plant is illustrated in Fig. 26. These plants were not light sensitive. Chromosome counting, performed for 15 plants of the 4 clones, demonstrated that all of the plants were polyploid and aneuploid, with from 70 to more than 100 chromosomes and the chromosome number was variable in different cells of the same plant (Table 16). In Fig. 26, a metaphase plate with 88–92 chromosomes in a root cell of hybrid D6-1 is shown. Callus tissues were originated from petioles of five plants and after two subcultures they were grown on agar medium containing 10 mg l^{-1} of 5-methyltryptophan. The callus tissues of four plants (D6-1, D10-1, H9-1, H9-2) were found to be more resistant to 5-methyltryptophan as compared to cells of the sensitive parent (2GD; 2GH), but showed lesser resistance than cells of the parental strain with resistance to 5-methyltryptophan (Fig. 27). Self-pollination of 24 plants from clones of group A demonstrated a very low degree of self-fertility for 3 plants (0.01%–0.001%), while the remainder were completely self-sterile. The majority of the seedlings of the F1 progeny of plant D10-1 were light resistant (24 out of 29).

Group B included plants regenerated from 14 cells lines obtained from fusion products: H1, H2, H3, H6, H7, H8, H10, H13, D1, D2, D3, D4, D5, D7. All plants of this group were morphologically normal and similar to plants of var. Rosa (Fig. 26). They had sessile leaves typical for plants of this variety, whereas plants of var. Xanthi possess petiolated leaves. A few of these plants, however, revealed new characters, such as variegation and crimped leaf shape. All of the plants tested proved to be light sensitive. Chromosome counting was done for 34 plants from 14 clones and revealed either haploid or diploid chromosome sets in

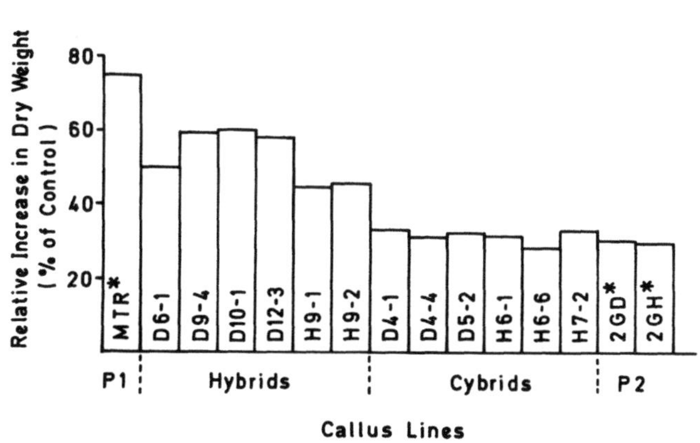

Fig. 27. Protoplast fusion and the fate of nuclear genetic determinants in the somatic hybridization of *Nicotiana tabacum*, var. Xanthi, resistant to 5-methyltryptophan (×) *N. tabacum*, var. Rosa, homozygous light sensitive. The relative increase in dry weight of various cell cultures in media containing 5-methyltryptophan was taken as a measure of resistance to this drug. The resistance of the hybrids is intermediate between the resistant parental type (*MTR*) and the sensitive parental type (*2GD, 2GH*)

Table 17. Combinations of different plant genera in which division of heterokaryons has been observed after protoplast fusion

Source of protoplasts		Ref.
Leaf mesophyll	Callus culture	
Hordeum vulgare	*Glycine max*	Kao et al. (1974)
Zea mays	*Glycine max*	Kao et al. (1974)
Pisum sativum	*Vicia hajastana*	Kao et al. (1974)
Pisum sativum	*Glycine max*	Constabel et al. (1975)
Melilotus alba	*Glycine max*	Constabel et al. (1975)
Medicago sativa	*Glycine max*	Constabel et al. (1975)
Caragana arborescens	*Glycine max*	Constabel et al. (1975)
Brassica napus	*Glycine max*	Kartha et al. (1974 b)
Hordeum vulgare	*Daucus carota*	Dudits et al. (1976c)
Nicotiana tabacum	*Glycine max*	Constabel et al. (1976)
Nicotiana glauca	*Glycine max*	Constabel et al. (1976)
Nicotiana rustica	*Glycine max*	Constabel et al. (1976)
Colchicum autumnale	*Glycine max*	Constabel et al. (1976)
Petunia hybrida	*Daucus carota*	Reinert and Gosch (1976)
Atropa belladonna	*Petunia hybrida*	Gosch and Reinert (1976)
Nicotiana glauca	*Glycine max*	Kao (1977)
Nicotiana tabacum	*Arabidopsis thaliana*	Gleba and Hoffmann (1978)
Nicotiana sylvestris	*Arabidopsis thaliana*	Gleba and Hoffmann (1978)
Brassica campestris	*Arabidopsis thaliana*	Gleba and Hoffmann (1978)
Brassica napus	*Arabidopsis thaliana*	Gleba and Hoffmann (1978)
Brassica nigra	*Arabidopsis thaliana*	Gleba and Hoffmann (1978)

all cases (Table 16). A typical metaphase plate with 48 chromosomes from the hybrid plant D4-4 is shown in Fig. 26. Calli were obtained from leaf petioles of five plants of group B clones and were grown on medium containing 5-methyltryptophan (Fig. 27). All of the plants tested were sensitive to 5-methyltryptophan at the same level as that of the sensitive parent (haploid and diploid cells of wild type: 2GD, 2GH). All 35 plants of group B studied were fertile and the F_1 progeny obtained by self-pollination were also light sensitive.

Group C consisted of the plants obtained from two fusion products, D9 and D12. Heterogeneity was observed among the regenerants of this group. Plants of group A type [abnormal morphology, polyploidy and aneuploidy, chromosome instability, light resistance, resistance to 5-methyltryptophan (Fig. 27), and self-sterile] and those of group B type (morphologically regular, diploid, light sensitive, and self-sterile) were found (Table 16).

A total of 28 variegated plants were found among the regenerants from a total of 19 clones from all three groups (Fig. 26). They represent 9% of the total of the regenerated plants. This variegation was transmitted uniparentally and maternally in control reciprocal crosses. Six albino plants were found in regenerants from two clones.

The 50 plants regenerated from clones of individual parental haploid mesophyll protoplasts were similar to haploid plants of the parental cultivar Rosa, with the one exception that 18 out the 30 plants analyzed were diploid.

Group A plants correspond to what can be expected in the case when protoplast fusion is accompanied by nuclear fusion. These forms are somewhat resistant to the tryptophan analog, aneuploid and display chromosome instability – which are all characteristics of the parental callus cell line. They also manifest nearly normal morphogenesis and have sessile leaf shape – which are characteristics of the cv. Rosa. Chromosome numbers of these plants correspond roughly to the simple additive sum of those of the parental cells. Thus, the plants of group A are true hybrids (i.e., hybrids in the common sense of the word) and both cell fusion as well as nuclear fusion of two different parental cells occurred in the protoplast fusion products which gave rise to clones H5, H9, D6, and D10.

Group B plants resemble plants of cv. Rosa in almost all characters (the exception is variegation, which is not a nuclear character). There are three important points to consider when trying to explain why these plants resemble only one of the parental types with respect to nuclear characters. First, it is improbable that a consistent error in judgment during the mechanical selection of the fusion products was made. Second, analysis of the variegation character demonstrates the heterozygosity of the cytoplasm of these clones, indicating that they arose from the fusion of two different cell types. (The variegation probably arose as a result of the mutant nature of the callus cell line cytoplasm.) Third, the proportion of cytoplasts contaminating the protoplasts used in these experimentals well below 1% of the total and thus the probability of fusion of a protoplast with a cytoplast is correspondingly low and cannot be responsible for all these plants. Keeping these points in mind, there are only two reasonable explanations for the sorting out of pure parental genotypes from heterokaryons. First, it cannot be totally excluded that nuclear fusion took place and was followed by selective and quantitative chromosome segregation. In our opinion, this seems quite improbable, since this chromosome segregation would have to cease with the production of strictly euploid plants. In addition, all the dominant alleles of the light resistance genes and all the codominant factors of resistance to 5-methyltryptophan should have been eliminated. Since these hybrids have an intraspecific combination of chromosomes, such a strict specificity in chromosome elimination of one of the parents cannot arise from the noncoordination of the mitosis mechanism, as occurs in hybrids between distantly related species (see Chap. 5). If such profound alterations in the mitosis mechanism are postulated to be a function of the long in vitro culturing of the callus cell line, it is incomprehensible why similar specific chromosome elimination did not occur in the progeny of all of the fusion products, including those of group A. Therefore, the sole satisfactory explanation of all these phenomena seems to be that the nuclei of the two parental cells did not fuse and afterwards segregated into different daughter cells. Then, only those cells with morphologically potent nuclei were selected at the stage of plant regeneration.

Analogous conclusions concerning the fate of nuclei in protoplast fusion products have also been drawn by Hungarian investigators from the results of their own experiments (Menczel et al. 1978). They examined the progeny from the fusion of *Nicotiana knightiana* mesophyll protoplasts with callus protoplasts of *N. sylvestris* (from a cell line with resistance to kanamycin, obviously as a result of extranuclear gene mutation; Dix et al. 1977; Maliga et al. 1977). The fusion was induced with polyethylene glycol at a high calcium concentration and high pH (Kao 1977; Maliga et al. 1977). The fusion products were identified visually on the basis of their morphology, were manually isolated with a micropipette 48–72 h after the beginning of culture, and were transferred into medium containing the parental protoplasts of the kanamycin resistant line of *N. sylvestris*. A total of 50 fusion products were introduced into culture. After 2 weeks, the colonies were transferred onto solid, organogenesis-inducing medium. The hypothetical hybrids were re-identified as green colonies (the cells of the parental line resistant to kanamycin are defective for photosynthesis and form albino colonies on this medium). A total of 45 green colonies were found and subsequently analyzed. Biochemical study of esterase and dehydrogenase isoenzymes revealed that 21 of the 25 cell lines analyzed contained only isoenzymes specific for *N. knightiana*. Four lines possessed a number of specific isoenzymes typical for both species as well as some novel isoenzymes. These four lines were capable of morphogenesis (both parental forms were defective for this character; Maliga et al. 1977) and their cells revealed resistance to kanamycin. Therefore, these lines were defined as true hybrids. The cells of the remaining 20 lines were defective for morphogenesis (with the exception of one line) and were sensitive to kanamycin. The investigators explained these results as follows. Cell hybridization is followed by nuclear hybridization in at least 15% of the cases (4 out of 24) whereas, in the remaining cases, only the exchange of cytoplasm and organelles takes place, giving rise to colonies with the nuclei of *N. sylvestris* and the green plastids of *N. knightiana*. It should be stressed that in this study cell hybridization of *different* plant species was analyzed and this might have affected the observed frequency of nuclear fusions. Also, the term "cloning" used by the authors seems to be inappropriate in this case, since cell colonies in mixed culture can aggregate with other colonies or themselves disassociate, leading to errors in the quantitative aspects of the experiment.

All of these experiments support a general conclusion that *the nuclear genetic determinants are inherited both bi- and uniparentally in plant hybridization by protoplast fusion. The uniparental inheritance appears to be a consequence of the nonfusion of the nuclei followed by their subsequent segregation during divisions of the heterokaryocytes*. Such a segregation is observed in both intra- and interspecific species combinations. The segregation is observed in heterokaryocytes obtained by fusion of mesophyll cells with callus cells as well as those obtained via fusion of two mesophyll cells.

The fate of nuclei and nuclear genes in hybridization of closely related forms in experimental systems that do not employ deliberate use of physical or chemical influences, which might affect the genetic processes in the fused cells, have been examined. The hybridization of distantly related species is discussed in Chap. 5 and the problems of transmission genetics in closely related parasexual crosses where X-ray and gamma ray irradiation or inactivation of one of the parents by iodoacetate have been utilized are discussed in the following section (4.4). Some of these experiments should be commented on here since their results are in apparent contradiction with the conclusions of this section. Menczel et al. (1982) and Sidorov et al. (1981 a) observed a low percentage of nuclear segregants (1.3% –1.4%) upon hybridization of mesophyll protoplasts of tobacco and *N. plumbaginifolia*, where one of the parents was treated with iodoacetate. After hybridization of callus and mesophyll protoplasts in the distantly related combinations *Arabidopsis* (×) *Brassica* (Gleba and Hoffmann 1978, 1979, 1980) and *Atropa* (×) *Nicotiana* (Gleba et al. 1982a–c), all of the mechanically isolated heteroplasmic fusion products (with one exception) from a total of 26 clones have given rise to clones of true nuclear hybrids. Such a high percentage of nuclear hybrids and the paucity of cybrids in a number of experiments might be explained as a peculiarity of the individual experimental system under investigation. However, in some experiments hybrids have demonstrated unequivocal growth heterosis, for example, in the case of *Arabidopsis* (×) *Brassica* system. It is evident that under these conditions, an overt selection for nuclear hybrids operates. Likewise, only a portion of the fusion products survive mechanical isolation and cloning and therefore a certain selection functions also, albeit poorly understood.

4.3.2 Chromosome Sets of Hybrid Cells

Prior to the progress of the last decade in the cultivation and fusion of isolated protoplasts and the actual hybridization of higher plants by somatic cell fusion, there were no assurances that the cell hybrids produced by such an "unnatural" method would be capable of regeneration of normal hybrid plants capable of subsequent sexual reproduction. This problem initially appeared to be very complicated since the natural process of sexual hybridization is so strictly regulated and is also accompanied by the complex events of gamete production for sexual fusion. Thus, it was especially surprising to discover that the first somatic hybrids produced were so apparently normal (as previously mentioned, the term "normal" refers to their resemblance to sexual hybrids between the same parents).

It is even now surprising how very little time has been required to elaborate techniques of parasexual hybridization that permit not only the formation of novel somatic hybrids, but also the regeneration of fertile hybrid plants (although admittedly for a limited number of species combinations).

An informative index of the nature of a parasexual hybrid is the characteristics of its chromosome set, namely, the number and type of chromosomes and the stability of the chromosome number. From the preceding chapters, it is clear that some of the hybrids produced by protoplast fusion are products of the fusion of more than two cells, while others bear the nucleus of only one parent and have a hybrid or mixed cytoplasm. This section deals with specific experiments concerning chromosome sets of true plant hybrids, which for the most part, have apparently originated from the fusion of two cells and the fusion of their nuclei. It is evident that the chromosome numbers of hybrid plants also provides information on the possible and/or probable number of cells that fused to give rise to the hybridization products that were regenerated into these plants. Those investigations dealing with intraspecific hybridization will be considered first.

Melchers and Sacristan (1977) performed a very extensive, and to date, the most complete study of chromosome numbers in hybrids obtained by fusion of haploid mesophyll protoplasts of tobacco (Melchers and Labib 1974). (It is known that cells of tobacco leaf mesophyll are homoploid and haploid in haploid plants; Nagata and Takebe 1971; Gleba et al. 1978.) Among the regenerants obtained from 55 different hybrid callus colonies, 30 were diploid, 5 were triploid, and 6 were tetraploid. Moreover, aneuploids with near diploid (four plants), triploid (eight plants), or tetraploid (two plants) chromosome numbers were also found. Thus, the majority of the hybrids were genetically regular, fertile amphiploids. This would be expected when cell fusion is accompanied by full integration of their nuclei and when the culture conditions or internal processes in the hybrid cells do not result in chromosomal changes. At the same time, a relatively large number of aneuploid plants was observed.

It is a well-known and generally occurring phenomenon that there is considerable variation in chromosome numbers of cells of tobacco tissues cultured in vitro. Thus, no special explanation is necessary to account for the formation of aneuploid hybrid plants, since the hybrid cells were grown in in vitro culture for a sufficient number of generations from the moment of fusion until plant regeneration to account for the aneuploidy.

Schieder (1977) reported data on the chromosome numbers of intraspecific hybrids of *Datura innoxia* (2n = 24). In this study, mesophyll cells of diploid plants were fused. (These protoplasts, being derived from diploid plants, were evidently euploid and most likely diploid.) The majority of hybrid plants obtained from 20 hybrid cell colonies were also euploid (5 tetraploids, 8 hexaploids, and 3 octoploids), while 4 were aneuploids. Similar data, although less extensive and therefore more ambiguous, on chromosome numbers of intraspecific hybrids have also been reported by other investigators (Gleba et al. 1975; Belliard et al. 1977a, b; Glimelius and Bonnett 1981; Glimelius et al. 1981; in the last two studies, callus cells were used as parental forms).

Unfortunately, these experiments provide no indication of to what extent the ratio of euploid to aneuploid regenerants reflects the ratio of these cell types in the hybrid calli from which they were obtained. This is important since the process of regeneration itself is known to screen for definite genotypes and in this situation, euploid cells with small chromosome sets seem to have some advantages. Moreover, cells with different chromosome sets surely experience different selective pressures in specific conditions of cultivation. For example, Schieder (1977) reported that colonies of aneuploid hybrid cells of *Datura innoxia* grew much more slowly than colonies of euploid (especially hexaploid) cells.

If the genetic regularity of the parental cells is inherited by at least some of the parasexual progeny, the question naturally arises of whether genetic instability is inherited as well. This problem still has not been seriously studied. However, a positive answer to this question seems to be provided by our experiments with cloning the fusion products of haploid and diploid mesophyll cells with aneuploid tobacco callus cells (Gleba and Berlin 1979; see Gleba 1980). Chromosome instability and hyperdiploid chromosome sets are characteristic for the tobacco callus line used. The nuclear hybrids obtained by the fusion of stable with unstable cells gave rise to hybrid plants with chromosome instability. Control hybrids containing the nuclei of stable cells and the cytoplasm of unstable cells did not exhibit chromosome instability. The quantitative aspects of this instability and whether or not the instability of the hybrid applies equally to the chromosomes from both parental cells is of obvious interest and is worthy of further investigation.

The interpretation of events that lead to the production of hybrids with different chromosome sets is even more complicated in those cases of interspecific somatic hybridization. Most investigations have dealt with just this sort of hybridization. Chromosome instability is a normal phenomenon for many species of plants commonly employed in interspecific hybridization.

A priori, it would be expected that the somatic hybrids in these same species combinations would also display this instability. A good example of this is found in two studies of hybridization of *Nicotiana tabacum* (×) *N. rustica*. Nagao (1978) fused diploid mesophyll protoplasts and regenerated 13 hybrid plants. Without exception, all of these plants were found to be aneuploid and possessed 60–91 chromosomes, instead of the expected 96. The variability of chromosome number among different cells of the same plant was not mentioned. Most of the 13 plants exhibited a high degree of fertility. In a separate study, Douglas et al. (1981 a–c) fused protoplasts from freshly initiated callus cultures (4-months-old for *N. tabacum* and 18-months-old for *N. rustica*). All but one of the 15 regenerants studied possessed less than 96 chromosomes (the range was from 68–96) and the chromosome numbers of different cells from the same plant did not differ significantly. In some cases, chromosome sticking and fragmentation were observed. Additionally, these hybrid plants were essentially sterile. In both experiments, aneuploid plants which had lost several chromosomes of one or both parents were produced. These results are in good agreement with the data of analyses of sexual hybrids between *N. tabacum* and *N. rustica* (Kostoff 1943; Murty and Swaminathan 1957). The sexual hybrids also display chromosome instability.

Similar chromosome instability, perhaps resulting from genetic noncoordination of mitotic separation of chromosomal material, is also found in the somatic hybrids *Nicotiana glauca* (×) *N. langsdorffii* (Smith et al. 1976; Chupeau et al. 1978). In the experiments of Smith et al. (1976), 23 plants regenerated from 19 different hybrid callus colonies were examined. The chromosome numbers of all of them varied from 56 to 64. The origin of 10 of these plants possessing exactly 60 chromosomes might conceivably be explained as a result of the fusion of two *N. langsdorffii* cells (2n = 18) with one cell of *N. glauca* (2n = 24), i.e., triple fusion. The remainder of the plants were aneuploid. Chupeau et al. (1978) studied chromosome numbers in regenerants and the corresponding callus tissues. They demonstrated that cells of the 16 callus lines studied were extremely variable in chromosome number, ranging from 28 to 143, and that most cells were aneuploid. However, two amphidiploids (2n = 42) as well as several plants with unstable chromosome sets containing from 55–80 chromosomes were found among the regenerants. The chromosome instability in culture and the tendency for the chromosome number to increase during unorganized cell growth in culture make highly probable the assumption that those regenerants with 60 chromosomes in the experiments of Smith et al. (1976) were a consequence of evolution of the chromosome number in in vitro culture, rather than as a result of their formation from the fusion of three cells. Chromosome instability has also been described for the sexual hybrids *N. glauca* × *N. langsdorffii* and is also characteristic for tumorous cells in vivo (Burk and Tso 1960).

Somatic hybrids between the sexually incompatible species *Nicotiana tabacum* and *N. repanda* have recently been produced by Nagao (1982). Although both species have 2n = 48 chromosomes, the 3 hybrids recovered had 52–64 chromosomes, which is considerably less than the sum of the parental chromosomes.

In the light of these observations, the chromosome instability in somatic hybrids *N. tabacum* (×) *N. knightiana* (Maliga et al. 1978) are quite explicable, even though we have no data about the genetic instability of sexual hybrids between these species.

In contrast to the number of interspecific somatic hybrids that exhibit chromosome instability, it is especially interesting and important that many interspecific somatic hybrids have also been produced that are quite stable amphieuploids. Examples of these stable hybrids include some species combinations of *Nicotiana* (Chupeau et al. 1978; Nagao 1979; Medgyesy et al. 1980; Evans et al. 1980, 1981; Menczel et al. 1981, 1982; Uchimiya 1982), *Datura* (Schieder 1978a, 1979), *Petunia* (Power et al. 1976, 1980), and *Daucus* (Dudits et al. 1977).

In conclusion, it may be stated that *the hybridization of somatic cells in closely related species combinations may result in the production of normal fertile plants containing the sum of the parental chromosomes sets*, at least in some of the cases. The divergence from amphieuploid values by a considerable percentage of the regenerants from hybrid cells may be due to several causes, including (1) aneuploidy and/or chromosome instability of one or both of the parents; (2) suboptimal culture conditions at the stage of unorganized growth of the hybrid cells in vitro; and (3) genetic noncoordination of processes in the hybrid cells.

4.4 Transmission Genetics in Hybridization Systems Utilizing Parental Cell Inactivation

4.4.1 Consideration of Animal and Bacterial Systems

A priori, it may be assumed that the process of parental cell inactivation may have far-reaching consequences on the subsequent genetic processes within the fusion product and its progeny. Thus, the use of specific inactivating agents might also be used to manipulate the fate of the parental genetic material. Such an approach was first formulated and practically demonstrated by Pontecorvo (1971) using animal somatic hybrids formed by fusion of mouse and Chinese hamster cells, both of which have morphologically distinguishable chromosomes. The hybrid cell clones of these species are usually stable and do not eliminate chromosomes in a selective, species specific manner. Irradiation of the cells of one of the parents with X-rays and gamma rays, as well as treatment with bromodeoxyuridine and exposure to visible light, led to the preferential loss of the chromosomes of this parent during subsequent replication of the hybrid cells. In particular, irradiation of Chinese hamster cells with a dose of 600 roentgen (r) resulted in the loss of six chromosomes and a decrease in the plating efficiency to 15% of the control. Doses of 1000 r led to increased asymmetry, a decrease in viable clones and an increase in the frequency of chromosome aberration. At doses of 1600 r, no viable hybrids were produced. In the experiments with bromodeoxyuridine, selective chromosome elimination was observed after treatment with 10^{-5} molar bromodeoxyuridine for 20–22 h and illumination with visible light 24 h after hybridization.

Conceptionally similar experiments have been performed with bacteria: treatment with streptomycin, heating and UV-irradiation have been used as inactivating agents (see the excellent review of genetic studies with bacterial protoplasts by Hopwood, 1981). While the heating and the treatment with streptomycin significantly decreased the production of recombinants, high doses of ultraviolet light irradiation were found suitable for hybrid selection, since the cell inactivation rate upon increase of the dose was considerably higher than the decrease in the formation of recombinants (hybrids). For example, when the survival of the irradiated cells approached 1%, the number of recombinants constituted about 25% when compared to the control and at a survival rate of 0.01%, the production of recombinants was still 10% that of the control. The ultraviolet light irradiation resulted in the elimination of nonselective characters and in the selection of recombinants with individual "donor" markers, whereas the treatment with streptomycin or heating did not markedly affect the pattern of recombination of parental traits.

Finally, it should also be mentioned that similar works have already been performed with higher plants without the use of somatic hybridization. Pandey (1975) reported that pollination of tobacco with pollen irradiated with high radiation doses results in lower frequency transfer of some characters from donor to recipient. Further work in this area has been reported (Pandey 1979). Such investigations, jokingly termed "genetic engineering for poor men," are currently underway in a number of laboratories.

The most systematic and complete studies of the influence of two different inactivating agents (iodoacetate and irradiation) on genetic processes and genetic constitution of the progeny of higher plant somatic hybrids have been performed in the laboratory of Maliga. Interspecific hybridization in *Nicotiana* was chosen as the experimental system to investigate these phenomena. In this chapter, only those studies concerning closely related hybridizations will be described. The single report of an intergeneric species combination, which used irradiation to affect the transfer of chromosome markers in *Petroselinum* (×) *Daucus* (Dudits et al. 1980), will be discussed in Chap. 5.

4.4.2 Treatment with Iodoacetate

The treatment of tobacco mesophyll protoplasts with 10 m*M* iodoacetate for 30 min at 25 °C leads to the inactivation of the protoplasts; however, this may be reversed by hybridization (Menczel et al. 1981). Fusion of streptomycin resistant tobacco protoplasts, inactivated with iodoacetate, with mesophyll protoplasts of *N. sylvestris* and subsequent selection for antibiotic resistance permitted Medgyesy et al. (1980) to isolate 137 resistant clones. Of these clones, 105 were nuclear hybrids, 12 were a mixture of hybrid cells of one of the parents, and 16 were cybrid cells of *N. sylvestris* with tobacco plastids. Thus, the inactivation process not only permits efficient screening of the somatic hybrids, but also does not significantly affect the production of true nuclear hybrids among the progeny (in this experiment, more than 85% of the clones analyzed were true nuclear hybrids).

Sidorov et al. (1981 a) fused mesophyll protoplasts of *Nicotiana plumbaginifolia*, inactivated with 3 m*M* iodoacetate for 20 min at 7 °C, with cells of a tobacco plastome mutant, irradiated with gamma rays (6 krad). Among the 40 clones obtained, 23 were nuclear hybrids, 2 contained both hybrid cells and cells of *N. plumbaginifolia*, and 15 were of *N. plumbaginifolia* type. The plastids from 31 of 34 of these clones were derived from the irradiated tobacco cells. No segregants containing only tobacco nuclei were isolated in these experiments. Therefore, it seems that the treatment with iodoacetate did not lower the production of true nuclear hybrids as compared to the nontreated control obtained by mechanical cloning, which had a frequency of nuclear hybrids close to 100%. In similar studies, Menczel et al. (1982) fused mesophyll cells of a streptomycin resistant tobacco mutant treated with 10 m*M* iodoacetate for 30 min at 25 °C with mesophyll protoplasts of *N. plumbaginifolia* as a control. The proportion of nuclear hybrids resistant to streptomycin among the regenerants was about 98%–99% and was thus similar to the results of Sidorov et al. (1981 a).

The inactivation of carrot callus cells with 30 m*M* iodoacetate for 30 min at 0 °C enabled Lazar et al. (1981) to select for nuclear hybrids.

Thus, these data indicate that cell inactivation by iodoacetate does not lead to qualitative nor visible quantitative changes in the spectrum of genetic diversity of the plant forms produced by parasexual hybridization.

4.4.3 Irradiation

Menczel et al. (1982) have studied the influence of various doses of gamma ray irradiation of streptomycin resistant (mutant) tobacco protoplasts on the fate of nuclear and plastome genes in fusion products of mesophyll protoplasts of *Nicotiana tabacum* (×) *N. plumbaginifolia*. Tobacco cells were inactivated by iodoacetate instead to irradiation for the selection of hybrids in the control. The control experiments demonstrated that without irradiation the proportion of true nuclear hybrids among the progeny in this system constitutes 98%–99%. The frequency of nuclear hybrids obtained after application of doses of 50, 120, 210, and 300 J kg^{-1} (5, 12, 21, and 30 krad) constituted 56%, 43%, 17%, and 30%, respectively. The segregants possessed tobacco plastids (i.e., from the irradiated donor) and were strictly euploid diploid or tetraploid. Thus, these experiments persuasively demonstrate that (1) gamma ray irradiation results in inactivation of the nucleus and elimination of the nuclear genes of the irradiated parent; (2) an increase in radiation dose leads to a decrease in the frequency of viable nuclear hybrids; (3) the plastids and the plastome are apparently not affected by the radiation doses employed; (4) the inactivated nuclei are probably completely rather than partially eliminated from the segregants; and (5) the irradiation does not result in visible mutations.

Data of other investigations employing similar doses of radiation to inactivate one of the parents for the selection of hybrids are in good agreement with the results of Menczel et al. (1982). Hybrids of this same species combination were isolated by Sidorov et al. (1981) after fusion of *N. tabacum* mesophyll protoplasts with *N. plumbaginifolia* callus protoplasts. The hybrids were isolated by both mechanical isolation and cloning of the individual heterokaryocytes and by inactivation of the tobacco cells by irradiation and of the *N. plumbiginifolia* cells by iodoacetate. All of the ten mechanically isolated clones were found to be nuclear hybrids. One clone segregated both hybrid and tobacco type plants and seemed to be the result of fusion of more than two protoplasts. Analysis of the progeny from the 40 cell lines obtained by double inactivation showed that 23 were nuclear hybrids, 15 were *N. plumbaginifolia* type with most possessing tobacco plastids, and 2 lines regenerated both nuclear hybrid and *N. plumbaginifolia* type plants. In this case, the gamma ray irradiation (60 J kg^{-1}) decreased the frequency of nuclear hybrids from 100% to about 60% and the irradiated nuclei were eliminated in the segregants. At the same time, the irradiation not only did not inhibit, but rather seemed to stimulate the primary transfer of plastids in 31 of 34 cases. These plants are further discussed in section 6.2.4.

In other studies (Zelcer et al. 1978; Aviv and Galun 1980), the inactivation of protoplasts was accomplished using X-rays. The treatment of tobacco protoplasts with a dose of 42 J kg^{-1} resulted mainly in the formation of nuclear segregants (13 of 15 recombinants) in the system *N. tabacum* (×) *N. sylvestris*. Since the use of iodoacetate as an inactivating agent in the same species combination produces recombinants, about 85% of which are nuclear hybrids (Medgyesy et al. 1980), one can state that the irradiation apparently suppresses the formation of true hybrids and increases the proportion of nuclear segregants.

To date, there is insufficient data for a rigorous consideration of the effect of irradiation on the screening of organelles of either parent and mutation processes arising as a consequence of the irradiation.

4.5 Genetic Diversity of Hybrid Plants Obtained by Protoplast Fusion

The investigations described in this chapter illustrate that parasexual hybridization by protoplast fusion makes posible the production of hybrid plant forms with new combinations of parental genes. A synthesis of these data permits a comparison of the transmission genetics of parasexual hybridization by protoplast fusion to that of sexual crossing in *Nicotiana* (Fig. 28). The nuclear genes are inherited both uni- and biparentally in parasexual hybridization, whereas they are inherited only biparentally in the sexual process. Extranuclear genes are inherited biparentally in hybridization by protoplast fusion, but maternal, uniparental inheritance of these genes takes place in the sexual process. Since the plant cell possesses at least two distinct cytoplasmic genophores (those in chloroplasts and mitochondria), the scheme of transmission genetics after parasexual hybridization is complex (Fig. 28).

An additional complication also arises as a result of mitotic segregation of extranuclear genes (genophores) upon cell division of the fusion products (Fig. 28). Taking this into consideration and assuming that the higher plant cell has only two extranuclear genophores, the fusion of protoplasts of two parental types may theoretically yield 27 different combinations of nuclear and extranuclear gene systems (the possible recombination of extranuclear genomes has been excluded in this consideration). This is in stark contrast to the sexual process where only two combinations may be obtained by reciprocal crosses.

Additional diverse forms might be produced by parasexual hybridization as a result of selection of genetically heterogeneous copies of parental mitochondrial

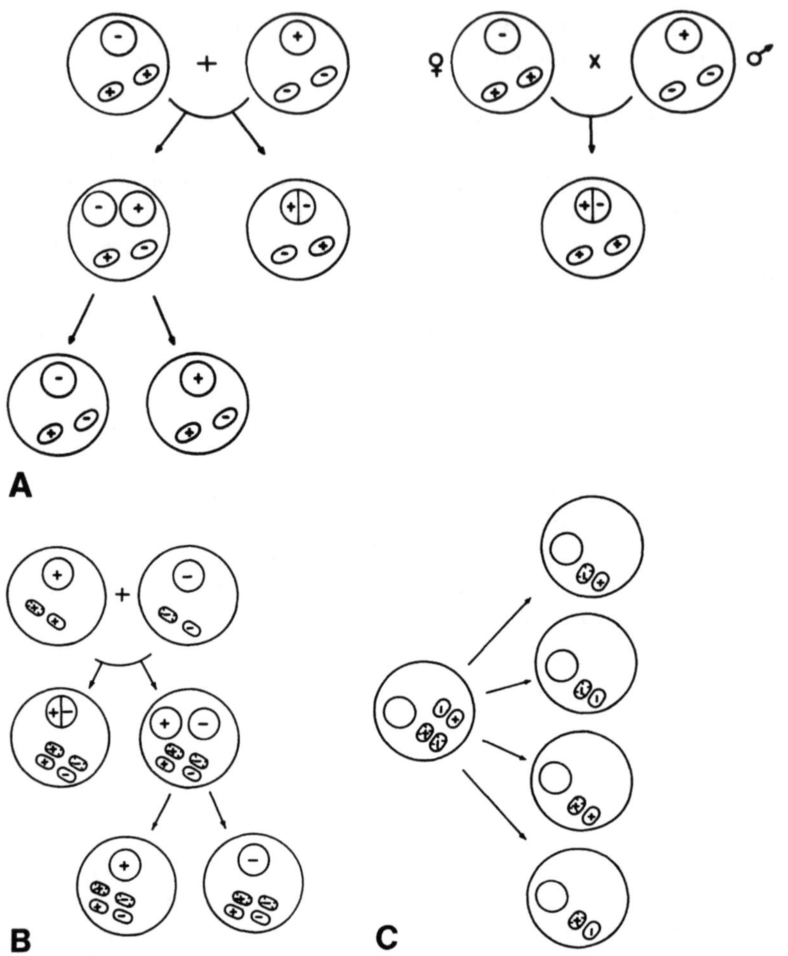

Fig. 28. Potentials for inheritance of determinants in parasexual and sexual hybridization. **A** The inheritance of parental nuclear (*circles*) and extranuclear (*ovals*) determinants in parasexual (*left*) and sexual (*right*) hybridization. **B** The same scheme as in **A** for parasexual hybridization, taking into account the inheritance of two independent extranuclear genophores. **C** The progeny of cells formed via parasexual hybridization may also be novel segregants for extranuclear genophores (this is in addition to the cytoplasmic heterozygotes shown in **B**)

DNA in hybrid cells and by genetic recombination of plasmagenes. In particular, protoplast fusion enables the synthesis of novel:

1. plants heterozygous for extranuclear genes;
2. plants bearing nuclei from one parent and cytoplasm from both or another parent;
3. plants bearing some extranuclear genes of one parent (for example, plastome) and the other extranuclear genes from the other parent (mitochondrion);
4. plants with reconstructed extranuclear genes as a result of genetic recombination.

5 Protoplast Fusion and Hybridization of Distantly Related Plant Species

5.1 Introduction

Parasexual hybridization of distantly related species of higher plants is the most interesting trend of the investigations involving the reconstruction of eukaryotic plant cells. The fate of a newly formed hybrid cell depends greatly upon the degree of coordination of its mechanisms. The subsequent potentiality of the hybrid is a measure of this coordination or compatibility. Such hybrid cells may reveal metabolic activity which was absent in both parental cells via genetic complementation as well as give rise to hybrid cell progeny via reproductive potentiality or even regenerate an entire hybrid plant via morphogenetic potentialiality. All of these potentialities of hybrid cells constructed by protoplast fusion may be experimentally studied. These studies may yield new data concerning the mechanisms of various cellular activities and in particular may give clues as to the phylogenetic similarities of these mechanisms. Analogous studies of animal hybrid cells have clearly demonstrated the broad potentials and perspectives of this type of genetic investigations (Ringertz and Savage 1976).

Moreover, the hybridization of distantly related plant species by protoplast fusion is of practical interest as a potential tool for bypassing the incompatibility barriers inherent in sexual crossing of agriculturally important plant species.

5.2 Initial Stages of Culture of Hybrid Cells of Distantly Related Plant Species

There are several reports on attempts to produce cell hybrids between distantly related plant species (Table 17). In many of the studies, direct visual observation of the hybrids was possible because morphologically different parental cell types, generally callus and mesophyll protoplasts, were fused. In most cases, however, the observations were limited to the first few divisions of the fusion products and in only a few cases were attempts made to isolate the hybrid cell lines. The use of this type of observation is obviously rather limited, but nonetheless is sufficient to illustrate the important fact that hybrid cells in some interfamily combinations are able to subsequently divide several times.

Unfortunately, cytological analysis of hybrid cells have not provided precise information concerning the fate of parental genetic material in the process of cell hybridization of distantly related species. Although cytological studies are quite numerous, most of them are not rigorous and thus permit the formation of assumptions rather than conclusions. The inferences of a number of investigators have frequently been based only on the visual observation of divisions in heterokaryocytes, whereas the conjugation of nuclear material and the formation of common metaphases has been demonstrated only for a few cases (Kao et al. 1974; Constabel et al. 1975a, 1977; Gosch and Reinert 1976; Kao 1977).

The behavior of individual fusion products during the initial stage of cultivation seems to be determined by the state of the parental cells before fusion. The fact that the fusion of a callus cell with a mesophyll cell induces the mesophyll cell nucleus to undergo mitosis during one to three successive divisions still provides no evidence for the capacity of the fusion product for multiple reproduction later, when the molecular milieu within the cell reflects the action of genes of the hybrid cell rather than that of the parents. Electron microscopic studies have shown that hybrids produced by protoplast fusion experience mixing of their cytoplasm within a few hours. This has been observed in different species combinations (Fowke et al. 1975a, 1976, 1977; Sidorov et al. 1978).

Mitosis in intergeneric heterokaryons is frequently synchronous; in other cases, it is asynchronous (Kao et al. 1974; Constabel et al. 1975a, 1977; Gosch and Reinert 1976; Kao 1977). The number of di- and multinuclear heterokaryocytes as a rule decreases during the first few days of cultivation. There is one report on the potentiality of hybrid cells *Vicia* (×) *Petunia* with two nuclei for dividing and forming daughter cells with two nuclei (Binding and Nehls 1978a). Apparently, multinuclear fusion products perish in most cases. Early studies in which fusions of interphase nuclei were observed (Dudits et al. 1976; Constabel et 1975a; Fowke et al. 1977; Binding and Nehls 1978) have recently been critically evaluated (Constabel 1978) and it is believed that the observed conjugation took place mostly in dying heterokaryocytes. The principal mode of conjugation is the formation of a common metaphase plate by the parental chromosomes and the utilization of a two pole spindle. Multipolar mitosis in heterokaryocytes is observed more rarely (Kao et al. 1974) (Figs. 29 and 30).

After five or six cell generations, the two parental chromosomal sets are not completely integrated in intergeneric hybrid cells. As a rule, the chromosomes of each parental type are grouped together and a separate localization of each type is also observed within the common interphase nucleus.. These inferences were made by Constabel et al. (1977), who studied the hybrids produced by fusion of *Glycine max* and *Vicia hajastana* callus cells (Fig. 31). A similar, separate grouping of parental chromosomal material in interphase nuclei and in the mitotic cycle was also characteristic for products of the first divisions of the interfamily hybrids *Glycine max* (×) *Nicotiana glauca* (Kao 1977) (Figs. 30 and 32). The mechanisms of these phenomena are still not known. Disturbances in mitosis from the first division of hybrid cells, resulting in the formation of anaphase bridges and reconstructed chromosomes (i.e., super long, ring, or fragmentary), are typical for interfamily cell hybrids, in contradistinction to intergeneric cell hybrids. These observations are in good agreement with the result of analysis of cell clones of a number of intertribal and interfamily hybrids several months after hybridization, as discussed later. However, since the data concerning the cytological analysis of processes in reconstructed systems are still insufficient, the drawing of general conclusions is somewhat premature.

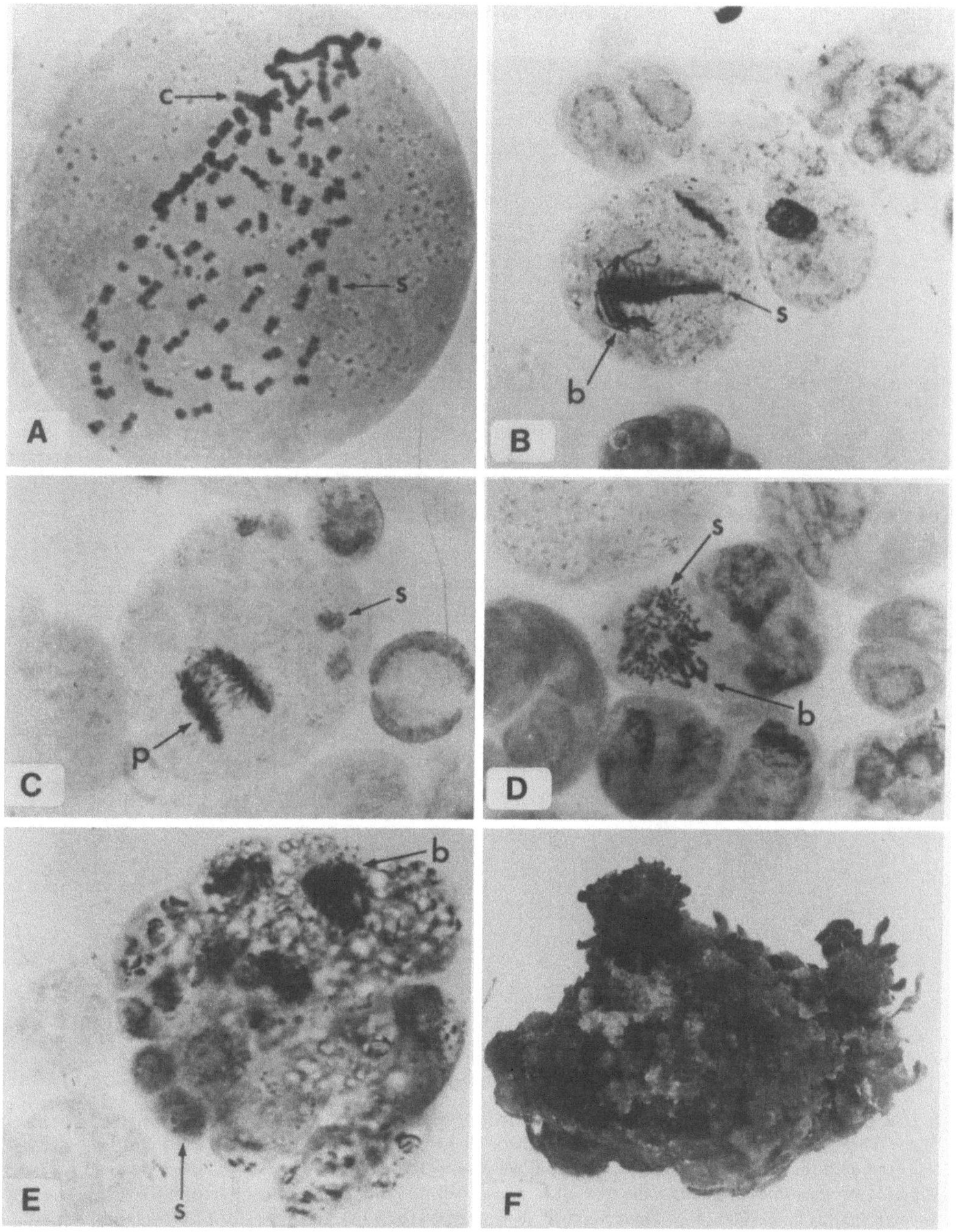

Fig. 29. Cytological analysis of dividing intergeneric hybrid cells. **A** Synchronized nuclear divisions in a corn (×) soybean heterokaryocyte after 5 days of culture, × 1 100. Both corn (*c*) and soybean (*s*) chromosomes are visible. **B** Slightly unsynchronized nuclear divisions in a barley (×) soybean heterokaryocyte after 5 days of culture, × 500. Barley (*b*) and soybean (*s*) chromosomes are present. **C** A pea (×) soybean heterokaryocyte after 3 days of culture. A synchronous anaphase of the pea (*p*) and soybean (*s*) regions with multiple poles may be seen, × 500. **D** A soybean (×) barley hybrid undergoing its third cell division after 5 days of culture, × 500. Soybean (*s*) and barley (*b*) chromosomes are visible in different regions of this metaphase. **E** A chimeral cell colony of barley and soybean cells, probably derived from a heterokaryocyte, 5 days after culturing (*b* barley nucleus; *s* soybean nucleus), × 500. **F** A tumor-like callus of *Nicotiana glauca* and *N. Langsdorffii* cells derived from heterokaryotic protoplasts, × 4 (Kao et al. 1974)

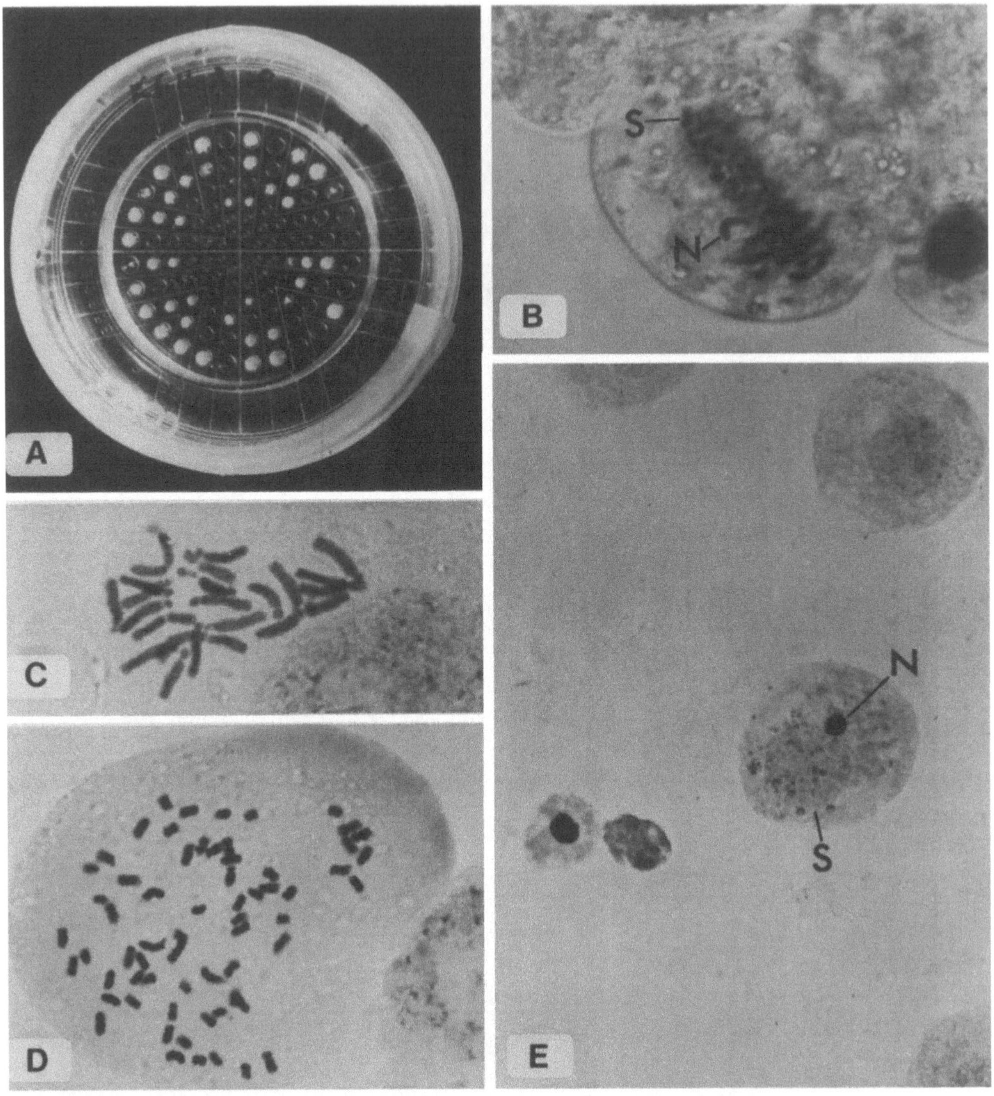

Fig. 30. Chromosomal behavior in somatic hybrids of *Glycine max* (×) *Nicotiana glauca*. **A** A Cuprak dish with prolific growth of the hybrid cells in the wells of the inner chamber. **B** A hybrid cell during the first mitosis, with soybean (*s*) and *N. glauca* (*n*) chromosomes, 48 h after fusion, ×1200. **C** Chromosomes of a colchicine treated *N. glauca* cell, ×1200. **D** Chromosomes of a colchicine treated *G. max* cell, ×1200. **E** A heterokaryocyte with a soybean (*S*) and *N. glauca* (*N*) nucleus. Parental protoplasts are also visible. (Kao 1977)

Only a few studies have dealt with the fate of cytoplasmic organelles in the process of fusion and culture of heterokaryocytes. These experiments have been performed primarily with the aid of the electron microscope (Fowke et al. 1975a, 1976, 1977; Sidorov et al. 1978; Kajita et al. 1980) (Fig. 33). The results of these studies, taken as a whole, suggest that there are no pronounced incompatibility processes during the first several days of cultivation of these hybrids. An excep-

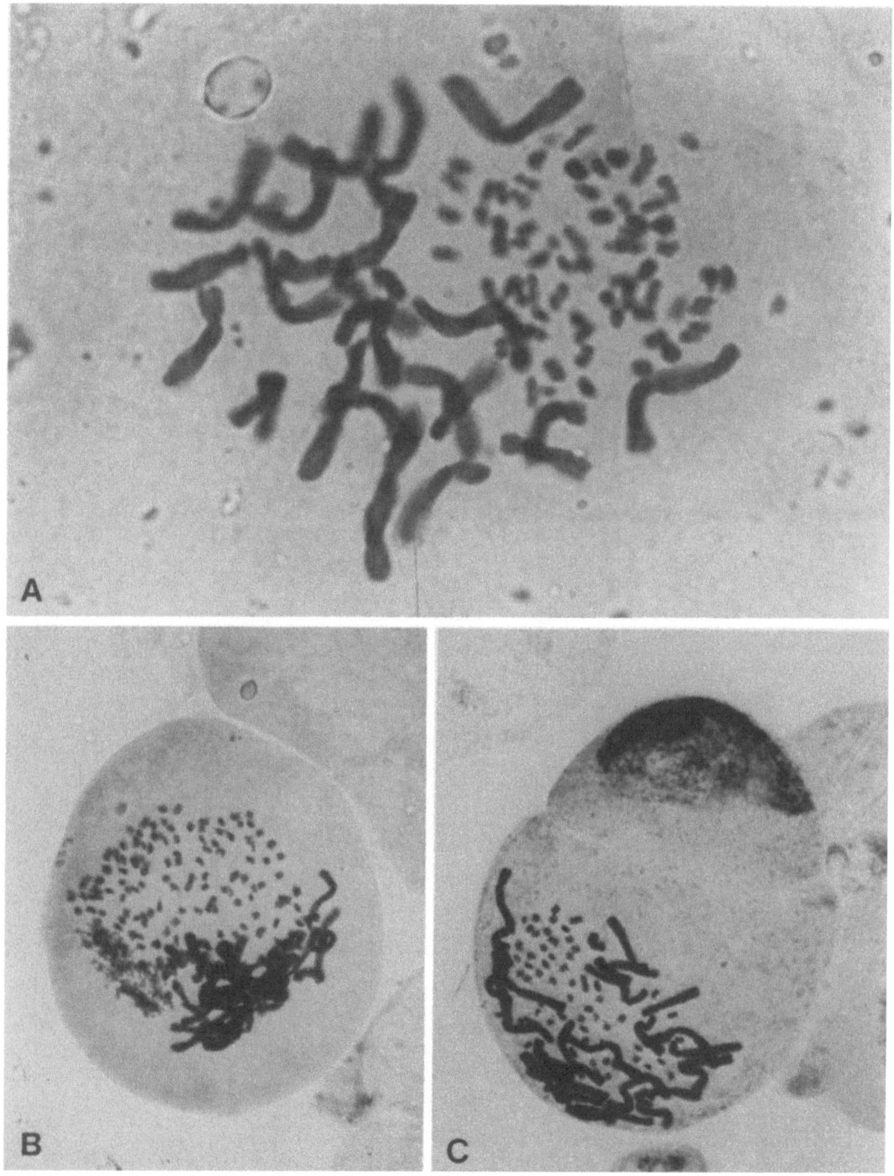

Fig. 31. Behavior of chromosomes in somatic hybrid cells of *Glycine max* (×) *Vicia hajastana*. **A** Asynchronous behavior of soybean and *V. hajastana* chromosomes in metaphase, × 1 500. **B** This hybrid cell also exhibits mitotic asynchrony, × 1 000. **C** A metaphase in a hybrid cell after 14 days in culture, × 1 000. (Constabel et al. 1977)

tion to this rule is the cessation of chloroplast divisions and the degradation of mature chloroplasts following the fusion of callus and mesophyll chloroplasts. This may be explained by a process of dedifferentiation of the mesophyll components.

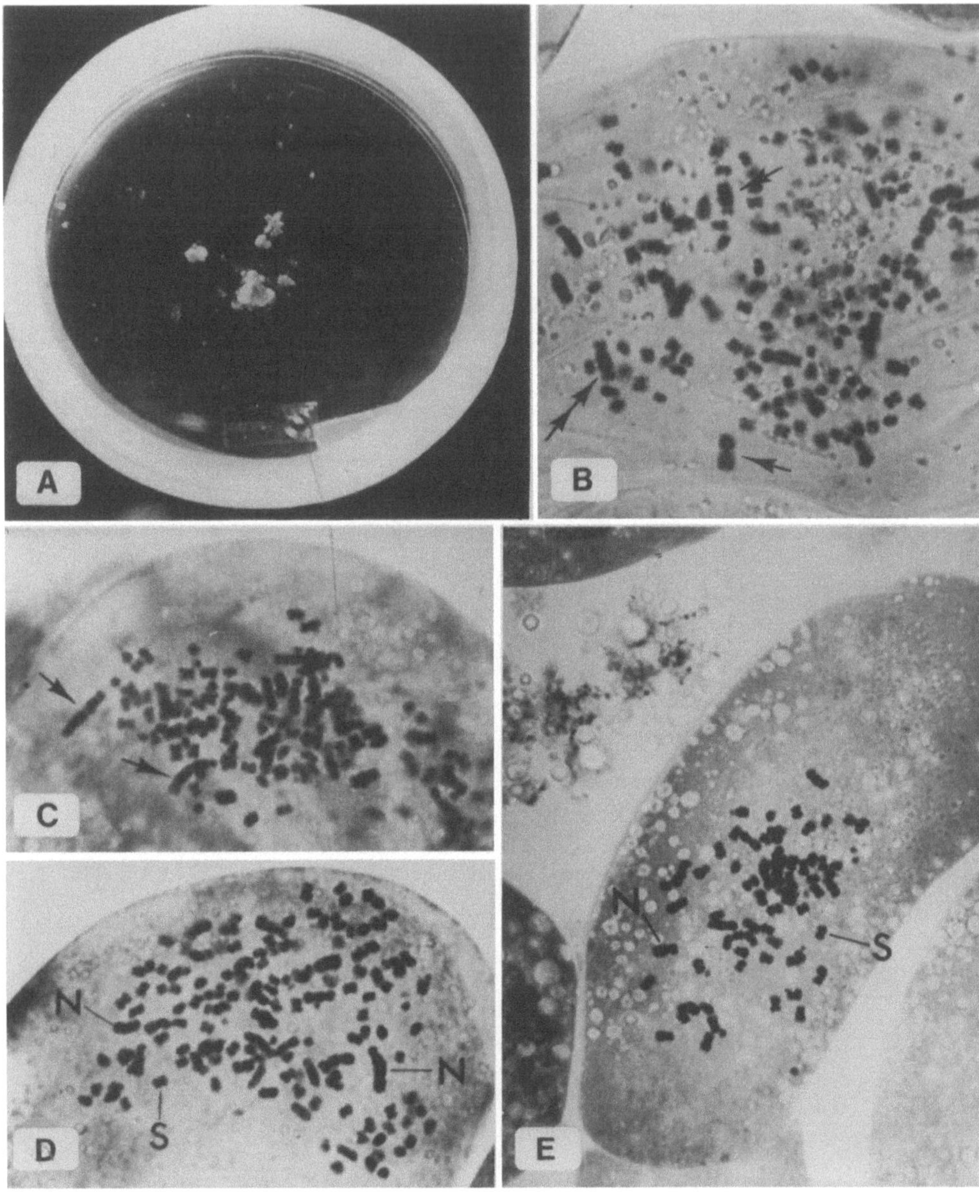

Fig. 32. Chromosomal behavior in somatic hybrids of *Glycine max* (×) *Nicotiana glauca*. **A** Calli of a soybean (×) *N. glauca* hybrid. **B** Chromosomes of *N. glauca* type with one (*arrow*) or two (*double arrow*) constrictions in a hybrid cell of a 4-month-old culture treated with colchicine, ×1200. **C** Triconstrictional chromosomes of *N. glauca* type (*arrows*) in a hybrid cell similar to **B**, ×1200. **D–E** Chromosome figures of two different somatic hybrids after 6 months of culture, with soybean (*S*) and *N. glauca* (*N*) chromosomes, colchicine treated, ×1200 (Kao 1977)

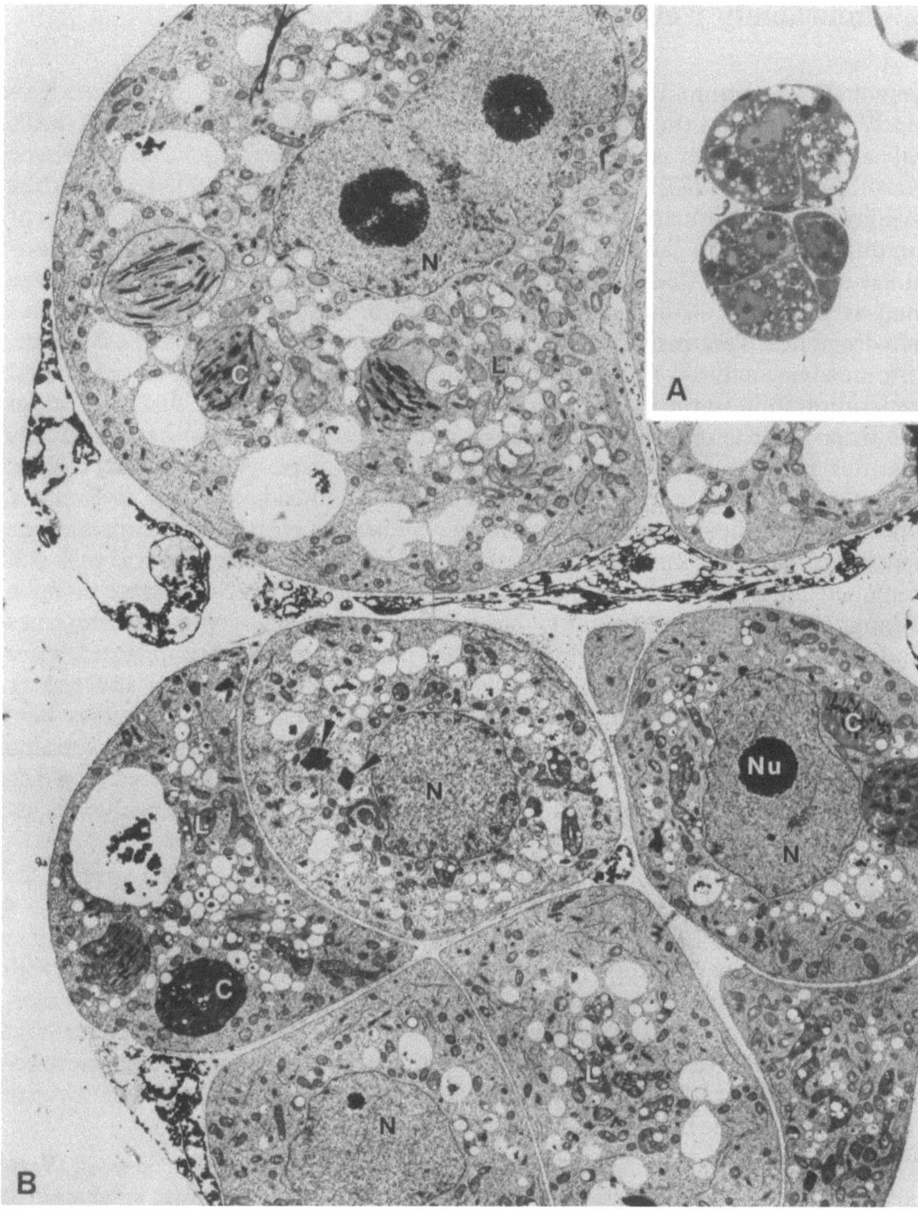

Fig. 33. A multicell cluster developed from a single fusion product of pea and soybean protoplasts, as seen in the light microscope **A**, × 500. Electron micrograph **B** of the same cells. Serial sections revealed that all of these cells contained at least one pea chloroplast **C** and many soybean leucoplast **L**, as well as a nucleus N and nucleolus Nu. Note the pseudocrystalline inclusions (*arrows*). (Fowke et al. 1977)

5.3 Interfamily Cell Hybrids of Higher Plants

First of all, an account of the successful work of the Canadian experimenters Kao and Wetter aimed at the isolation and study of clones of interfamily cell hybrids will be presented. They isolated the cell hybrids *Glycine max* (×) *Nicotiana glauca* (Kao 1977; Wetter 1977; Kao and Wetter 1977, Wetter and Kao 1980). Individual dividing fusion products were mechanically isolated and cultured in droplets of conditioned medium. Twenty cell lines were obtained and their growth was maintained during more than 8 months (about 100 cell generations). Chromosome analysis and study of the isoenzymes of aspartate aminotransferase and alcohol dehydrogenase were performed from the second to the eight month of culture. Chromosome analysis of the first cell divisions was also performed on the general preparation (not on the clones). In a 48 h culture, the premitotic fusion of nuclei was observed. The replication and division of genetic material in the hybrid cells was nonsynchronous and the tobacco chromosomes displayed a tendency to fuse. Chromosomal bridges were observed in anaphase. During the first divisions, a super long chromosome as well as ring chromosomes with several constrictions were observed (Fig. 32). After 1–2 months of individual culture, hybrid cells still contained long chromosomes, chromosome fragments, and anaphase bridges. During the subsequent months, the number of tobacco type chromosomes gradually declined. Standard type tobacco chromosomes were detected more rarely, whereas chomosomes with two or more constrictions and short thick chromosomes that might be reconstructed or fused tobacco chromosomes were frequently observed. After 6 months of culture of these cell hybrids, some tobacco chromosomes were still preserved in 5 of the 20 hybrid cell lines studied (Fig. 32). In contrast, the chromosome morphology was invariable in the parental callus tissue of *N. glauca* and soybean cultured for the same time as a control.

Biochemical study confirmed the presence of isoenzyme forms specific for both tobacco and soybean in all of the cell lines at the early stages of culture (2–4 months). However, the tobacco specific enzyme forms were no longer detected in the majority of the cultures by the eighth month. This is in good agreement with the data of the chromosome analysis. The authors did not report any type of morphogenesis (it may be inferred from the data contained in the different papers that the soybean cells were incapable of morphogenesis and the absence of morphogenesis in the hybrid cell lines was associated with the predominance of soybean chromosomes in the hybrids from the very onset of culturing).

The same authors then performed an elegant continuation of their work (Wetter and Kao 1980). Hybrid cells of *Glycine max* (×) *Nicotiana glauca,* which had lost all or almost all *N. glauca* chromosomes, were "back-crossed" twice by somatic hybridization with mesophyll protoplasts of *N. glauca*. The first "recurrent" protoplast fusion was carried out 27 months after the start of the initial hybrid lines and the second one was performed 7 months after the first "back-fusion". As a reminder, the original hybrid line had almost completely lost the parental tobacco and recontituted (hybrid?) chromosomes and no longer displayed the isoenzyme forms of aspartate amino-transferase characteristic for tobacco. After the back-crossings, a considerable increase was observed in tobacco

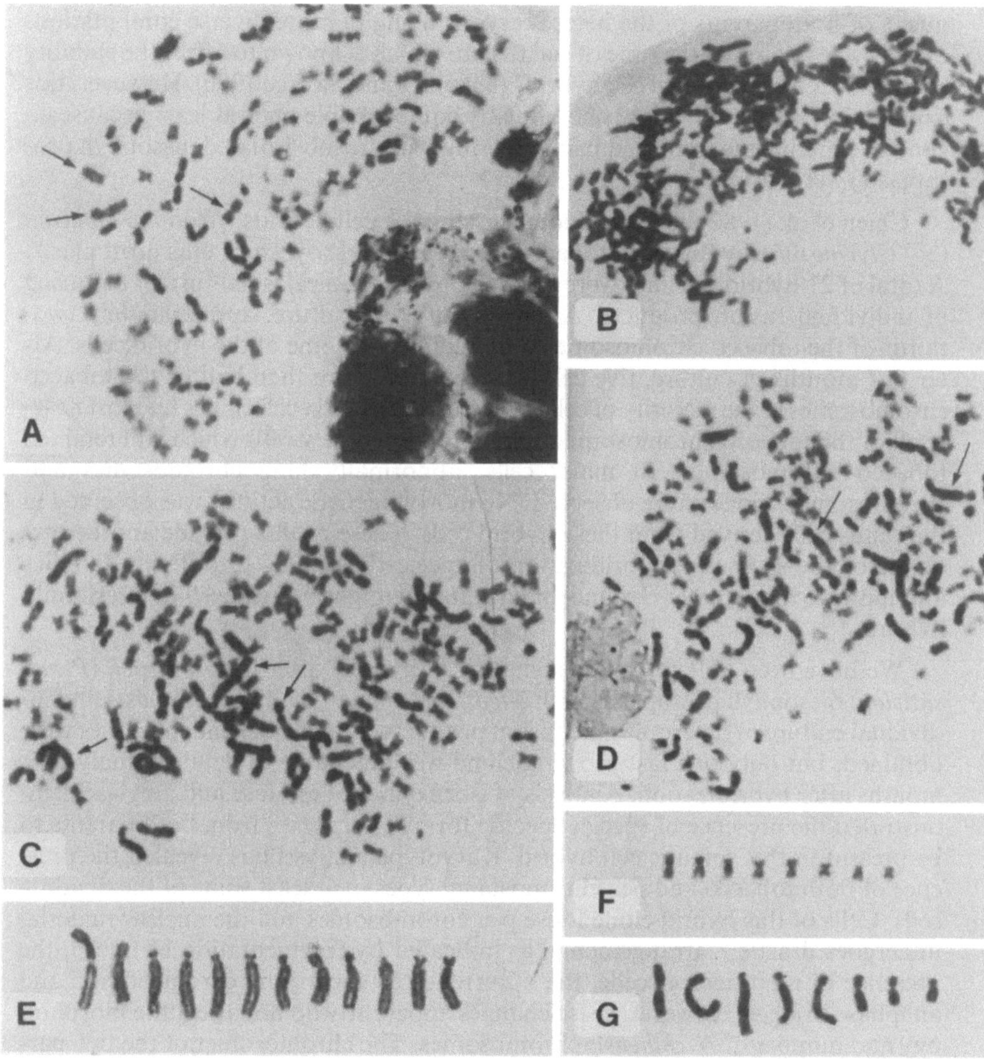

Fig. 34. Chromosomes of various hybrid cell lines of soybean (×) *Nicotiana glauca*. **A** After 3 years culture in a liquid medium. **B** Three months after the second "back-fusion" of the soybean (×) *N. glauca* hybrid to *N. glauca*. **C–D** Six months after the second back-fusion. *N. glauca* type chromosomes and their derivatives are indicated by *arrows*. **E** Karyotype of *N. glauca* (n = 12). **F** Some typical soybean chromosomes (n = 20). **G** Typical *N. glauca* type chromosomes in the somatic hybrid lines 6 months after the second back-fusion. (Wetter and Kao 1980)

chromosome content, in the number of large, reconstructed chromosomes (derivatives of tobacco chromosomes) and the isoenzyme forms of *N. glauca* were detected again, even 6 months after the second recurrent fusion of the cells (Fig. 34). The authors reasoned that since the mesophyll protoplasts of *N. glauca* used for the recurrent crosses were cultured for 24 h prior to fusion, the increase of tobacco chromosome material in the hybrids after back-crossing was a result of the cell nuclei of both parents of the heterokaryons being in more diverse combinations of cell cycle phases by the time of cell fusion, which is known to affect the stability of chromosomes in animal cell hybrids (Ringertz and Savage 1976). However, this explanation seems to be only one of many equiprobable ones, at least because no control experiments (repeated back-crossing with freshly isolated mesophyll protoplasts) were carried out.

Chien et al. (1982) have produced the somatic cell hybrids *Nicotiana tabacum* (×) *Glycine max* by fusion of tobacco mesophyll and soybean callus protoplasts. A total of 21 hybrid cell lines were obtained by mechanical isolation and culturing of individual fusion products. After 3 months of culture, approximately two-thirds of the tobacco chromosomes were retained in some of the hybrid cells. After 6–7 months of culture, five cell lines retained more than half of the tobacco chromosomes in the majority of cells, whereas another six cell clones had lost nearly all of the tobacco chromosomes, except for a very few cells which still retained tobacco chromosomes. In many cells, abnormally long chromosomes and chromosomal bridges were observed. No morphogenetic activity was observed in the callus cells derived from these hybrid cells. These results provide another example of the successful hybridization of species belonging to different families and indicate that greater chromosomal stability might be achieved in interfamily hybrids.

We have been able to isolate somatic cell hybrids after fusion of pea (*Pisum sativum*) mesophyll protoplasts with *Nicotiana chinensis* callus protoplasts and individual culture of heteroplasmic fusion products. Three independent clones were obtained, but only one fast growing clone was amenable to detailed analysis 6 months after hybridization. Analysis of isoenzymes of esterase and amylase demonstrated the presence of species specific forms of enzymes from both parents to be present in this somatic cell hybrid. Karyotype analysis has revealed the presence of both tobacco and pea chromosomal types in at least some of the dividing cells. Cells of this hybrid clone loose pea chromosomes and the nuclear material undergoes drastic rearrangements, as indicated by fragmentation of nuclei, the presence of multinuclear cells, the occurrence of super long chromosomes, and anaphase bridges. However, these changes apparently do not affect the morphology and number of *N. chinensis* chromosomes. The chromosomes of the two parental types tend to group separately in common metaphases, indicating a spatial separation of the parental genomes in hybrid nuclei. To date, no morphogenetic activity has been observed in this cell line, even though efforts have been made to induce morphogenesis (Gleba et al. 1983b).

These experiments are of great significance. The initial data show that parasexual hybridization of somatic cells enables the production of viable lines of cell hybrids between species belonging to different families and even different orders. Interfamily plant cell hybrids demonstrate chromosome reconstruction and elimination of chromosome material of one of the species, as observed in animal somatic cell hybrids (Ringertz and Savage 1976). It is still an open question whether morphogenesis can be induced in cell cultures of interfamily hybrids. A more detailed discussion of the properties of distantly related hybrids will be reported in connection with intertribal hybrid plants (Sect. 5.4.3).

A brief report of experiments on interfamily cell hybridization by protoplast fusion of *Vicia faba* (×) *Petunia hybrida* has been published (Binding and Nehls 1978a). However, these experiments did not result in the production of viable hybrid cell lines.

5.4 Intertribal Hybrids of Higher Plants

5.4.1 Hybridization of Cruciferae Species

The tribe is a taxonomic unit intermediate between family and genus. To the best of our knowledge, no successful sexual crosses between higher plant species belonging to different taxonomic tribes have been reported. The successes of the parasexual technique in hybridization of distantly related plant species and the production of intertribal hybrid plants are limited to a few experiments (Gleba and Hoffmann 1978, 1979, 1980; Krumbiegel and Schieder 1979, 1981; Dudits et al. 1979, 1980a; Gleba et al. 1982a, b, c; Skarzhynskaya et al. 1982). The purpose of our work (Gleba and Hoffmann 1978, 1979, 1980) was the production and analysis of hybrid cell lines and hybrid plants of *Arabidopsis thaliana* (×) *Brassica campestris*. This work was the first investigation that contained a complete and well-grounded demonstration of the possibilities and limitations of the parasexual technique for the hybridization of distantly related plant species. This work also provided the first evidence for the potential to produce genetically distinct plant hybrids, i.e., symmetric and asymmetric hybrids, from the same initial material.

Arabidopsis thaliana (L) Heynh cultured cells were used as a source of callus protoplasts. Callus tissues were initiated from chlorophyll deficient seedlings of an albino recessive genome mutant. Leaves from axenically grown plants of *Brassica campestris* (L) were used as a source of mesophyll protoplasts. Mixed suspensions of protoplasts were fused by a combination of agglutination with polyethylene glycol (according to Kao and Michayluk 1974) followed by a wash with a high pH and high Ca^{+2} solution (Keller and Melchers 1973; Kao 1977). The fusion products proliferated vigorously in Nagata and Takebe B medium. Fusion products 3–5 days old contained four to eight cells and were able to grow independently as isolated cultures in microdroplets. Within 3 weeks after transfer to microdroplets in Cuprak dishes, 54 out of 113 single fusion products developed into small colonies, each consisting of several hundred cells. Thus, a total of 31 hybrid cell clones, each derived from a single protoplast fusion event, were obtained (Fig. 35).

Neither *Arabidopsis thaliana* nor *Brassica campestris* are favorable species for chromosome analysis. Chromosomes of *Arabidopsis* ($n = 5$; designated a) are very small and uniform. The largest chromosome, a 1, is about twice as long as the smallest chromosome, a 4. Chromosomes a 1, a 2, and a 3 have submedian centromeres; chromosomes a 4 and a 5 have subterminal centromeres. Cytological analysis of the *Arabidopsis* callus cell line used in our experiments showed that the majority of the cells observed in metaphase were octoploid. *Brassica* chromo-

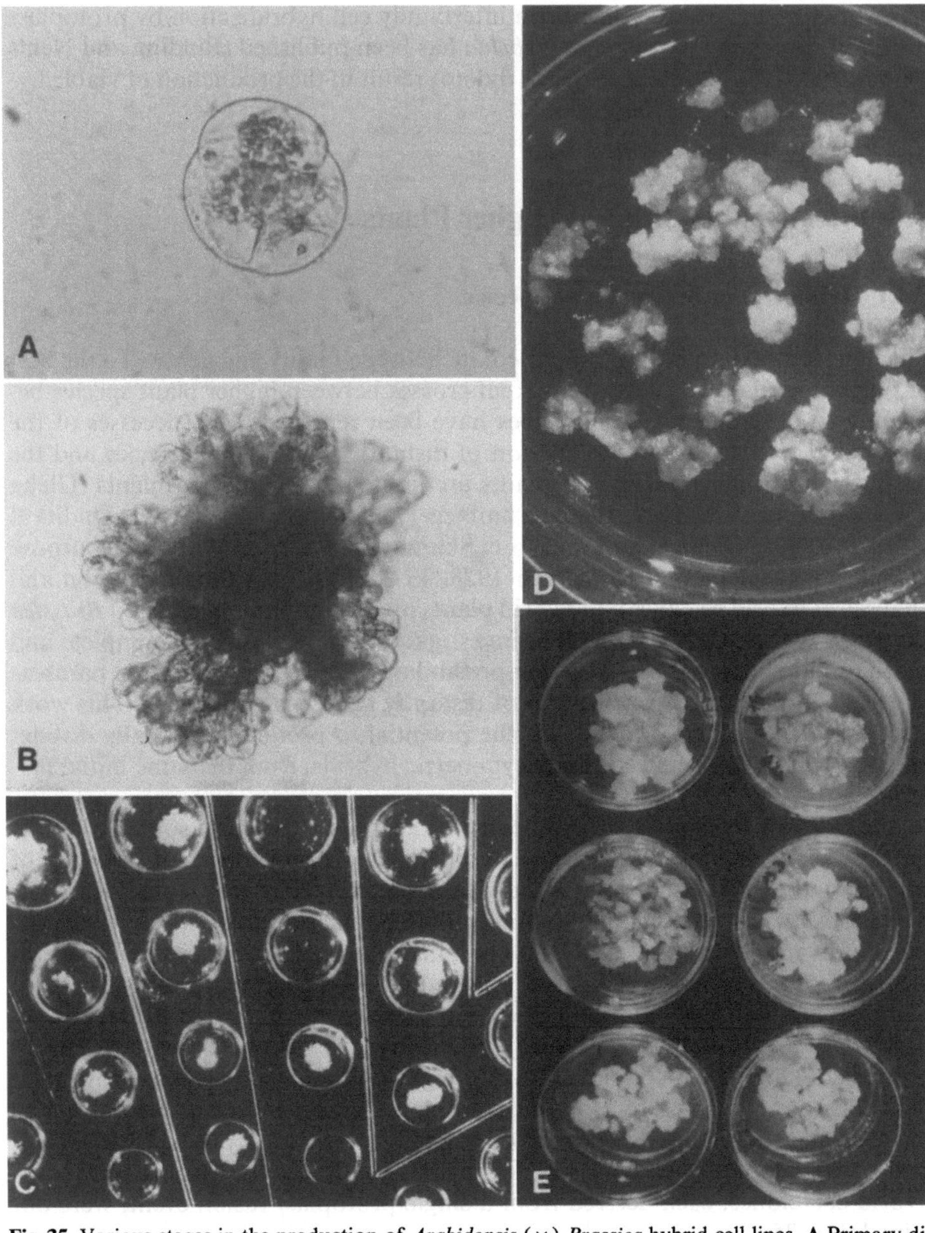

Fig. 35. Various stages in the production of *Arabidopsis* (×) *Brassica* hybrid cell lines. **A** Primary divisions in a heteroplasmic protoplast fusion product after 5 days of culture, ×400. **B** Cell colony after 25 days in culture, ×100. **C** Cell colonies similar to that in **B** can be seen in the individual wells of the Cuprak dish after 25 days of culture, ×3. **D** Callus tissue of line AB 1 after 75 days of culture. **E** The six actively growing hybrid lines AB 1, 2, 7, 10, 15, and 16. (Gleba and Hoffmann 1979)

somes (n = 10; designated b) are quite heterogenous in mophology and include two long chromosomes, b 1 with a submedian centromere and b 2 with a subterminal centromere. Comparative analysis indicates that except for chromosome b 10, all of the *Brassica* chromosomes are longer and thicker than *Arabidopsis* chromosomes a 2, a 3, a 4, and a 5 and thus may be readily differentiated. A metaphase of *Brassica campestris* chromosomes in a root tip cell is presented in Fig. 36. Chromosome analysis of the six fast growing hybrid cell lines of *Arabidopsis thaliana* (×) *Brassica campestris* (lines 1, 2, 7, 10, 15, and 16), each isolated as single fusion products, was conducted 4–7 months after isolation. In the 4–5 month old cultures, cells of all six lines were found to contain both *Arabidopsis* and *Brassica* chromosomes in all metaphases studied. Cell lines 1, 2, 7, and 16 had 60 chromosomes, while cell lines 10 and 15 both had nearly 80 chromosomes. Judging by the chromosome morphology, cell lines 1, 2, 7, and 16 had 40 a-type chromosomes and 20 b-type chromosomes, including two b 1 chromosomes. Cell lines 10 and 15 had 40 a-type chromosomes and 40 b-type chromosomes. Thus, cell lines 1, 2, 7, 10, and 16 probably each originated from the fusion of one octoploid *Arabidopsis* cell with one diploid *Brassica* cell and cell lines 10 and 15 probably each originated from the fusion of one octoploid *Arabidopsis* cell with either one tetraploid or two diploid *Brassica* cells.

Cytological analysis revealed some chromosome rearrangements in the hybrid cells. Chromosomal bridges were frequently observed in anaphase. In metaphase, the chromosomes revealed a tendency to stick together and to form multichromosomal chains. The b 1 chromosome was found to be unchanged in all of the cell lines. To the contrary, the b 2 chromosome, with a subterminal centromere, was modified in the majority of the cell lines to contain a small thinner fragment and thus appeared as a chromosome with two constrictions. Another smaller b chromosome was also modified in the same way. In line 2, a ring chromosome was observed during the fourth and fifth months of culture, but it later disappeared.

Analysis of the chromosomes in 6–7 month old cultures revealed changes in the chromosome numbers, but only few, if any, additional chromosome rearrangements. Most cells of lines 1, 2, 7, and 16 had 65–75 chromosomes, although some cells still possessed 60 chromosomes. Most cells in lines 10 and 15 had 65–85 chromosomes. Analysis of the individual chromosome morphology indicated that all of the morphological types of chromosomes were still present in the cells of all six hybrid cell lines studied (Fig. 36).

A biochemical study of the hybrid cell lines was conducted 6–7 months after fusion. This analysis included electrophoresis of the total protein on polyacrylamide gel slabs and subsequent staining for the specific enzyme activities of esterase, lactate dehydrogenase, alcohol dehydrogenase, and peroxidase, as well as thin-layer chromatography of low molecular compounds. A test for the presence of myrosinase, a crucifer obligate enzyme, was also performed. Electrophoretic patterns and a diagrammatic representation of esterase isoenzymes from the parental cells and hybrids are presented in Fig. 37. *Arabidopsis* cells possess two fast moving major bands as well as another major and two or three minor slow moving bands. *Brassica* cells possess two fast moving major bands, three intermediate bands, and two or three slow moving bands. All esterase isoenzyme bands of these

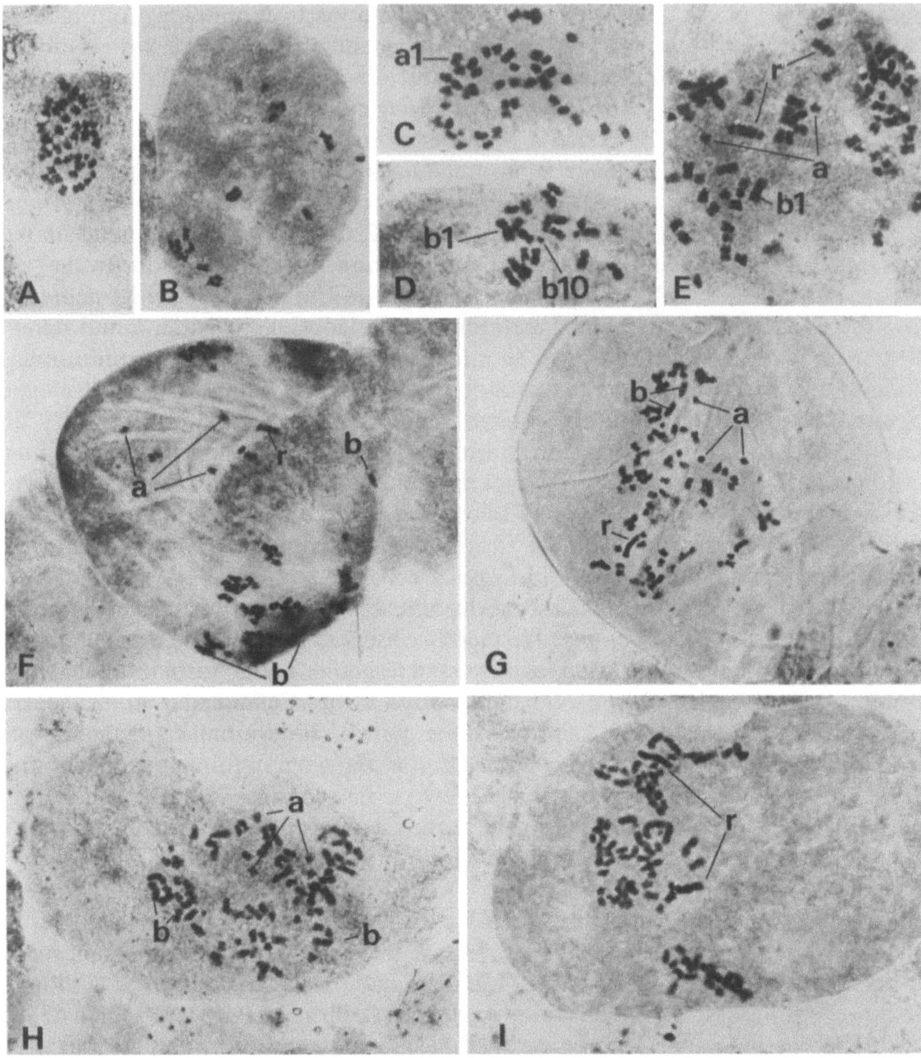

Fig. 36. Chromosome analysis of *Arabidopsis thaliana* (×) *Brassica campestris* hybrid cell lines. **A** Metaphase plate in cell culture of *Arabidopsis* with 8n = 40 chromosomes. **B** Metaphase plate in root cell of *Brassica* with 2n = 20 chromosomes. **C** A portion of a metaphase illustrating the morphology of *Arabidopsis* chromosomes, including the long, submedian *a1*. **D** A portion of a metaphase illustrating *Brassica* chromosome morphology, including the long, submedian *b1* and the shortest subterminal chromosome, *b10*. **E** The chromosomes of hybrid cell line ab 2, with both parental (*a, b1*) and reconstructed (*r*) chromosomes. **F** Metaphase plate of cell line ab 2 four months after hybridization. **G** Cell line ab 2 seven months after hybridization. Hybrid cell lines ab 7 (**H**) and ab 1 (**I**) 7 months after hybridization. (**A, B, F–I** are × 1 000; **C–E** are × 1 500.) (Gleba and Hoffmann 1978)

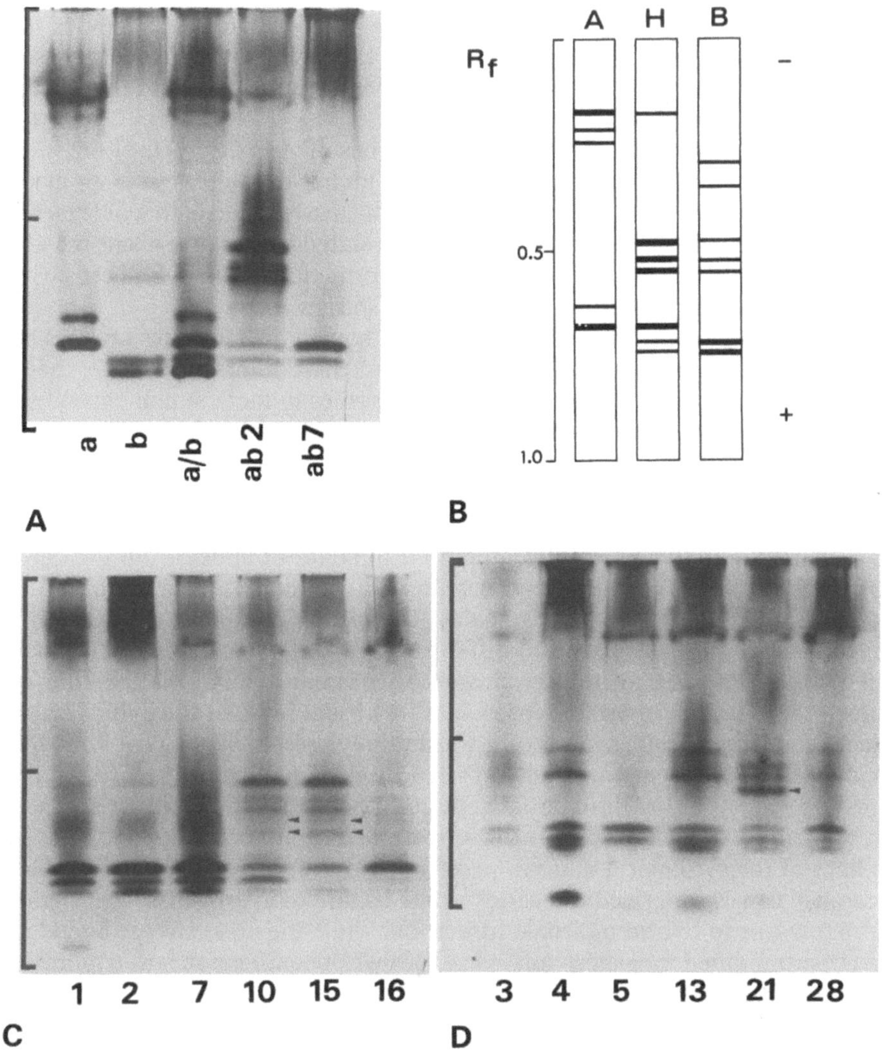

Fig. 37. Electrophoretic analysis of isoenzymes of esterase of *Arabidopsis* (×) *Brassica* hybrid cell lines. **A** Isoenzyme spectra obtained for *a Arabidopsis* callus, *b* Brassica callus, *a/b* mixture of parental extracts, and hybrid lines *ab 2* and *ab 7*. **B** Zymograms of esterases for callus tissues of (*A*) *Arabidopsis*, (*B*) *Brassica*, and (*H*) hybrid. **C** Isoenzyme spectra for the indicated hybrid lines with active growth. **D** Isoenzyme spectra for the six indicated slowly growing lines. The new isoenzymes which were not found in the parental forms are indicated by *arrows*. (Gleba and Hoffmann 1978)

two species differ from each other in their electrophoretic mobility. In cells of hybrid line 2, two a bands and five b bands can be clearly resolved. In all 12 hybrid cell lines studied, only one fast and one slow moving band from the four or five parental *Arabidopsis* bands and two fast and three intermediate bands from the seven or eight parental *Brassica* bands are expressed (Fig. 37). The different cell lines displayed different relative intensities of these bands, but these differences were not reproducible. New isoenzyme bands were regularly seen in lines 10, 15, and 21 and were sometimes observed in lines 1, 2, and 3 also (Fig. 37). These new bands of enzyme activity are not intermediate in their mobility and thus are probably not hybrid enzymes.

To the best of our knowledge, these cell lines of *Arabidopsis* (×) *Brassica* are the first hybrid cell lines of distantly related higher organisms with no evidence for specific chromosome elimination. The data clearly indicate that at least within one family, genetic material of two distantly related higher organisms may be integrated and organized for the successive reproduction of the hybrid cells without elimination of the chromosomes of one of the species.

These experiments have also shown that some chromosomes are less stable and are more likely to undergo rearrangement than others. Of course, the question arises whether these rearranged chromosomes in fact contain chromosomal material from both species – that is, are they hybrid chromosomes? One interesting approach with potential to answer this question is the use of differential chromosome staining, which has been successfully applied in the analysis of chromosomes in hybrid animal cells (Bobrow and Cross 1974).

Isoenzyme analysis has fully supported the data obtained from the chromosome studies. All 12 lines analyzed were found to contain isoenzymes of both parents. Although the results were unequivocal and reproducible, the expression of several isoenzymes was found to differ greatly from one experiment to another. Several parental isoenzymes were invariably present in all of the cell lines, while some others, such as esterase band a 3, were absent in all of the hybrid cell lines. Since each *Arabidopsis* chromosome was initially present in eight copies in the hybrid cells, the apparent uniformity of all of the cell lines in selective isoenzyme expression reflects either differential gene activity or selective (but not species specific) chromosome elimination or inactivation. The latter possibility seems likely in light of the cytological data on selective chromosome reconstruction. The appearance of new intermediate bands in the lactate dehydrogenase zymograms is probably due to hybrid molecule formation, while the new esterase bands might be either modified molecules or isoenzymes whose expression was suppressed in the parental cells.

Isoenzyme and chromosome analyses were continued for hybrid cells lines 1, 2, 7, 10, 15, and 16 during the interval of 6–15 months after hybridization. It was found that lines 1, 2, 10, and 16 retained all marker chromosomes and all esterase isoenzymes during this period. Line 7 gradually lost one esterase b band, but retained all marker chromosomes. Line 15 lost almost completely both esterase b bands and long b-type chromosomes by the end of 15 months of culture.

However, at the time of the morphogenetic studies (7–10 months), the elimination process in these lines had evidently only just begun. Callus cells of the cytologically and biochemically studied hybrid lines 1, 2, 7, 15, and 16 were grown on various media, including those with a high cytokinin to auxin ratio. Root formation was induced in three of the lines. Active rhizogenesis was observed on B5 medium containing 6-benzylaminopurine (1.5 mg l^{-1}) as the sole hormone. Cells of line 2 responded most actively with root formation when this medium was additionally supplemented with charcoal (1 g l^{-1}). The responses of all three lines were general: most cells displayed organized growth. Nonetheless, growth of these roots was very slow and they ceased to grow longer than 2–3 cm. Transfer onto B5 medium with 1-naphthylacetic acid ($0.5\,mg\,l^{-1}$) and coconut milk ($10\,ml\,l^{-1}$) resulted in extensive secondary root formation.

Chromosome analysis of the root cells was significantly hampered by the slow growth of these cultures; nevertheless, metaphases of lines 1 and 2 were examined. Marker chromosomes of both parents were found in the same metaphases in all cases. Biochemical study included esterase isoenzyme analysis of the roots and the corresponding callus cells from all three lines. No differences were observed in the isoenzyme patterns of cells growing in either organized or unorganized forms. Isoenzymes of both species were equally expressed in callus and root cells.

The anatomy of the structure produced by these three lines clearly indicated the organization typical of a root. A longitudinal section demonstrated a root tip with its characteristic regions: the rootcap, the precursor tissues of the epidermis, the cortex, and the central vascular cylinder. A transverse section revealed the anatomy typical of a young root with a differentiated central cylinder. Whether some anomalies in the cell patterns of the apical meristem and the inability to distinguish an endodermis or a distinct pericycle were due to the in vitro origin of the roots or to their hybrid nature must be examined further.

Shoot-bud formation was induced in 2 of the 6 cell lines tested on B5 medium (Gamborg et al. 1968) containing 6-benzylaminopurine ($1.5\,mg\,l^{-1}$) as the sole phytohormone. This same culture medium formula had been shown to be effective in inducing extensive root formation in these lines. However, shoot morphogenesis appeared as a very rare event. To date, a total of six independent shoot regeneration events have been discovered. Two have been in line 1, three in line 15, and one in line 7 and each has been responsible for the regeneration of hundreds of plants (Fig. 38 and 39). Persistent efforts to induce shoot formation in the other four lines have been unsuccessful.

Regeneration lines 1-1 and 1-2, obtained from hybrid cell line 1, both gave rise to plants which appear to be morphologically intermediate between *Arabidopsis* and *Brassica*. The leaves are mostly petiolar, thick and wide, round or lobed, with no hairs (both parents have trichomes). The leaves of most of the plants are red because of anthocyanin, an *Arabidopsis* character. All of the plants are green even though the *Arabidopsis* parental cell line originated from an albino seedling homozygous for a recessive genome mutation. These plants are very homogeneous and morphologically regular, but only a few of them have produced roots. In addition, they have never been observed to flower.

Chromosome analysis of cells from young shoot tips of the regenerants of line 1-1 revealed both long b-type chromosomes and short, submedian a-type chromosomes to be present in all metaphases examined (Fig. 40). In contrast, we were not able to identify with certainty any of the reconstructed, biconstrictional chromosomes found earlier in the callus tissue. The number of chromosomes in these regenerants was between 55–60; however, this variation might simply be due to uncertainty in the counting.

Esterase isoenzyme analysis of the plants from lines 1-1 and 1-2, as well as of calli originated from these plants, displayed both *Arabidopsis* and *Brassica* specific bands to be present in all cases. No reproducible differences have been observed between isoenzyme spectra for the initial hybrid cell line 1 on the one hand

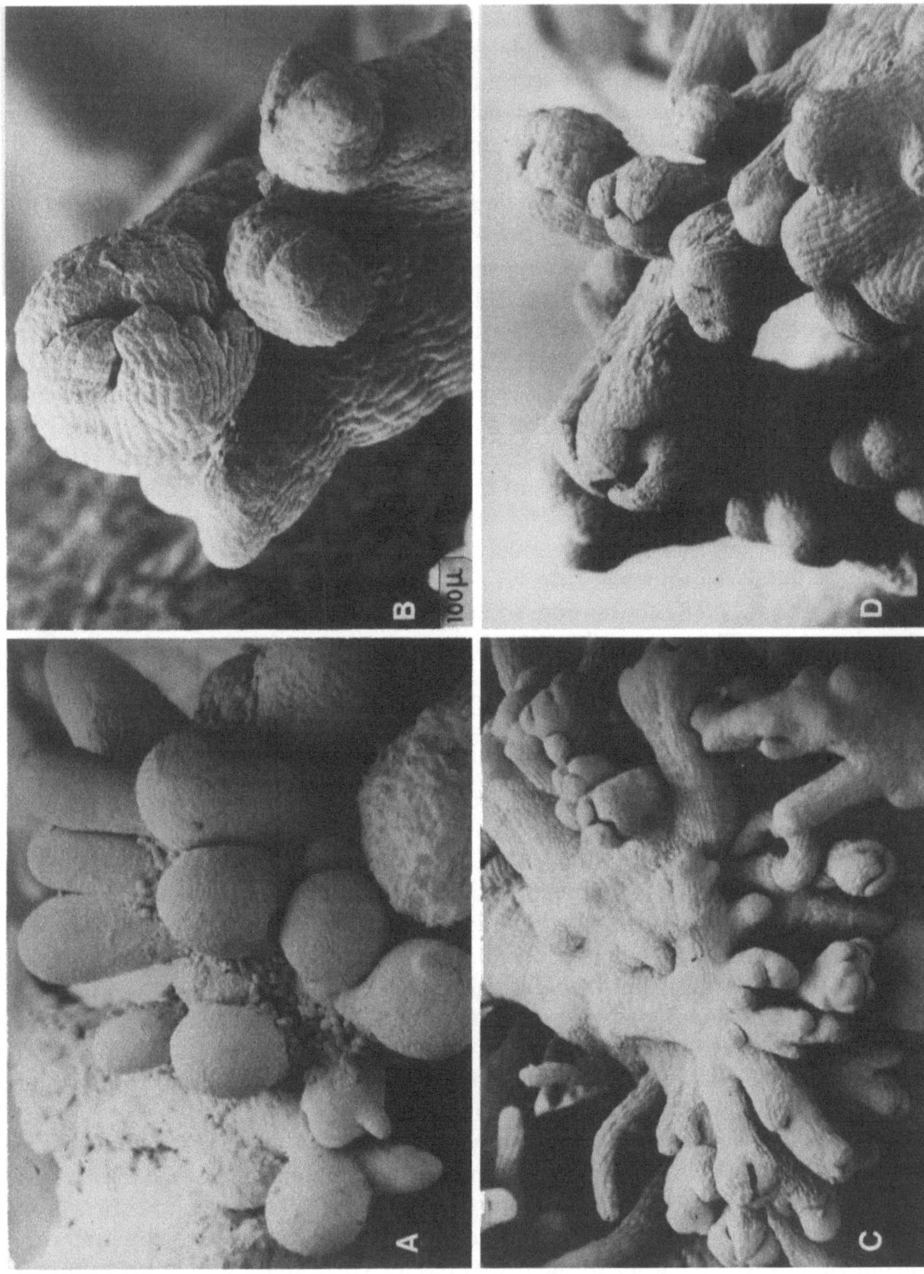

Fig. 38. Scanning electron microscopy of shoot organogenesis in a culture of hybrid cell line ab 15 of *Arabidobrassica*. **A** Primary nidi of organization. **B** Leaf and **C**, **D** flower primordia, × 50–100. (Gleba 1980)

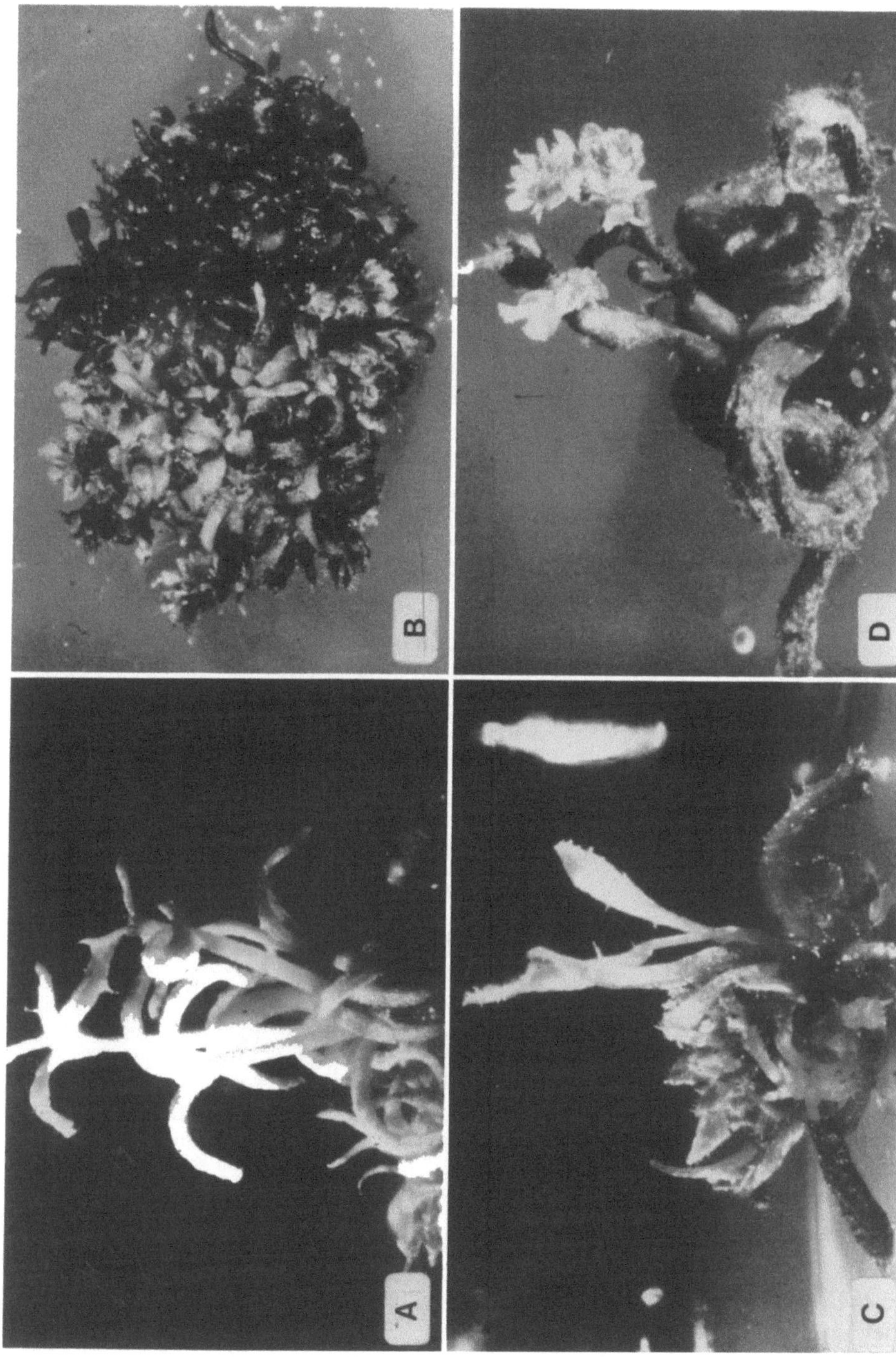

Fig. 39. The plant regenerants obtained from cells of *Arabidobrassica*, line ab 15. **A** Abnormal teratoma forms from ab 15-3. **B** Regenerants with a tendency to flower. **C** Regenerants possessing normal leaves (line ab 15-1). **D** A "grown-up," flowering monster from line ab 15-1. (Gleba 1980)

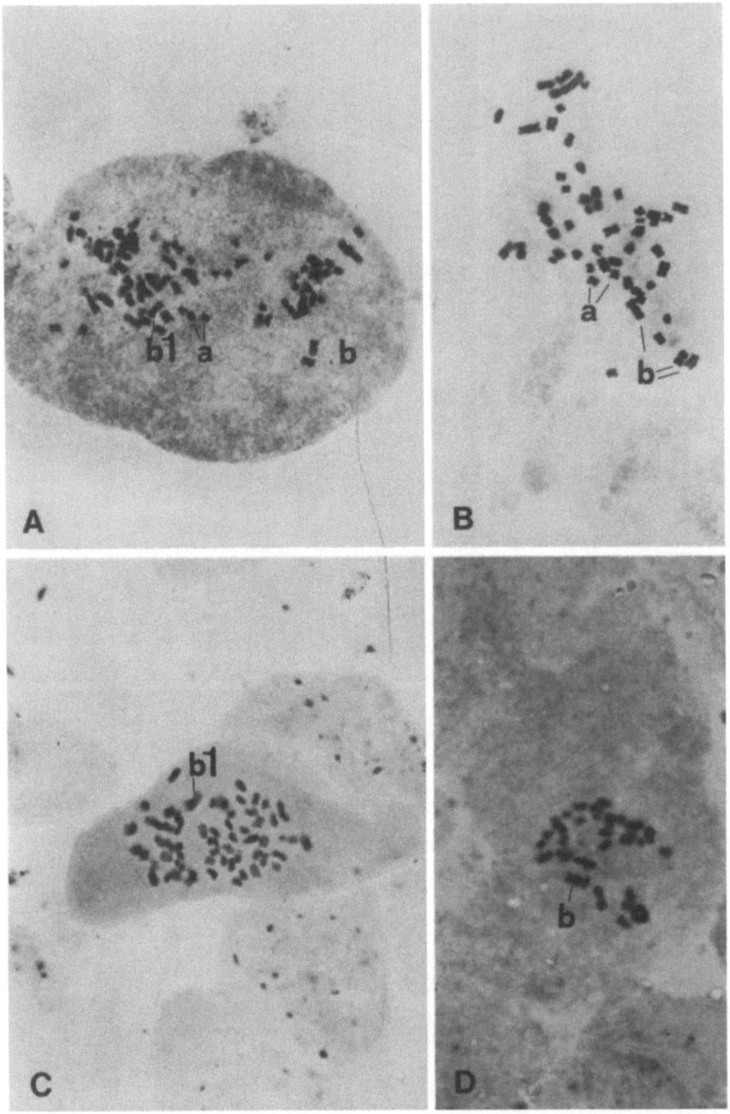

Fig. 40. Chromosome analysis of *Arabidopsis* (×) *Brassica* somatic hybrid plants. Metaphase chromosomes in a **A** callus cell of line ab 1; **B** shoot tip cell of a plant regenerant from the same cell line, ab 1-1; **C–D** flower primordia cells from regenerant line ab 15-1. *Arabidopsis* (*a*) and *Brassica* (*b*) chromosomes, including the longest (*b 1*), are indicated. All enlargements are ×1 000. (Gleba and Hoffmann 1980)

and the regenerated plants and secondary calli derived from these plants on the other hand (Fig. 41).

Regenerants from hybrid cell line 15 (regeneration lines 15-1, 15-2, and 15-3) are extremely heterogenous (Fig. 39, 42), in contrast to those from hybrid cell line 1. Only highly abnormal shoot-like and leaf-like structures have been observed so

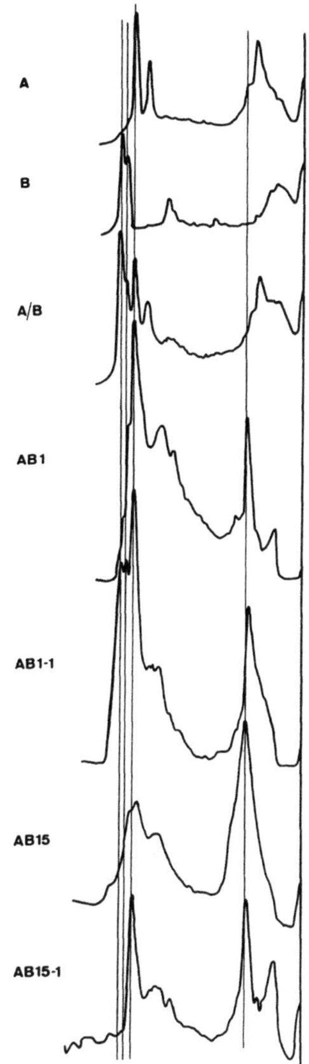

Fig. 41. Densitograms of esterase spectra obtained by scanning the electropherograms for *Arabidopsis* callus (*A*), *Brassica* callus (*B*), mixture of parental extracts (*A*/*B*), cell line ab 1 (*AB 1*), regenerants from cell line ab 1 (*AB 1-1*), cell line ab 15 (*AB 15*), and regenerants from cell line ab 15 (*AB 15-1*). (Gleba and Hoffmann 1980)

far in lines 15-2 and 15-3. These teratoma never really become normal and tend to brown and die after reaching some stage in development. Cytological analysis of these lines was not possible for technical reasons. Analysis of the esterase isoenzymes revealed only *Arabidopsis* specific bands to the present (Fig. 41). However, line 15-1 produces shoots with a sometimes perfect organization that more closely resemble the *Arabidopsis* parent with respect to size and gross morphology (Fig. 38 and 39). This line also produced the first flowering plantlets from this hybridization. Although these plants tend to flower profusely, none of them have yet formed roots. Since trichome analysis is extensively used in Cruciferae taxonomy, hybrids were compared with the parental species with respect to their hair structure. The trichomes of *Arabidopsis* have three branches, while *Brassica campestris*

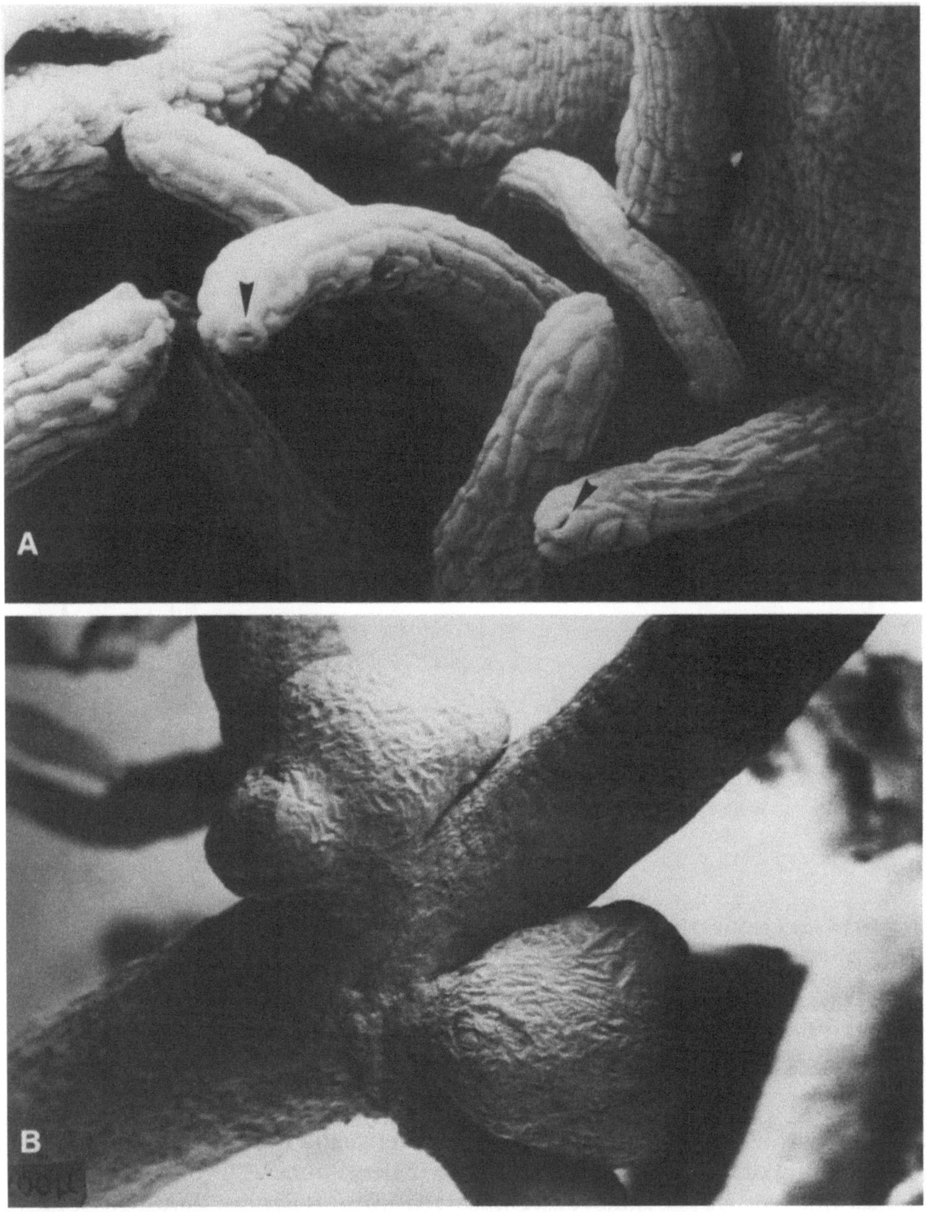

Fig. 42. Morphological abnormalities of the *Arabidobrassica* parasexual hybrid cell line *ab 15-1*. **A** Leaves with almost radial symmetry. The stomata are indicated by *arrows*. **B** Abnormal development of axillary buds, × 100. (Gleba 1980)

leaves have only simple, nonbranching hairs. Leaves of the hybrid plants are usually covered with different hair types having one, two, three, and sometimes even four branches (Fig. 43). Hairs with two branches prevail in most plants. A few plants have been observed with one or the other parental type of trichomes only. The fine morphology of the trichomes of the hybrid plants, even those of the nonbranching *Brassica-type*, resembles that of the *Arabidopsis* parent more than that of the *Brassica* parent (Fig. 44).

Considerable heterogeneity was also observed in flower morphology. A large number of the plants have filled, monstrous flowers, singular or in bunches of a few; in some extreme cases, bunches with dozens of small flowers were observed. The flowers are of the same size as the *Arabidopsis* parent in all cases. Large, filled flowers are usually completely nonfunctional, although a partially transformed pistil and ovary are sometimes observed (phyllody). On the other hand, smaller, morphologically perfect flowers were found with the classical crucifer formula. All possible forms intermediate between these two extreme types were also present. For example, otherwise morphologically normal flowers were observed to be lacking petals or anthers (usually the outer circle of smaller types was reduced) or flowers were found in which both petals and anthers were transformed into leaf-like structures. The anatomy of the more-or-less regular flowers were studied in greater detail. In some cases, there were indications of the initiation of the microsporogenesis process; nonetheless, the pistil and ovary were completely nonfunctional. In other flowers, nearly perfect pistil and placenta with ovules were found, but the anthers produced no pollen (Fig. 45). *Arabidopsis* flowers have white petals and *Brassica* flowers have yellow petals. The flowers of the hybrids are predominantly pure white, although light yellow and sectorially yellow types were also found. All of the hybrid plants appear to be self-sterile.

A chromosome study of the flower primordia of the plants from hybrid line 15-1 revealed a greatly reduced chromosome number as compared to the callus line 15. Most of the plants possessed 30–45 chromosomes with the chromosome numbers varying within the same plant. The reduced chromosome numbers suggest that chromosome elimination and/or segregation occurred at a step prior to morphogenesis. A great majority of the chromosomes are very small and can be described as *Arabidopsis*-type; in addition, at least one or two long *Brassica*-type chromosomes, but never the b 1 chromosome, were present in most metaphases (Fig. 36). Reconstructed, biconstrictional chromosomes were not observed. Despite repeated attempts, no correlations could be made between the presence or absence of *Brassica*-type chromosomes and morphological peculiarities such as trichome type and pattern or degree of flower organization.

Analysis of the esterase isoenzymes of the plants regenerated from line 15-1 (Fig. 37) and of the callus tissue derived from these plants revealed only one or two *Arabidopsis* specific bands and virtually no more *Brassica* specific bands than were found in the corresponding callus line, hybrid line ab 15, before morphogenesis. The observed appearance of new bands as well as deviations in the pattern intensity may also be a consequence of the hybridization, although the new bands are not intermediate in position.

A study of the subunits of ribulose-1,5-bisphosphate carboxylase (Fraction I protein) was performed using the various symmetric and asymmetric hybrids of

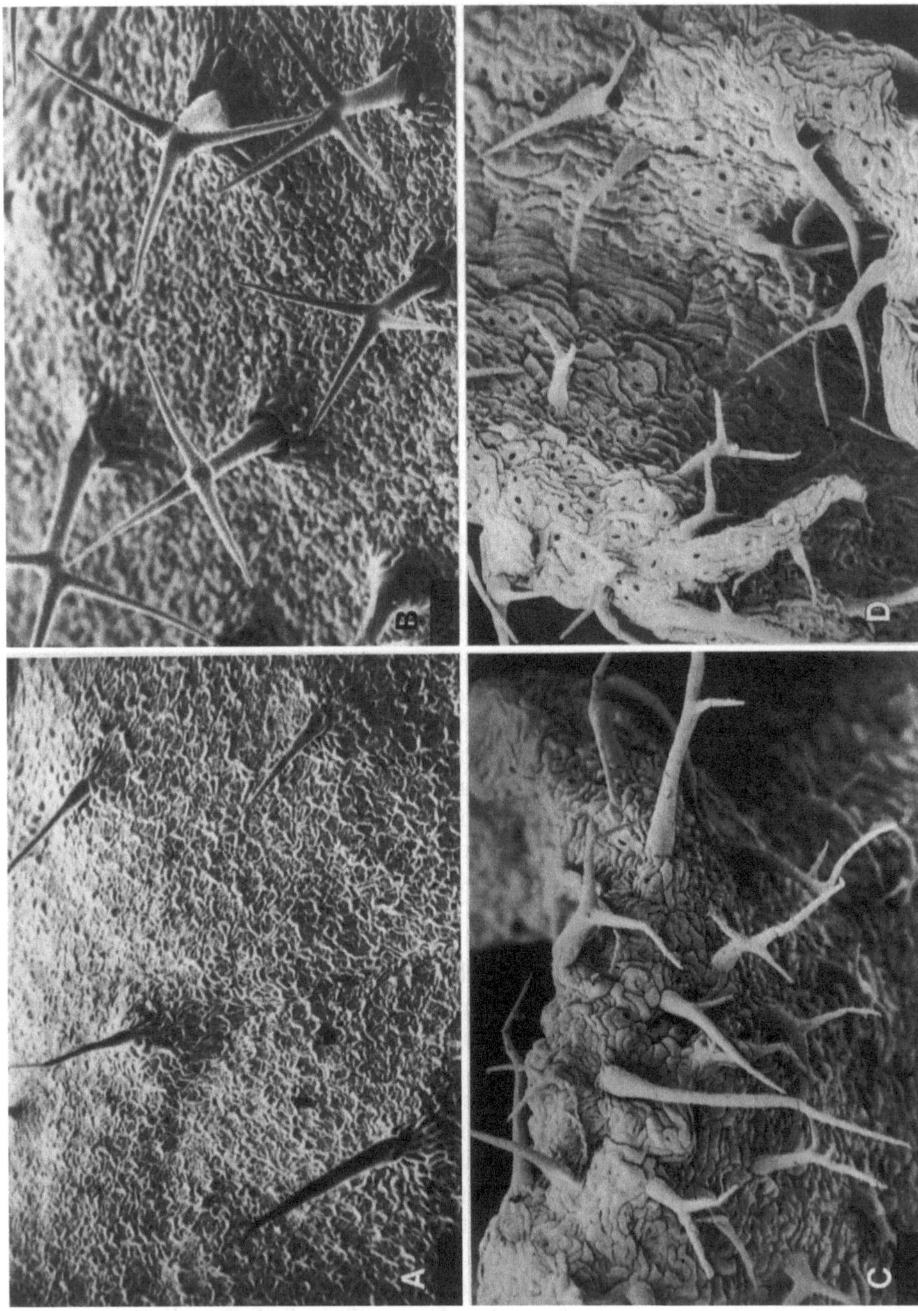

Fig. 43. Trichome analysis of the parental forms and hybrids of *Arabidopsis* (×) *Brassica* by scanning electron microscopy. **A** Trichomes of the lower surface of a *Brassica* leaf. **B** Trichomes of the lower surface of an *Arabidopsis* leaf. **C–D** Trichomes on the leaf surface of a regenerant of *Arabidobrassica* line ab 15-1, ×200. (Gleba and Hoffmann 1980)

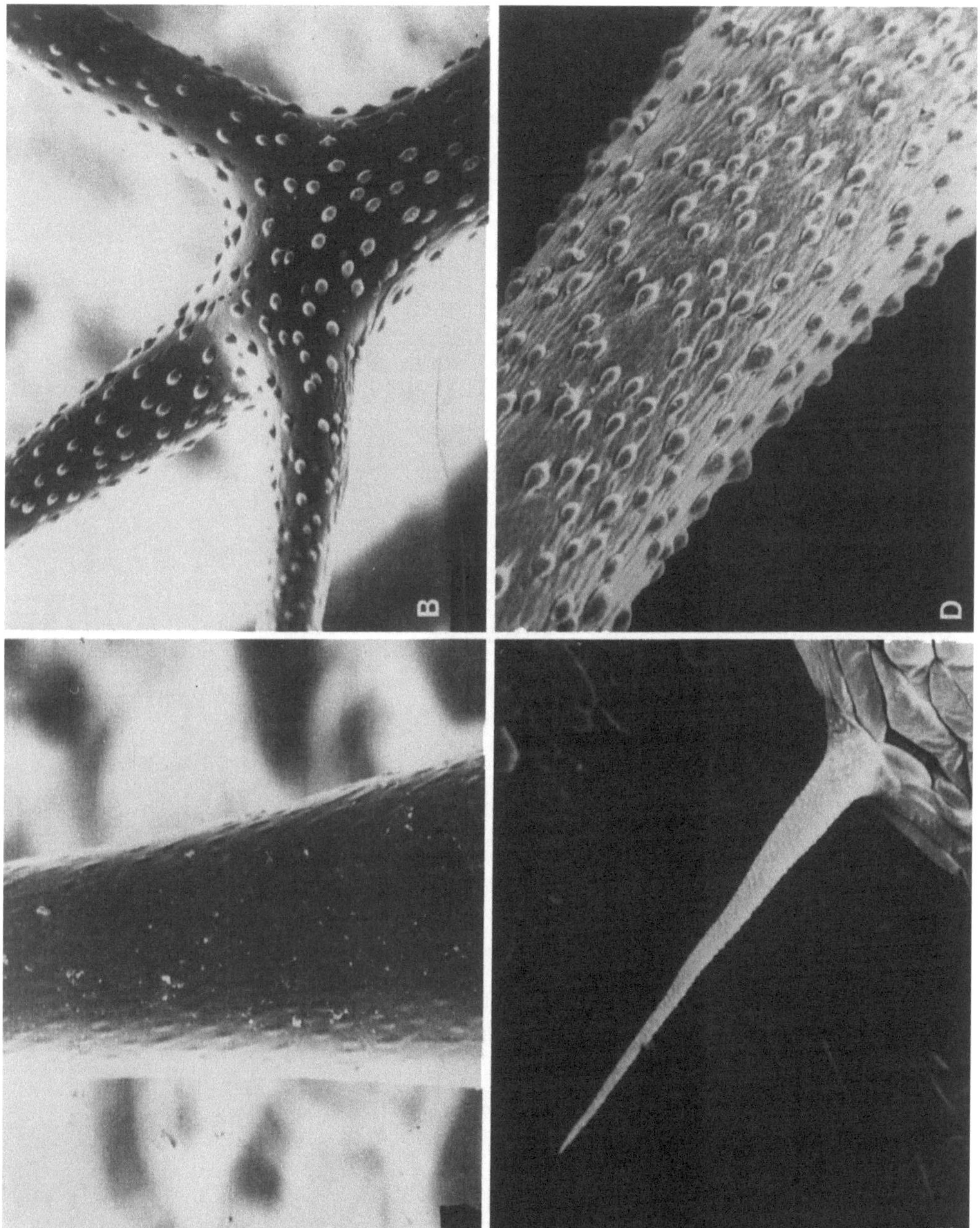

Fig. 44. Ultrastructure of trichomes of *Arabidopsis* **A**, × 5 000, *Brassica* **B**, × 5 000, and of a trichome with the macromorphology of *Brassica* type from the regenerant *Arabidobrassica* line ab 15-1, **C** × 1 000 and **D** × 5 000. (Gleba 1980)

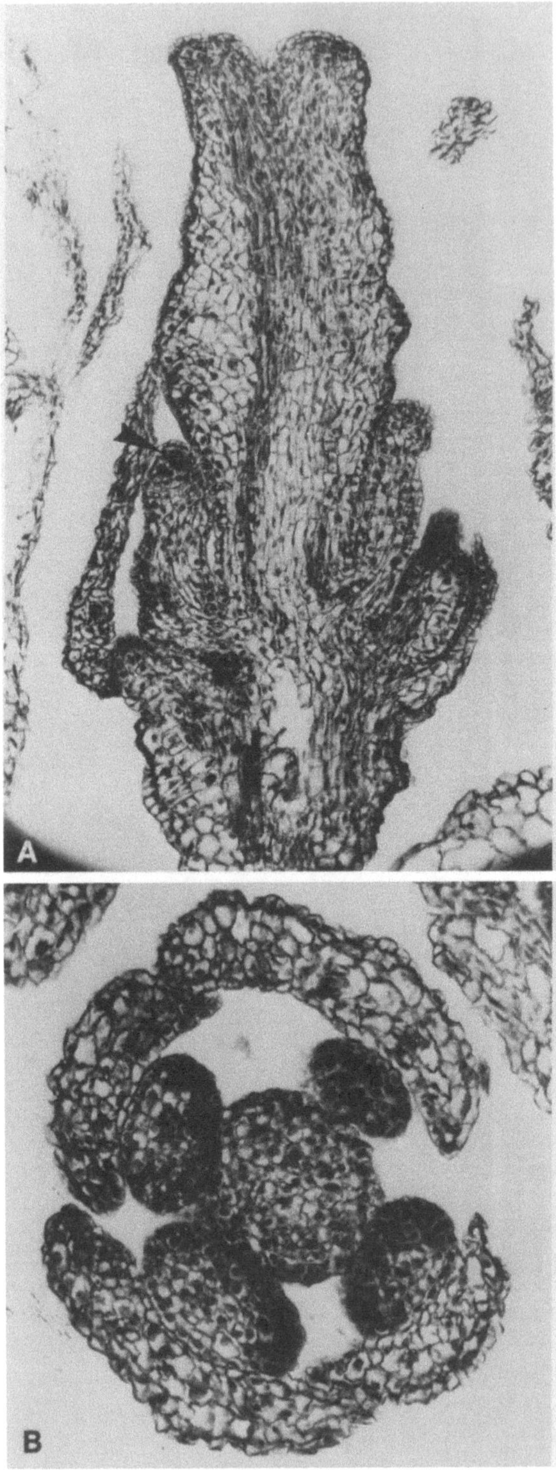

Fig. 45. Anatomy of flowers of asymmetric hybrids of *Arabidopsis* (×) *Brassica*. **A** Longitudinal and **B** transverse sections of flowers stained with saphranin, × 150. The initiation of microsporogenesis is indicated by an *arrow*. (Gleba and Hoffmann 1979)

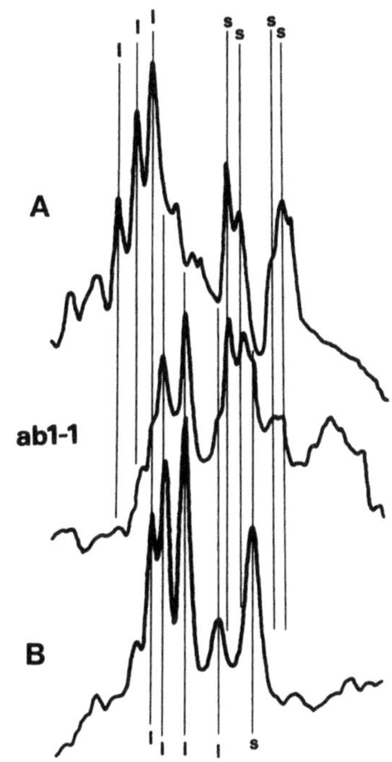

Fig. 46. Analysis of polypeptides of Fraction I protein (ribulose-1,5-bisphosphate carboxylase) in hybrid plants of *Arabidobrassica*. Densitograms of the electrophocusing separation of the subunits of Fraction I protein from *Arabidopsis* (**A**), *Brassica* (**B**), and hybrid plants of line ab 1-1 (*ab 1-1*). Both large subunits (*l*) and small subunits (*s*) of this enzyme are resolved. (Komarnitsky and Gleba 1981)

Arabidobrassica. The name *Arabidobrassica* was given to the somatic hybrids *Arabidopsis thaliana* (×) *Brassica campestris* for scientific and honorary reasons (Gleba and Hoffmann 1979, 1980). The significance of the analysis of the subunit structure of this enzyme for the elucidation of the origin of parasexual hybrids is discussed in Sect. 3.4. Densitograms of the isoelectric focusing patterns of the subunits of the Fraction I protein of both parents and a symmetric hybrid are shown in Fig. 46. The complete analysis showed that the large subunit of this enzyme and thus the chloroplasts of both symmetric and asymmetric plants were from the *Brassica* parent. Likewise, the small subunits of this enzyme and thus the nuclear genes were inherited from both parents in all cases. These data constitute independent evidence of the hybrid nature of the forms obtained and are in perfect agreement with the data of the previous investigations.

In summary, cytological, biochemical, and morphological analysis of the calli, roots, and plants which differentiated from individual *Arabidopsis* (×) *Brassica* protoplast fusion products indicate that the genetic material of both species has combined and is active in these isolates. The experiments show that, in four of the six hybrid cell lines investigated, no specific elimination of genetic material of either of the species occurred during 15 months of unorganized cell growth. In line ab 7, one of the *Brassica* specific esterase isoenzymes disappeared during this time. In line ab 15, isoenzyme and chromosome studies both indicated the elimination of most, but not all, *Brassica* chromosomes. Morphogenetic experiments demonstrated that at least some hybrid cell lines were able to regenerate roots (lines ab 1, ab 2, and ab 15) and even whole plants (lines ab 1 and ab 15). The

Arabidopsis cell line used as one of the parents was defective in morphogenesis and cell cultures originating from *Brassica campestris* mesophyll protoplasts seldom regenerate roots and, to date, have never regenerated shoots. The regeneration of plants by these hybrid cells was evidently due to some restoration of the morphogenetic potential as a result of the hybridization. The demonstration of plant regeneration from not only the nonsegregating line ab 1, but also from line ab 15, which lost chromosomes and isoenzyme bands, is an excellent indication of the diversity of the profound and long-term genetic rearrangement processes switched on by parasexual hybridization. These processes evidently lead to a genetic and perhaps even epigenetic variability within the population of cells derived from a single fusion event. Despite these processes, *Arabidobrassica* cell lines, unlike animal cell hybrids, do not show the tendency for specific elimination of chromosomal material of one of the parents. The different cell lines illustrate that very different genetic situations within hybrid cells can successfully result in organized growth. Plants regenerated from the nonsegregating cell line ab 1 are genetically and morphologically homogeneous and display characteristics of both species. The relatively large chromosome numbers of these plants indicate that the process of plant regeneration probably did not require or involve chromosome elimination. Despite evidence of good growth and more-or-less normal morphology, these plants do not flower. These plants appear to be hybrids in the common sense of the word. On the other hand, plants regenerated from line ab 15, which looses *Brassica* chromosomes, were extremely heterogeneous in morphology. They encompass a full spectrum of forms from morphologically nearly perfect flowering types to severely malformed shoot-like or leaf-like teratomas. Flowering plants from this line have been found to have a greatly reduced chromosome number. Nevertheless, at least a few long, *Brassica*-type chromosomes can be identified in most cases. Morphological analysis and study of the subunits of Fraction I protein demonstrate that some *Brassica* specific characters are present. This type of hybrid is termed an "assymmetric hybrid" to indicate that one of the parental types is represented by less than a full set of chromosomes.

5.4.2 Hybridization of Distantly Related Members of the Solanaceae and Other Families

Krumbiegel and Schieder (1979) reported the regeneration of hybrid shoots and teratomata in the intertribal hybridization of two Solanaceae species: *Datura innoxia* and *Atropa belladonna*. These two species possess chromosomes of greatly differing size (those of *Datura* are about twice as long as those of *Atropa*), making the chromosome analysis easier. Protoplasts were isolated from the leaves of a chlorophyll deficient mutant of *Datura* and from wild type *Atropa*. After fusion, cell colonies were regenerated and the hybrid calli were identified at the stage of shoot regeneration as green forms (an *Atropa* character) covered with thick trichomes (a *Datura* character). A total of 8 independent fusion experiments were performed and 13 purported hybrids were isolated from 5 of them. Chromosome analysis demonstrated the presence of chromosomes of both parental species in the common metaphase plates of all 13 hybrids. One presumably amphidiploid line was also found; however, it was incapable of shoot morphogenesis. Grossly abnormal shoots were produced by other lines carrying polyploid and apparently aneuploid chromosome sets. These data unequivocally demonstrate that the hybridization of these two species did indeed occur.

Additional investigations (Krumbiegel and Schieder 1981) have demonstrated that more or less normal, albeit sterile, plants can be selected from the teratoma after a year and a half. These hybrid plants are similar to *Datura innoxia,* but possess a few *Atropa* chromosomes, including the chromosome responsible for the restoration of chlorophyll synthesis. This line of regenerants frequently sorts out albino areas, thus indicating the segregation and elimination of *Atropa* chromosomes. Shoots possessing not more than six chromosomes of *Atropa* and a complete set of *Datura* chromosomes have been observed.

In other experiments, clones of hybrid cells of the intertribal species combination *Atropa belladonna* (×) *Nicotiana chinensis* were obtained using mechanical isolation and individual cloning. Mesophyll protoplasts of *Atropa belladonna* were fused with callus cells of *Nicotiana chinensis.* A total of 12 cell clones were produced and analyzed 4–6 months and 10–12 months after hybridization. These cells lines were analyzed by chromosome size and morphology, cytophotometric study of nuclear DNA content, isoenzyme forms of esterase, amylase and ATP-ase, as well as by the comparative study of the mutation frequency for resistance to parafluorophenylalanine and 5-methyltryptophan. Abnormal shoots and plants were regenerated 4–5 months after hybridization (Fig. 47). Most of the tests performed on the hybrid cell lines were also performed on these regenerants. Additionally, the polypeptide composition of Fraction I protein, restriction endonuclease mapping of chloroplast DNA and analysis of alkaloids typical for *Atropa belladonna* and *Nicotiana chinensis* were investigated in these hybrid plants.

The chromosome analysis of this combination of species was particularly informative since the chromosomes of *Atropa belladonna* are much smaller and thinner than those of *Nicotiana chinensis* (Fig. 48). Thus, the comparison of chromosome types permitted not only the confirmation of the hybrid origin of the cell lines, but also a rather accurate evaluation of the ratio of the chromosomes of both parental types. The study of six clones after 4–6 months in culture demonstrated the presence of chromosome types characteristic for both parents in all of the clones. In some cases, reconstructed, chain or ring chromosomes were observed. Clones were found with a near diploid set of *Atropa belladonna* chromosomes (2n = 72) and a near triploid set of *Nicotiana chinensis* chromosomes (70) and other clones possessed a near diploid number of *Atropa* chromosomes and 150–200 chromosomes of *Nicotiana.* These clones may have been derived from the fusion of parental cells of different ploidy or by the fusion of various numbers of cells. In cell line 9, the species specific elimination of *Atropa* chromosomes was detected and the number of *Atropa* chromosomes in metaphase plates of hybrid cells did not exceed 10–15. Biochemical studies of the isoenzymes of amylase and esterase and the electrophoretic analysis of soluble proteins in the 12 cell clones *Atropa belladonna* (×) *Nicotiana chinensis* after 4–6 months of culture revealed the presence of species specific forms of these enzymes and proteins from both parents in all but one hybrid cell line, line 14 (Fig. 49).

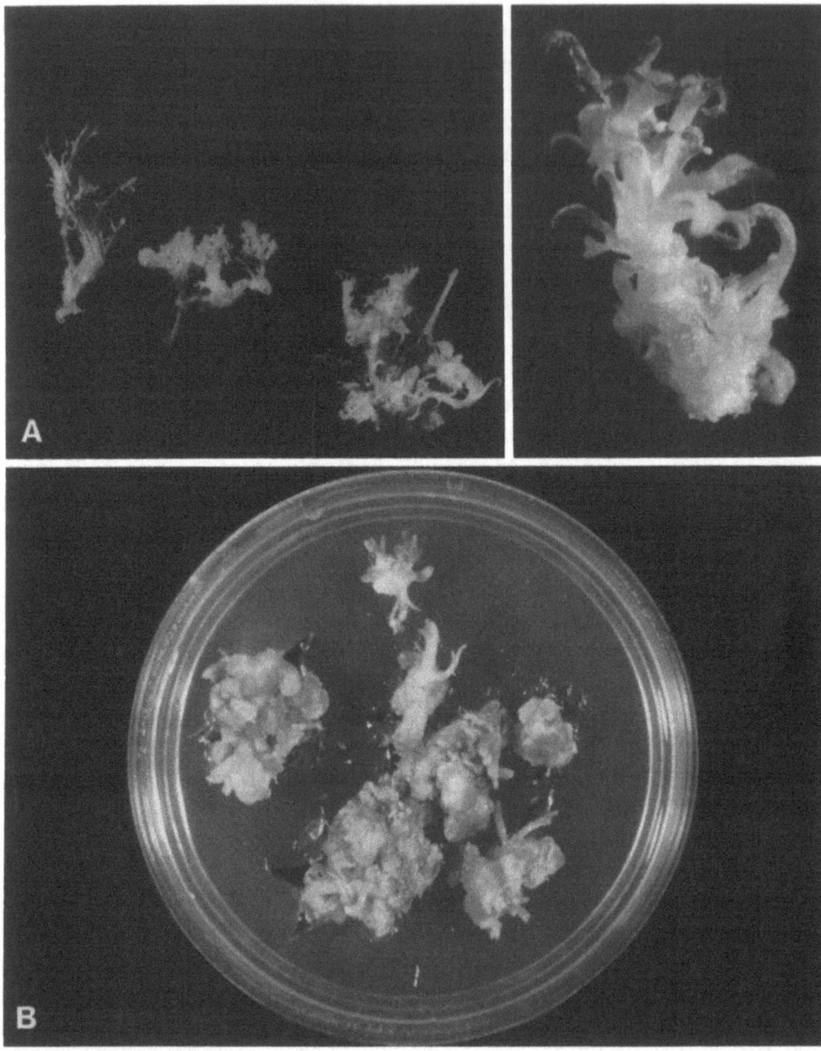

Fig. 47. Intergeneric, intertribal hybrid plants obtained by protoplast fusion of **A** *Solanum tuberosum* (×) *Nicotiana tabacum* (Skarzhynskaya et al. 1982) and **B** *Atropa belladonna* (×) *Nicotiana chinensis*. (Gleba unpublished)

Chromosome analysis of these cell lines 12 months after hybridization was more complete and included all of the clones (Fig. 50). It was found that most of the hybrid cell lines were relatively stable and retained chromosomes of both parental types. However, almost complete, selective elimination of *Atropa* chromosomes took place in cell lines 1 and 9. Nonetheless, stable chromosome numbers was a common characteristic for all of these hybrid cell lines. Cytophotometric studies of the nuclear DNA content also demonstrated the regularity of these cell lines and confirmed the genetic diversity of the independently derived clones (Fig. 51). Thus, similar to the results obtained with *Arabidobrassica,* these results have

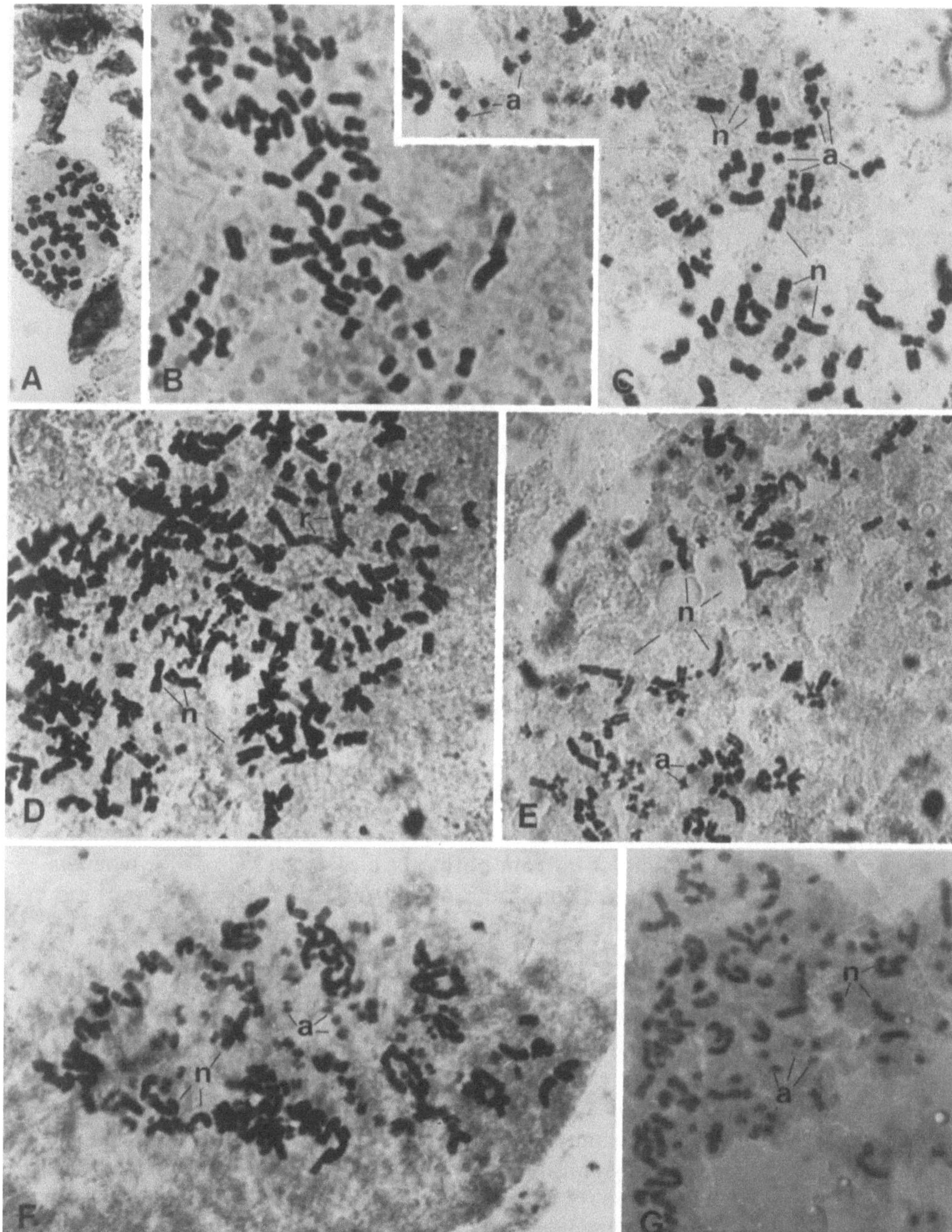

Fig. 48. Chromosomal analysis of somatic hybrids of *Atropa belladonna* (×) *Nicotiana chinensis*. **A** Metaphase chromosomes in a cell of *A. belladonna* root tip. **B** Metaphase plate in a callus cell of *N. chinensis*. Chromosomes 5 months after hybridization in (**C–D**) hybrid cell line 11 and (**E**) hybrid cell line 5. (**F–G**) Metaphase plates in primary callus cells derived from the shoot apex of a plant regenerant from clone 13, × 1 300. Chromosomes typical for *A. belladonna* (*a*), *N. chinensis* (*n*), and reconstructed chromosomes (*r*) can be seen. (Gleba et al. 1982a)

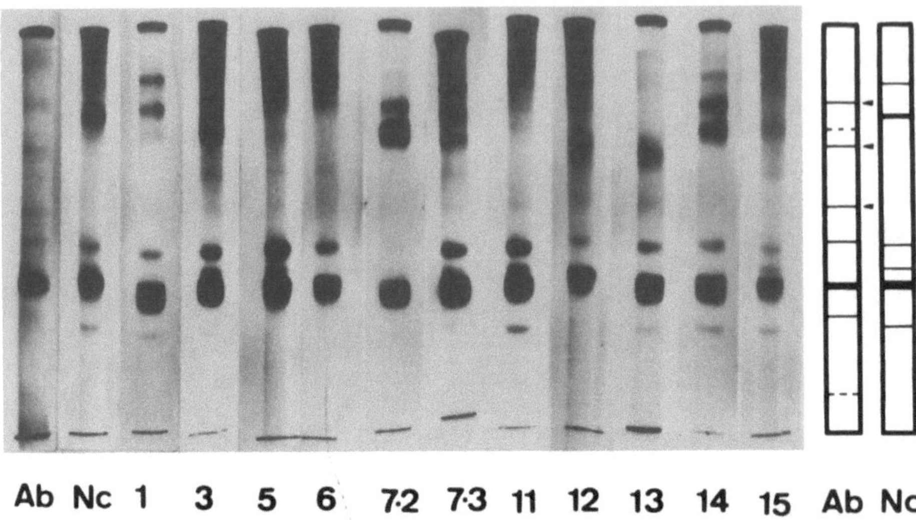

Fig. 49. Electrophoretic analysis of esterase isoenzyme spectra in cells of hybrid clones of *Atropa belladonna (×) Nicotiana chinensis*. Cells analyzed included callus tissues of *A. belladonna* (*Ab*) *N. chinensis* (*Nc*) and 10 clones of hybrid cells (*1, 3, 5,* and *6*; two subclones of *7, 7·2,* and *7·3; 11, 12, 13, 14,* and *15*). On the *right* is a schematic diagram of the zymograms of esterase from the parental material. (Gleba et al. 1982c)

demonstrated that intertribal cell hybrids are relatively stable upon in vitro propagation and only a minor proportion of the cell lines is subjected to species specific elimination of the chromosomes of the mesophyll parental cell type in the hybridization.

Analysis of the shoots regenerated from *Atropa* (×) *Nicotiana* confirmed the presence of chromosomal material from both parental species (Fig. 48), whereas the search for alkaloids (nicotine, anabasine, and atropine) revealed only the presence of small amounts of nicotine. Cytophotometric studies of the nuclear DNA content of leaf cells of these regenerants has also confirmed their unique and hybrid nature (Fig. 52). By the time the analysis of hybrid plants regenerated from four clones was performed, mitotic segregation of the plasmagenes had already taken place and the plastids of either *Atropa* (two clones) or *N. chinensis* (two clones), but not of both species, were retained in the regenerants (as shown by chloroplast DNA and Fraction I protein analysis).

Since it was known that chromosomes of each parent were found in a separate location in metaphases during several successive cell cycles in the initial stages of culture of hybrid cells *Glycine* (×) *Vicia*, it was of interest to examine whether such specific localization of chromosomes might persist for even longer times. The distribution patterns of *Atropa belladonna* and *Nicotiana chinensis* chromosomes in different hybrid cell clones during mitosis in the absence of colchicine revealed that the chromosomes of both parents are mixed after many successive mitoses. That is, their separate localization is not preserved over 6–12 months of culture. However, metaphase plates frequently contained zones with predominant localization of chromosomes of one parent (Fig. 48, 50). Additional ex-

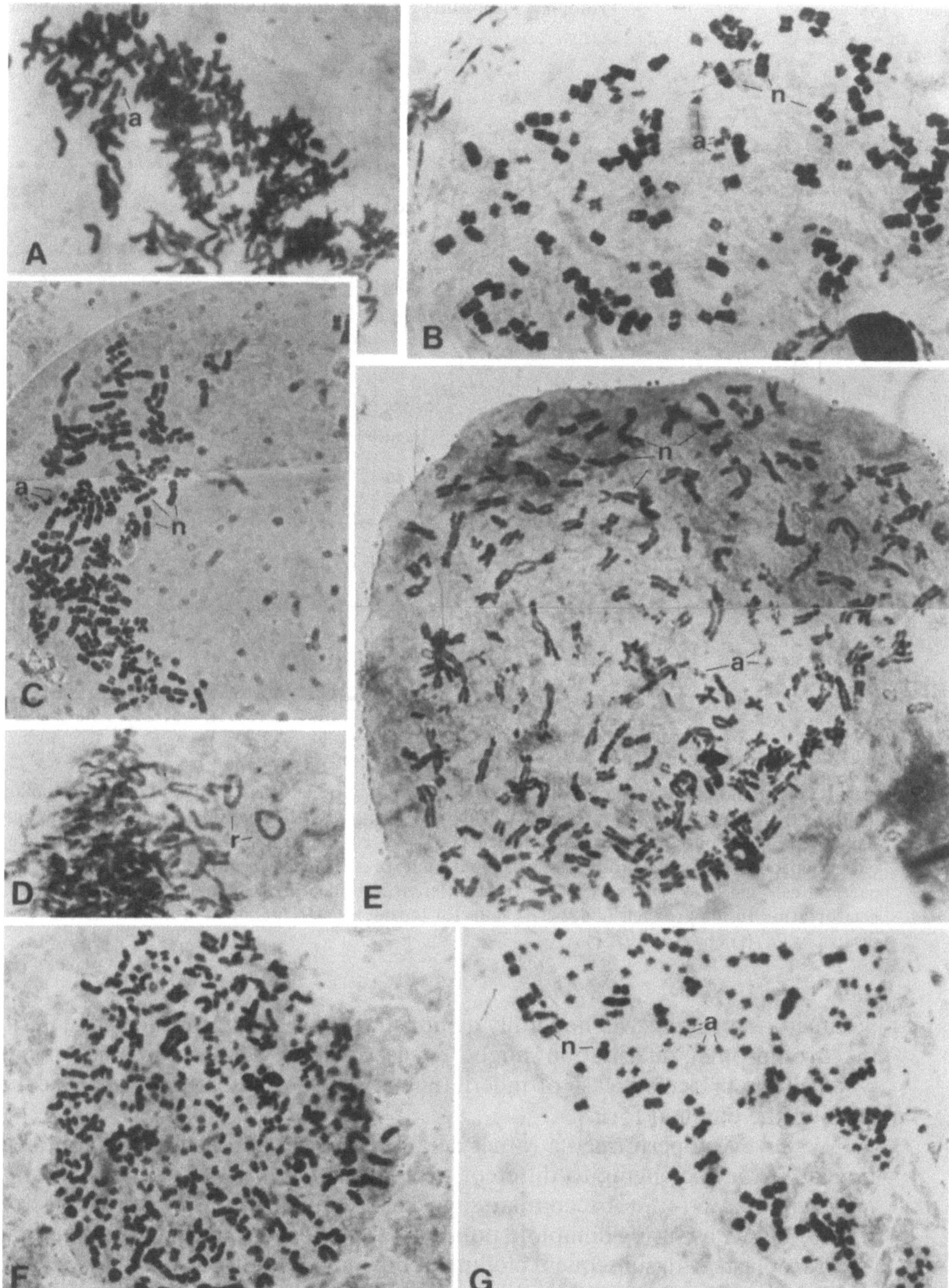

Fig. 50. Chromosomes in cells from different hybrid cell clones of *Atropa belladonna* (×) *Nicotiana chinensis* after 12 months of growth in vitro. **A** Metaphase of a callus cell of clone 1, which loses *Atropa* chromosomes. **B** Metaphase of clone 6, which has a low number of chromosomes of which at least 35 are from *Atropa*. **C** Metaphase of clone 7-2. **D–E** Metaphase of clone 11, which has a high chromosome number and a predominance of *N. chinensis* chromosomes and regularly possesses two ring chromosomes. (**F–G**) Metaphase of clone 12, which has a high chromosome number and a predominance of *Atropa* cromosomes, × 1 300. Chromosomes typical for *Atropa* (*a*), *N. chinensis* (*n*), and reconstructed chromosomes (*r*) are indicated. (Gleba et al. 1982c)

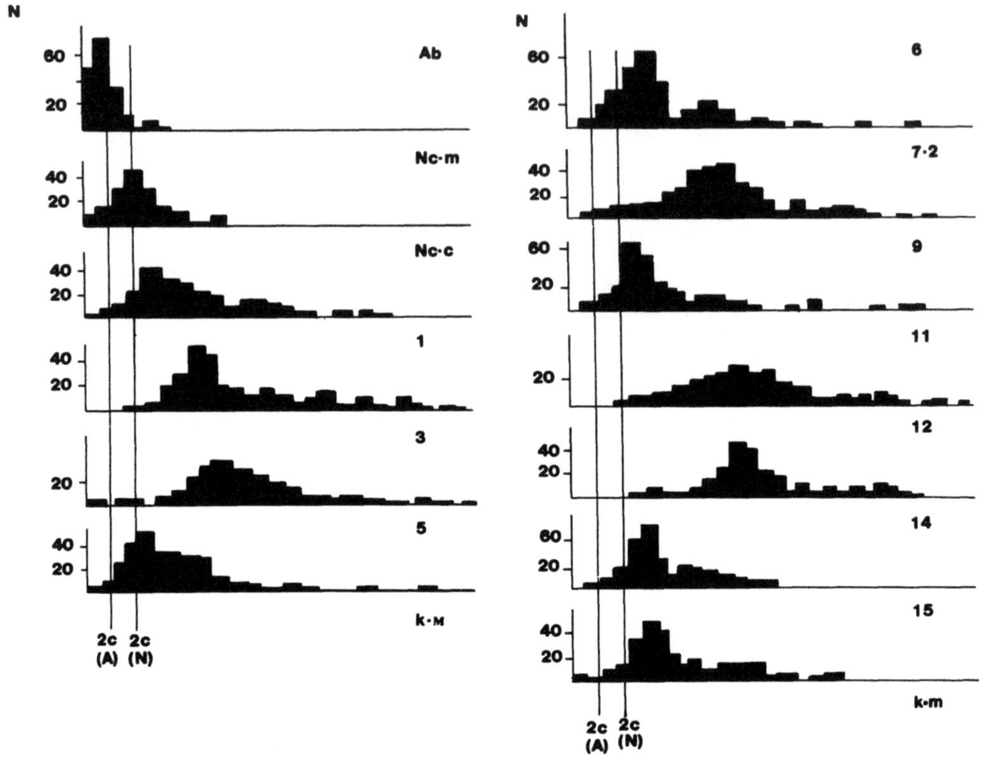

Fig. 51. Cytophotometric analysis of nuclear DNA content in callus cells of different clones of somatic hybrids of *Atropa belladonna* (×) *Nicotiana chinensis* 12 months after hybridization. The histograms represent cell distributions (N = number of cells) for DNA content ($k \cdot m$ = nuclear DNA content in arbitrary units) for: mesophyll protoplasts of diploid plants of *A. belladonna* (*Ab*) and *N. chinensis* ($Nc \cdot m$), parental cell line of *N. chinensis* ($Nc \cdot c$) and different hybrid cell clones obtained from individual heteroplasmic fusion products (*1-15*). The value for the nuclear DNA content of the diploid cells is indicated by *2C*. (Gleba et al. 1982c)

periments in progress will hopefully demonstrate to what extent such irregularity in chromosomal distribution might be ascribed to random fluctuations or whether it represents a case of inheritance of the specific chromosomal localization of the parental forms.

We have also performed a rather detailed study of another hybrid combination of species belonging to different tribes, *Solanum tuberosum* (×) *Nicotiana tabacum* [potato (×) tobacco, Skarzhynskaya et al. 1982]. The hybrid cell lines were obtained by fusing dihaploid potato mesophyll protoplasts and tobacco callus protoplasts with subsequent cloning of the individual fusion products. A total of 30 clones were obtained and the hybrid origin of many of them was confirmed using biochemical analysis of the isoenzymes of amylase and esterase of the initial cell lines and cells obtained by repeated cloning to exclude the possible chimaerization of the callus material. Morphogenesis was induced in nine cell lines 3 months after hybridization and resulted in the formation of abnormal shoots and plantlets. The study of esterase and amylase isoenzymes and the nicotine content

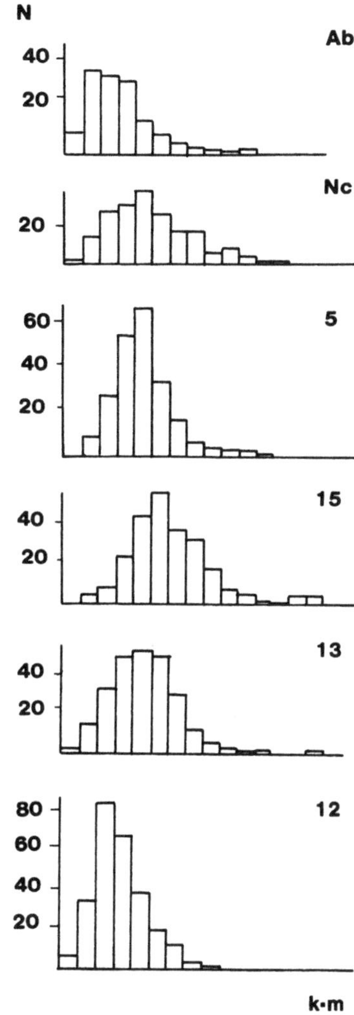

Fig. 52. Cytophotometric analysis of nuclear DNA content in leaf cell of somatic hybrid plants of *Atropa belladonna* (×) *Nicotiana chinensis*. The histograms represent cell distributions (N = number of cells) for DNA content ($k \cdot m$ = nuclear DNA content in arbitrary units) for: mesophyll cells of diploid *A. belladonna* (*Ab*) and *N. chinensis* (*Nc*) and from hybrid plants regenerated from the different hybrid cell clones *5*, *12*, *13*, and *15*. (Gleba et al. in preparation)

in the regenerated plants confirmed the hybrid origin of the abnormal shoots from three clones, while the plants regenerated from the rest of the clones did not differ from the potato parent for these characters. Cytospectrometric analysis of nuclear DNA content was performed on three lines of hybrid cell callus tissue and six lines of regenerated plants. The extreme heterogeneity observed for the callus cell lines significantly exceeds that which might be predicted by superimposition of the histograms of the populations of the parental cells (Fig. 53). The extreme heterogeneity of nuclear DNA content and the chromosome rearrangements observed in cytological studies seem to be a direct consequence of the noncoordination of metabolic processes within the hybrid cells. Moreover, a lowering of the ploidy of the regenerated plants to a level lower than the sum of the parental cells was also observed. Thus, these studies demonstrate that the process of regeneration is accompanied by the selection of genotypes with lower nuclear DNA content and the formation of asymmetric hybrids (Fig. 53).

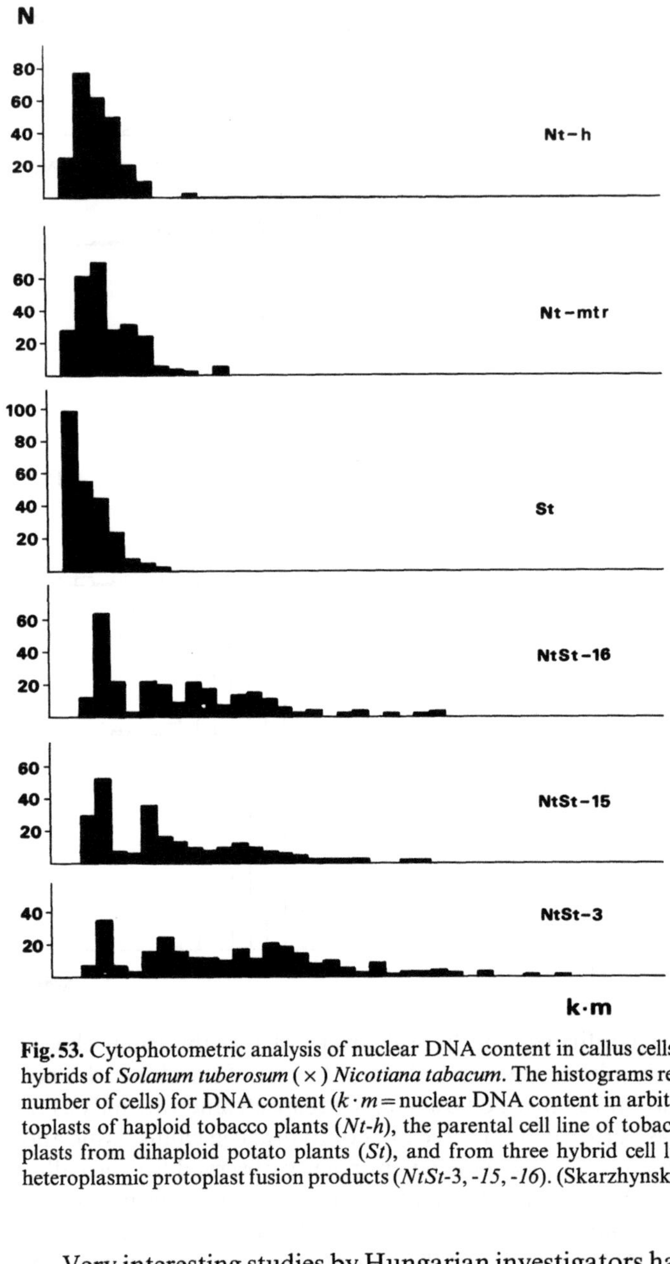

Fig. 53. Cytophotometric analysis of nuclear DNA content in callus cells of different clones of somatic hybrids of *Solanum tuberosum* (×) *Nicotiana tabacum*. The histograms represent cell distributions (N = number of cells) for DNA content ($k \cdot m$ = nuclear DNA content in arbitrary units) for: mesophyll protoplasts of haploid tobacco plants (*Nt-h*), the parental cell line of tobacco (*Nt-mtr*), mesophyll protoplasts from dihaploid potato plants (*St*), and from three hybrid cell lines obtained from individual heteroplasmic protoplast fusion products (*NtSt*-3, *-15*, *-16*). (Skarzhynskaya et al. 1982)

Very interesting studies by Hungarian investigators have produced parasexual hybrids between carrot (*Daucus carota*, 2n = 18) and goutweed (*Aegopodium podagraria*, 2n = 42) (Dudıts et al. 1979). Callus protoplasts from a suspension culture of a carrot chlorophyll deficient mutant (8.9×10^5 cells) were fused with mesophyll protoplasts of goutweed (6×10^6 cells). In the cell suspension treated with polyethylene glycol, the hybrid fusion products and the parental carrot cells were observed to divide, while the parental goutweed mesophyll protoplasts did not divide and were apparently eliminated from the culture at this stage. Ten photosynthesizing colonies were obtained after 2 months of unorganized growth. No green colonies were found in a control experiment in which more than 10^8 carrot cells were treated with polyethylene glycol. After an additional 3 months of unorganized growth, plants were regenerated from three of the hybrid cell lines. These regenerated plants were morphologically abnormal. One of them formed flowers, but they were devoid of anthers and pistils. Chromosome analysis revealed 18 chromosomes similar to carrot chromosomes in their morphology and heterochromatin distribution. The degree of molecular hybridization of nuclear RNA from the regenerated plants with the total cell DNA of goutweed was higher than the hybridization of carrot nuclear DNA with goutweed DNA (it is incomprehensible why the "reciprocal" hybridization of goutweed nuclear RNA and the newly obtained hybrid nuclear RNA with carrot DNA was not performed). An analysis of carotenoid pigments demonstrated the presense of neurosporene, a carotinoid typical for goutweed and absent in carrot. The restoration of chlorophyll synthesizing activity cannot be explained by a simple transfer of goutweed chloroplasts upon fusion, since the carrot form utilized was a nuclear genome mutant. As an interpretation of these seemingly contradictory data, the authors proposed that the protoplast fusion caused the premature chromosome condensation in nuclei of *Aegopodium* mesophyll cells (as observed in animal systems, Ringertz and Savage 1976) and thus caused the elimination of the chromosomal material of this species upon division. However, these processes were accompanied by fragmentation of the *Aegopodium* chromosomes and by partial integration of these fragments into the carrot chromosomes. Similar processes seem to take place in animal somatic cell hybrids in which the transfer of isoenzyme markers is sometimes observed without the transfer of visible chromosome fragments (Schwarz et al. 1971; Boyd and Harris 1973).

It is premature to classify the plants produced by Dudits et al. (1979) as hybrids. However, this study must be regarded as an important experiment that might clear the way to a new field of investigation in plant somatic hybridization.

In one extension of this study, Dudits (in press) has performed recurrent fusion, somatic backcrossing (Wetter and Kao 1980), of the hybrids *Daucus carota* (×) *Aegopodium podagraria* with an albino strain of carrot. Since the mesophyll protoplasts of the hybrids do not divide under the conditions employed, the backcrossed hybrids were selected as the green regenerants. A chromosome study of the plants obtained by backcrossing revealed that they possess 35 chromosomes, whereas the initial hybrid plants had 18 chromosomes. These new plants were phenotypically similar to carrot and produced normal flowers and were capable of growth in soil, in contradistinction to the initial hybrids. At present, attempts are being made to obtain the sexual progeny of these plants. To date, there are no biochemical data concerning the nature of these backcrossed somatic hybrids.

In addition to the experiments on hybridization of species belonging to different genera and different tribes already mentioned, we have also performed investigations of a limited extent to elucidate the possibilities of production of somatic hybrids in other intertribal combinations of Solanaceae species. These investigations consisted of the production of cell clones by cloning the individual heterokaryocytes and the study of their origin. Hybrid cell clones were obtained from *Atropa belladonna* (×) *Nicotiana tabacum* (2 clones; Gleba et al. 1982b), *Solanum sucrense* (×) *N. tabacum* (4 clones; Skarzhynskaya et al. 1982), *S. sucrense* (×) *N. chinensis* (14 clones), and *Lycopersicon esculentum* (×) *N. tabacum* (2 clones). In all cases except the last one, biochemical analysis of esterase and amylase isoenzymes confirmed the hybrid nature of these various cell lines. This conclusion was further substantiated by the study of chromosome size and morphology in the case of *Atropa belladonna* (×) *N. tabacum*. Special attempts to induce morphogenesis and to regenerate intact plants were not made, but morphogenetic activity and the formation of green, shoot-like teratoma were observed in one *Atropa belladonna* (×) *N. tabacum* clone.

Nagao (1982) has probably produced hybrid plants in the intertribal species combinations *Nicotiana tabacum* (×) *Salpiglossis sinuata,* but to date no biochemical analysis of the material obtained has been performed. The morphological peculiarities displayed by the presumed hybrids and cited as arguments for hybridization are not completely convincing and might also be the result of aneuploidy.

There are additional references concerning the production of intertribal cell hybrids; however, no detailed reports on these have been published. Gosch and Reinert (1976, 1978) fused protoplasts of *Atropa belladonna* and *Petunia hybrido* and selected hybrids on the rather inconclusive basis of their abnormal morphology (cit. after Krumbiegel and Schieder 1979). Constabel (1978), in his review of the presentations of the 1978 IAPTC Congress in Calgary, presented a list of interfamily and intertribal combinations in which division of the heterokaryons were reported [tobacco (×) tomato, tobacco (×) *Datura,* tobacco (×) potato, *Atropa belladonna* (×) tobacco, carrot (×) *Angelica archangelica*].

5.4.3 General Conclusions of Results Obtained with Intertribal Hybrids of Higher Plants

A summary of the many experiments mentioned in this section reveals that intergeneric and intertribal cell hybrids of higher plants behave in a rather similar way and are characterized by the following parameters. (1) *Relative stability of the hybrid state.* No complete elimination of the genetic material of either of the parents took place after months or even years of culturing for most of the species combinations studied. (2) *Genetic rearrangements.* Reconstruction and partial elimination of chromosomes gave rise to an increase in the heterogeneity of the populations for nuclear DNA content and also changes in the spectra of isoenzymes. (3) *Diversity of genetic constitution.* The hybrid cell lines were derived from different hybridization events and from different fusion products and thus reflect the varied genetic and epigenetic characteristics of the parental cells at the moment of hybridization. (4) *Limited morphogenetic potential.* The hybrid cells possessed a reduced capacity for morphogenesis and gave rise to morphologically and functionally abnormal plants, which were sterile in all cases.

It is central to our scientific dogma that species which are taxonomically distantly related can not be crossed with each other. Additionally, there are no extensive records concerning such attempts in the literature. However, there are a great number of reports of both successful and unsuccessful sexual crosses of plant species which are taxonomically closed related. These include intrageneric-interspecific and, in a few cases, intergeneric species combinations.

The entire classical problem of remote hybridization of plants has been a sort of sport conducted by plant breeders and recently by "tissue culturists," also. The taxonomists are burdened with the unenviable job of deciding how far apart the species in question should be placed according to taxonomic guidelines. Although taxonomy can hardly claim to be based only on objective criteria and be free of mistakes, its conclusions are nevertheless in good agreement with the results of plant hybridizers. In only a few cases has it been possible to sexually cross plants from different genera and no successful cross has been reported between plants which belong to different tribes or families.

The hybrid plants of *Arabidobrassica* produced in our experiments were the first plant hybrids between higher plant species belonging to different tribes. Can their production be explained by a mistake in the taxonomic classification of these two species or have new possibilities for hybridization via protoplast fusion been discovered? The familiy of Cruciferae has been classified by at least eight botanists and they all grouped *Arabidopsis* and *Brassica* into different tribes (Vaughan et al. 1976). Therefore, it is clear that distinctly new possibilities have been opened up by somatic hybridization of plant cells and incorrect taxonomy is not the explanation for the successful intertribal hybridization. Some of these possibilities for the bypassing of incompatibility barriers have also been demonstrated by the production of hybrids of *Datura* which can not be sexually crossed without the aid of embryo culture (Schieder 1978a) or the production of tomato (×) potato hybrids (Melchers et al. 1978), which can not be produced by sexual techniques (see 5.5).

When the experiments on *Arabidopsis* and *Brassica* protoplast fusion were undertaken, it was not expected that intact hybrid plants would be produced. A priori, it seemed that genetic material from two distantly related species could not be organized to the extent of supporting morphogenetic activity. Perhaps the most important consequence of all these experiments is the fact that hybrid cells of distantly related plant species are demonstrably capable of morphogenesis and this process is not obligatorily associated with the elimination or inactivation of genetic material of one of the parental species. This has been confirmed by new experiments and by the production of hybrid plants from other species combinations. It is now possible to state that parasexual hybridization is an efficient and promising tool for the crossing of distantly related species. The unsuccessful attempts to cross distantly related plant species using sexual techniques apparently reflects the noncoordination of various cellular mechanisms in these species rather than a morphogenetic impotency of the hybrid zygote.

All of the hybrid plants produced by parasexual techniques have been abnormal to some extent and appear "wretched" in comparison to other plants which were produced within the natural barriers of genetic exchanges inherent in sexual crosses. Theoretically, it intuitively seems that most of the anomalies of these hybrid plants result from chromosome unbalance, which is a situation similar to aneuploidy. But is it reasonable to expect a balanced chromosome situation in a cell containing chromosomes of an *Arabidopsis* cell line and a *Brassica* plant or even in the true amphidiploid when both genomes are represented by 2n sets of chromosomes? Current data suggest that mixtures of parental chromosomes, although initially euploid, are not compatible enough to persist after repeated mitotic cycles and are subjected to chromosomal rearrangement during unorganized cell growth. Although it is a bit premature to generalize, there is not much hope that the genetic material of distantly related species will remain unchanged once mixed. On the contrary, the complete variability of both chromosomes and nuclear encoded characters at the cell and plant level indicates that the hybridization process itself automatically enables genetic changes of all possible sorts. It remains to be seen whether the amplification of genetic variability due to parasexual hybridization is greater that that due to known mutagens. One can expect nearly every sort of organized form or plant to arise from the hybridized cells, with the probable exception of true alloeuploids (e.g., plants with two complete euploid sets of genetically unchanged parental chromosomes). It is difficult to imagine how such plants can be morphologically perfect and fertile. Do we need such monsters? What is their relevance and importance for society?

The hybridization of distantly related plant species gives the experimenter the unique possibility to combine various cellular mechanisms that are capable of only partial cooperation. Thus, a disorder may be produced in a plant cell that paradoxically permits the analysis of the mechanisms determining cell activity. One can enumerate several concrete questions which embody the sense of this line of experimentation. In our opinion, the answers to the following questions are important for the understanding of the organization of genetic material in plant cells and may be conveniently studied in distantly related parasexual hybrids.

Does such a fundamental parameter as the DNA content of a chromosome change after cell hybridization, especially in combinations of species with widely diverging values? Is the localization of chromosomes random in interphase nuclei and during metaphase? If not, what is the nature of the inheritance of chromosome organization in the nucleus? Are the chromosomes and nucleolar organizers of the two parental species located together in the interphase nucleus; that is, does a common or "hybrid" nucleolus form in the hybrid nucleus? What is the degree of expression of genetic material from each parent in cell hybrids of species with dramatically different nuclear DNA content? Is such a hybrid cell phenotypically distinguishable from the parent with the higher DNA content per nucleus? (Plant species differ considerably in nuclear DNA content and it is quite easy to find species differing by a factor of 100 in the mass of DNA per diploid nucleus.)

Aside from somatic cell hybridization, no other experimental approach seems suitable for answering these and similar questions.

From the point of basic research interests, a cell line with permanently changing genetics and which oscillates between various morphogenetic patterns is of general importance. With such a system, one can finally test for correlations between specific chromosome content and corresponding morphogenetic or differentiation patterns. Our experience with *Arabidobrassica* is that it is very easy to recover forms with morphogenetic blocks. Such forms are unable to pass certain stages in development and usually die after reaching this stage. Several patterns were observed in *Arabidobrassica* that are unusual for crucifer morphology and can presumably be attributed to some sort of atavism. However, thorough studies using species with chromosomes better suited to cytological investigation are required.

The hybrids *Atropa* (×) *Nicotiana* might serve as a model species combination for the elucidation of numerous questions concerning the mechanisms of cell reproduction and morphogenesis by using parasexual hybridization of distantly related plant species. The chromosomes of these species differ in size and the correct analysis of the relative content of chromosomes from each of the parents in the hybrids is possible. Additionally, the hybrids of this species combination can be relatively easily produced, they grow actively in vitro, are relatively stable, and are capable of morphogenesis (Gleba et al. 1982a, 1982b, 1982c).

Another fundamental question is whether the hybridization of distantly related species promises something interesting to plant breeders. The potential of this technique for breeding applications should be cautiously interpreted. In our opinion, interfamily and intertribal hybridization probably can not give rise to true alloeuploid plants. Of special interest, however, are the asymmetric hybrids. In the not too distant future, it will most likely be possible to produce fertile asymmetric hybrids. Assuming that the conditions for the selective elimination or retention of interesting chromosomes or even genes are developed, this chromosome/gene transfer will probably be of primary importance in practical plant breeding. Today, rather than calculating and theorizing on their potential value and applications, it is more appropriate to list the main questions that have been raised by these experiments.

1. What genetic processes occur in hybrid cells with unbalanced chromosome sets and how are they correlated with the processes of differentiation and morphogenesis?
2. What is the fate of the epigenetic programs of the parental cells upon hybridization?
3. Is it possible to influence the symmetry of the hybrids to the level single chromosome transfer, single gene transfer, or cybrid production?
4. Can such assymmetric hybrids be fertile?

5.5 Intergeneric Hybrids

The production of intergeneric hybrids via traditional sexual crossing is often difficult or impossible. The parasexual hybrid potato (×) tomato, produced by Melchers (Melchers et al. 1978) is of interest since this hybrid has never been produced sexually. Taxonomists are not unanimous in their attribution of both species to the *Solanum* genus and tomato is often classified as a separate genus, *Lycopersicon*. In Melchers' experiments, protoplasts from dihaploid potato were fused with mesophyll protoplasts of an aurea type chlorophyll mutant of tomato using the standard technique of polyethylene glycol, high pH and high Ca^{+2} treatment. Under the culture conditions employed, the mesophyll protoplasts of tomato were capable of division, but their callus tissues were not capable of shoot regeneration. Four morphologically abnormal forms with some of the characters (leaf

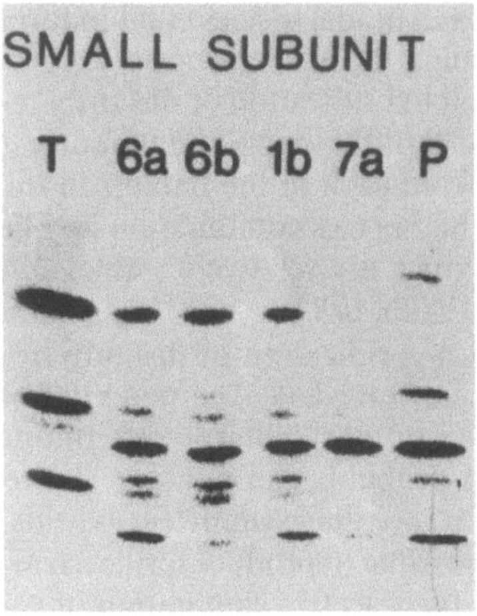

Fig. 54. Isoelectric focusing of S-carboxymethylated ribulose-1,5-bisphosphate carboxylase small subunits from tomato *T*, potato *P*, and different somatic hybrids potato (×) tomato. (Melchers et al. 1978)

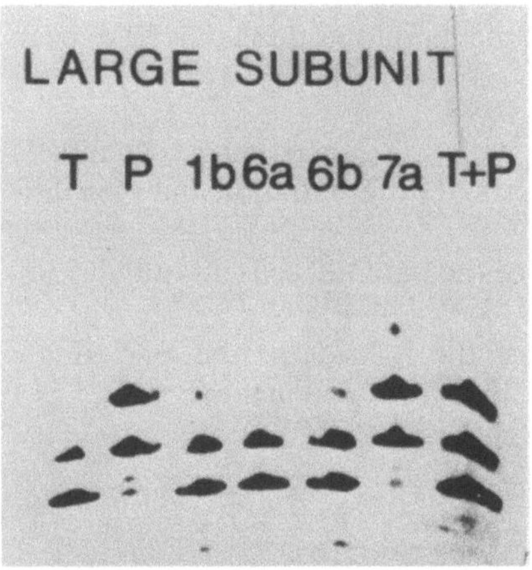

Fig. 55. Isoelectric focusing of S-carboxymethylated ribulose-1,5-bisphosphate carboxylase large subunits from tomato (*T*), potato (*P*), a mixture of tomato and potato (*T*+*P*), and different somatic hybrids potato (×) tomato. (Melchers et al. 1978)

shape and flower coloration) of both parents were observed among a large number of potato regenerants. Chromosome analysis has demonstrated that three of the plants had a hyperamphiploid chromosome number (50–56 instead of 48) and the other hybrid plant possessed nearly 72 chromosomes. The chromosomes of potato and tomato are morphologically indistiguishable and thus no conclusion can be drawn about the ratio of chromosome types. The analysis of the small subunit of the Fraction I protein, a marker for nuclear genes, has revealed the presence of polypeptides of both potato and tomato types (Fig. 54). The large subunit polypeptides of Fraction I protein, a marker for chloroplast genes, consisted of tomato type in three of the four plants and potato polypeptides and thus potato chloroplast genes were found in the hybrid form with 72 chromosomes (Fig. 55). These potato (×) tomato hybrids were abnormal in morphology with double flowers, thickened roots, and no stolons. They were sterile, which may be a consequence of either their aneuploid constitution or their hybrid nature. The visual selection of hybrids employed in this study may have been inappropriate to identify true allodiploids if these were morphologically more normal and thus similar to potato.

In other studies on these same potato (×) tomato hybrid cell lines (Melchers et al. 1978), the restriction endonuclease patterns obtained by digestion of chloroplast DNA with the enzymes Eco RI, Kpn I, and Pst I were examined (Schiller et al. 1982). Of the 12 hybrid lines screened, 8 contained exclusively potato chloroplast DNA and 4 contained only tomato chloroplast DNA, at a detection level of 0.1%–3%.

In a more recent publication, Melchers (1982) fused mesophyll protoplasts of a tomato chlorophyll deficient mutant with those of a potato chlorophyll deficient mutant. Euploid plants with 48 chromosomes were recovered among the regenerants (Fig. 56). However, these plants grow outside in vitro culture only when grafted onto tomato. Some of these hybrid plants flower and form parthenocarpic fruits (Melchers, personal communication).

From the results of both sets of experiments, one can probably conclude that aneuploid and euploid hybrid plants in the species combination *Solanum tuberosum* (×) *Lycopersicon esculentum* may be expected to be sterile and the last resort to produce fertile hybrids is to generate highly asymmetric hybrids that contain only a few nuclear characters or genes from one of the parents.

It is not out of place to recall here the optimistic expectations of scientists hoping to produce economically valuable hybrids by protoplast fusion. Cocking has fantasized and drawn a picture of a hypothetical hybrid between tomato and potato that is a plant with branches bent under the weight of tomato fruits and with stolons bearing tubers of impressive size (Cocking 1971). There is certainly some irony in the fact that the first parasexual hybrids of tomato (×) potato possess neither fruits nor tubers. Melchers has recovered dozens of tomato (×) potato somatic hybrids, a few of which are probably alloeuploids, but no fertile hybrid plant has been found among them (Melchers, personal communication).

Most experiments with somatic hybridization imply a practical, utilitarian aspect, at least indirectly. This aspect was alluded to in the work of Melchers et al. (1978), as well as in the other works analyzing potato (×) tomato hybrids. Unsuccessful attempts have been made by a number of investigators to obtain the sexual hybrids between these plants to transfer the cold resistance of potato to the cultured tomato forms. Smillie et al. (1979) studied the cold resistance of four somatic hybrids potato (×) tomato. The leaves of these hybrids and of the parents were maintained at 0 °C for different periods and the subsequent rate of recovery of photosynthetic electron transfer activity was determined by measuring the restoration of cytochrome f. The cold resistance of the hybrid forms was intermediate compared to the parental forms in all of the experiments. Similarly, the heat resistance of these hybrids was also intermediate as determined by changes in chlorophyll fluorescence in leaves heated at a rate 1 °C/min (Smillie and Gibbons 1979). However, the practical application of these hybrids for plant breeding is hampered by the complete sterility of the somatic hybrids potato (×) tomato.

Fig. 56 A–C

Ninnemann und Jüttner (1981) studied the low molecular weight volatile compounds produced by cultured cells of potato (×) tomato hybrids (Melchers et al. 1978) by gas chromatography. Among the 30 or so molecules that were identified, ethylvalerate, 1-pentanol, 1-hexanol, and ethyl benzoate were found to be specific for potato, while 3-methyl-1-butanol, methyl salycilate, benzyl alcohol, and benzaldehyde were specific for tomato. Unfortunately, since all of these substances are widely spread in plants, the "specificity" claimed by the authors must be regarded as quantitative rather than qualitative. The gas chromatographic analysis revealed a higher content of these substances in most of the hybrids as compared to the parental forms and, furthermore, compounds typical for both parents as well as novel molecules were found in the hybrids. For example, large amounts of terpenoid, which is absent in both potato and tomato, were found in some hybrid cell lines. The increased diversity of volatile molecules in these hybrid cell lines confirms the data of numerous investigations of sexually obtained intergeneric hybrids. However, the appearance of "novel" molecules might be explained not only as a result of new gene combinations resulting in the formation of products which are not synthesized in the parental cells, but also as a consequence of modification of regulatory mechanisms and derepression of biosynthetic pathways normally shut off in the parental cells.

Following the advice of Melchers (1980) to treat with ease the taxonomic categories which are "created by man" and therefore open to mistakes, it is appropriate to consider the experiments on hybridization of herbaceous (*Datura innoxia*) and tree (*D. sanguinea* and *D. candida*) *Datura* species by Schieder (1980a). According to some taxonomic classifications, the tree *Datura* species are attributed to the separate genus *Brugmannsia*. Thus this study is of interest as an independent measure of the relationship between the herbaceous and tree forms. Mesophyll protoplasts from chlorophyll deficient diploid plants of *D. innoxia* and diploid plants of *D. sanguinea* were fused and cultured under conditions where the cells of *D. sanguinea* did not divide. Forty-seven hybrid plants were obtained. Most of them were tetraploid, but hexaploid and aneuploid forms were also found. However, all of these hybrid plants were morphologically abnormal, incapable of root formation and flowering and were similar to the teratoma arising from tumorous transformation. The cells of these hybrids were also able to grow on hormone-free media. Two hybrid plants of *D. innoxia* (×) *D. candida* were produced in the same way. These plants were found to be morphologically normal and a chromosome study has demonstrated that both plants were hexaploid. Both plants were sterile. As mentioned by Schieder, these data differ significantly from the results of hybridization between herbaceous *Datura* species [*D. innoxia* (×) *D. discolor* and *D. innoxia* (×) *D. stramonium*] where the hybrids obtained were mainly tetraploid and were morphologically and functionally normal.

Fig. 56. Fruit **A** and "tubers" **B** of the somatic hybrid plant potato (×) tomato. The fruit was obtained after pollination with *Solanum stenotomum*. The wooden label measures 11 cm (Schiller et al. 1982). **C** Production of euploid potato (×) tomato hybrids after fusion of chlorophyll deficient mutants. From left to right, *1* tomato form from which the chlorophyll deficient mutant gilva *2* originated, *3* hybrid plant regenerated from the fusion of cells from plant *2* (×) plant *4*, *4* mutant of dihaploid potato line H_3703, and *5* dihaploid potato H_3703. (Melchers 1982)

5.6 Influence of Physical and Chemical Factors on the Fate of Genetic Material in Distantly Related Species Combinations

5.6.1 Introduction

Parasexual hybridization of distantly related plant species produces genetically unstable progeny and the subsequent propagation of hybrids is accompanied by a number of genetic processes, such as chromosome elimination and chromosome rearrangements. Although the production of genetically stable forms was the purpose of many investigations, some of the most successful and significant experimental results have been not only descriptions of various processes of genetic rearrangements as a result of hybridization, but also attempts at intensifying them so as to utilize them as instruments of investigation and means for specific cell reconstruction.

Some of the genetic processes observed in plant somatic cell hybrids that are of interest in hybrid manipulation include genetic recombination, genetic segregation, and directed asymmetry of the hybrid genome. Various physical and chemical agents have shown promise in attempts at manipulating the genetic makeup of hybrids in distantly related species combinations. The influence of physical factors (X-ray and gamma ray irradiation) and chemical factors (treatment with iodoacetate) upon the fate of genetic material in closely related species combinations is discussed in 4.4.

5.6.2 Irradiation and the Induction of Genetic Asymmetry

The elimination of large parts or even all of the genetic material of one parent has been observed after cell hybridization of distantly related plant species in all cell clones of *Glycine* (×) *Nicotiana* and *Aegopodium* (×) *Daucus* and in some clones of *Arabidopsis* (×) *Brassica, Atropa* (×) *Datura, Solanum* (×) *Nicotiana* and *Atropa* (×) *Nicotiana*. This phenomenon is termed asymmetrization. As was demonstrated in animal hybrid cells by Pontecorvo (1971), such an asymmetrization can be induced or its direction reversed by irradiation with X-rays and gamma rays (0.6–1.6 krad) of the parental cells whose genetic material was to be eliminated. A similar effect was also obtained after treatment with 5-bromodeoxyuridine. Irradiation was employed by several authors in experiments with somatic hybridization of closely related plant species to completely eliminate the nuclei of one of the parents and thereby produce cybrids (Zelcer et al. 1978; Aviv and Galun 1980; Menczel et al. 1982; Sidorov et al. 1981a, see 4.4.3). In the most detailed study (Menczel et al. 1982), an increase of the radiation dose from 50–300 J kg^{-1} (5–30 krad) caused the proportion of nuclear hybrids among the progeny to decrease from 56%–30%. However, the authors were not successful in recovering asymmetric hybrids containing only *a portion* of the chromosome set of the irradiated parent.

Dudits et al. (1980) employed irradiation to study the regulation of genetic processes in hybrid cells of distantly related plant species. The authors attempted to produce asymmetric hybrids by fusing *Petroselinum hortense* mesophyll protoplasts (2n = 22) irradiated with X-rays (9 krad) with callus protoplasts of a chlorophyll deficient genome mutant of *Daucus carota* (2n = 18). Among the green regenerants were plants with 19 chromosomes, which means they contained only a part of the genome of one of the parents. These hybrids possessed the isoenzymes specific for carrot of four different enzymes and displayed at the same time the band of glucose-6-phosphate dehydrogenase activity specific for *Petroselinum hortense*.

The transfer of genes accompanied by the presence (Goss and Harris 1975, 1977; Rodgers 1979) or absence (Rodgers 1979) of cytologically visible chromosomes or their fragments has also been demonstrated for animal cell hybridization after the use of irradiation.

5.6.3 Induction of Mitotic Chromosome Segregation by Chemical Agents

The induction of mitotic chromosome segregation in cell hybrids of different species is an interesting problem and holds great promise in the genetic engineering of somatic hybrids. Since few attempts have been made in this direction with higher plant somatic hybrids, results of analogous studies on hybrids of fungi (obtained by protoplast fusion) and sexual hybrids of higher plants will be discussed. The original idea to use chemical agents for this purpose can be traced to the work of Lhoas (1961), who demonstrated chromosome elimination and haploidization of vegetative diploid cells of fungi after treatment with para-fluorophenylalanine, an analog of phenylalanine. This approach is now widely applied in the study of fungal genetics, particularly for segregation analysis. In addition to para-fluorophenylalanine, chloral hydrate (Singh and Sinhar 1976), benomyl (benlate, methyl-1-(butylcarbamoyl)-2-benzynidazol carbonate; Hastie 1970) and other compounds are also known to induce chromosome segregation. Kevei and Peberdy (1979) grew the interspecific hybrids *Aspergillus nidulans* (×) *A. rugulosus*, obtained by protoplast fusion, in the presence of benomyl and chloral hydrate on complete medium and demonstrated random segregation of linkage groups as a result of these chemicals. Knight et al. (1963) observed aneuploid and haploid chromosome sets in the roots of *Ribes* sexual hybrids treated with para-fluorophenylalanine. In a similar way, Nitzsche (1973) demonstrated mitotic reduction of chromosomes in the intergeneric hybrids *Festuca pratensis* × *Lolium multiforum* (2n = 4x = 28). Treatment with para-fluorophenylalanine resulted in cells with decreased euploid numbers of chromosomes (21, 14, or 7). A reasonable interpretation of the mechanism of action of these chemicals on the process of segregation is not possible from these data. Preliminary conclusions by Gupta and Carlson (1972) about more efficient suppression of growth of high ploidy cells by para-fluorophenylalanine have been confirmed by Matthews and Vasil (1975); however, other authors were not successful in attaining similar results (Dix and Street 1974).

We were able to isolate a number of cell lines capable of active growth on para-fluorophenylalanine by cultivating cells from different clones of *Atropa* (×) *Nicotiana* intertribal hybrids on nutritive media containing either 0.33 m*M* or 1 m*M* para-fluorophenylalanine. The growth of cells on these media resulted in a decrease in the average nuclear DNA content in the hybrid cell lines. A more pronounced effect was observed in those clones with initially high ploidy. However, preliminary chromosome studies in these cells show that no selective species specific elimination of chromosomes takes place (Gleba et al. in preparation).

6 Use of Somatic Hybridization

6.1 Genetic Analysis Using Parasexual Hybridization

6.1.1 Introduction and Perspective

Parasexual hybridization by protoplast fusion may be used for genetic analysis exactly like conventional sexual crossing. Of special interest are those cases where the sexual process can not be employed. Some of these special cases that may be readily analyzed by parasexual hybridization include the analysis of genetics in species combinations that can not be sexually crossed, the analysis of nuclear genes in plants or plant cells defective in morphogenesis, the analysis of extranuclear genes of plant cells, the study of the fate of epigenetic programs of parental cells in the hybridization process, and the analysis of the fate of different cell mechanisms during hybridization. A tremendous amount of data on these and similar problems has been accumulated in the genetics of animal hybrid cells (Ringertz and Savage 1976).

This discussion will be necessarily short because only a few experiments have been performed on higher plants with the aim of genetic analysis via parasexual hybridization. Additionally, much of the limited data has already been treated in Chaps. 4 and 5 in other contexts and therefore will not be needlessly repeated here. Instead, this section will focus on the analysis of plasmagenes since significant data on them have been accumulated. Again, the focus of this section is the concept of using protoplast fusion and parasexual hybridization as a tool for genetic analysis.

6.1.2 Analysis of the Nature of Inherited Traits

The potentiality of plant cells for morphogenesis and differentiation presumes the existence of a mechanism of epigenetic inheritance or "organization inheritance." Theoretically, the variability in plant cell cultures may be of either a mutational (genetic) or epigenetic nature. The analysis of the origin of the variability (genetic versus epigenetic) in cultured plant cells represents a special problem in itself, which in some cases can not be solved by conventional methods. The most obvious method for elucidation of the nature of variability is the study of the maintainance or loss of characters in the morphogenetic cycle from cultured cell to plant and back to cultured cell. During this cycle, it is supposed that all or almost all the epigenetic determinations are destroyed, in contrast to the mutational changes, which remain (Gleba and Gleba 1978a). Unfortunately, not all cultured cells are capable of morphogenesis. Therefore, we proposed the use of parasexual hybridization for both the annihilation of epigenetic programs and for the restoration of morphogenetic potential as an alternative method of genetic analysis (Gleba and Berlin 1979; Gleba 1980).

In these experiments, tobacco cells resistant to 5-methyltryptophan and which had lost the capacity for morphogenesis were fused with normal tobacco mesophyll protoplasts. The hybrid cells regained their capacity for morphogenesis and hybrid plants were regenerated. Callus tissues were produced from the organs of the hybrid plants and were found to be partially resistant to 5-methyltryptophan. Thus, in this experiment, the epigenetic determinations of the parental cells were "destroyed" twice: first, during the cell hybridization and second, in the morphogenetic cycle of callus to plant to callus, Nonetheless, the resistance to 5-methyltryptophan was preserved. Therefore, one can conclude that the specific resistance to 5-methyltryptophan is at least partially controlled by genetic factors (Figs. 24–27).

6.1.3 Analysis of Nuclear Genes

The plant forms regenerated after mutagenesis are often defective for morphogenesis and/or gametogenesis. In these cases, protoplast fusion and regeneration of intact plants can be used not only as a method for the rescue of recessive or codominant mutations and their transfer into the heterozygous state, but also as a method of genetic analysis. Several studies should be mentioned as good examples. In the work of Grimsley et al. (1977a, b), protoplast fusion was utilized for the complementation analysis of auxotrophic mutants of the moss *Physcomitrella patens*. Mutagenesis gave rise to mutants dependent on para-aminobenzoic acid and nicotinic acid. In most cases, these mutants displayed developmental abnormalities and were sterile as a result. The technique of hybridization by protoplast fusion demonstrated that most of the changes induced by mutagenesis were recessive nuclear mutations. The complementation analysis of six independent strains auxotrophic for nicotinic acid revealed the existence of three nonoverlapping complementation groups (Table 18). These results were in good agreement with the analysis performed by sexual hybridization of some mutant strains (Ashton and Cove 1977). In these experiments, fertile heterozygous forms were obtained by protoplast fusion, thus opening a new perspective for a further and more detailed genetic analysis of the nature of the mutation a changes. Complementation analysis and dominance tests for para-aminobenzoic acid auxotrophs and mutants with abnormal development were subsequently performed (Grimsley et al. 1980).

Another example is provided by the studies on tobacco cell lines defective for nitrate reductase activity (Glimelius et al. 1978a). A large number of cell lines defective in nitrate reductase activity had been produced by selecting for chlorate resistance. The cell lines retained their resistance in the absence of chlorate and most of them had lost the capacity for morphogenesis (Müller and Graffe 1978). Study of acceptor activity and selectivity of the enzyme and reassembly of active

Table 18. Genetic anlysis of the moss *Physcomitrella patens* using protoplast fusion: allocation of the nicotinic acid auxotrophs to complementation groups (Grimsley et al. 1977 b)

Mutant line	Mutant line				
	nic A 4	nic A 10	nic B 5	nic B 6	nic C 7
nic A 2	–[a]	–	+[b]	+	+
nic A 4		–	+	+	+
nic A 10			+	+	+
nic B 5				–	+
nic B 6					+

[a] A minus sign (–) indicates no complementing hybrids were produced following protoplast fusion

[b] A plus (+) indicates that complementing hybrids were produced following protoplast fusion

enzyme in vitro from subunits derived from different variants demonstrated the existence of mutants with different metabolic blocks in the activity of nitrate reductase. For example, the mutant of nia type were defective for the synthesis of the nitrate reductase apoprotein and mutants of the cnx type had a defect in the synthesis of a molybdenum-containing enzyme cofactor. Complementation analysis using protoplast fusion has fully confirmed the data of the biochemical analysis and provided evidence that the nia and cnx mutant types are controlled by recessive nonallelic genes (Müller 1978).

A more detailed study of cnx mutants using somatic cell hybridization has recently been performed in the laboratory of Maliga (Marton et al. 1982). Three groups of cnx mutations complementary to each other were found. This is not unexpected, since several complementation groups of cnx mutants have been demonstrated in other organisms.

In flowering plants, protoplast fusion and complementation analysis of three different auxotrophs for isoleucine, leucine, and uracil have demonstrated that each of these mutations is recessive (Sidorov and Maliga 1982).

There are several studies utilizing cell fusion to analyze the nature of different mutations giving rise to resistance to various antimetabolites.

6.1.3.1 Resistance to 5-Methyltryptophan

In our work (Gleba and Berlin 1979; Gleba 1980), protoplasts of a tobacco hyperdiploid callus line resistant to 5-methyltryptophan (an analog of tryptophan) were fused with haploid or diploid tobacco mesophyll protoplasts. It had been previously shown that this resistance was not due to reversal of inhibition due to high levels of tryptophan, which is overproduced in these mutant cells. The hybrid cells displayed an intermediate level of resistance to the tryptophan analog as compared to the parental cells. Thus, this resistance appears to be codominant (Fig. 27). Resistance to 5-methyltryptophan was used to screen cell hybrids of *Nicotiana sylvestris* and an intermediate level of resistance between the levels of

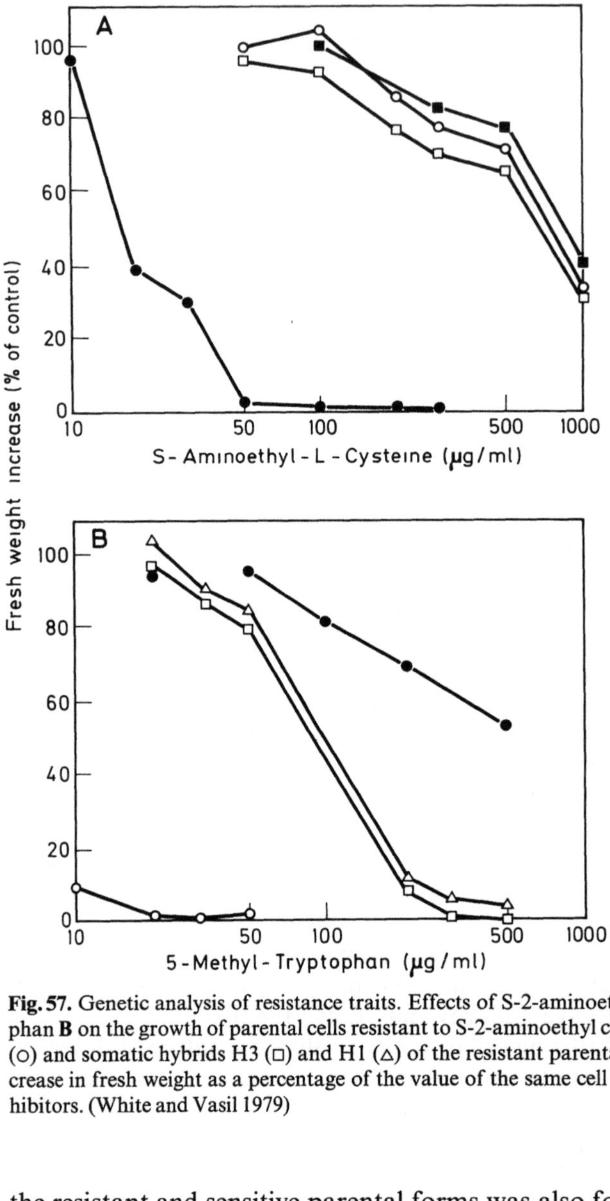

Fig. 57. Genetic analysis of resistance traits. Effects of S-2-aminoethyl cysteine **A** and 5-methyltryptophan **B** on the growth of parental cells resistant to S-2-aminoethyl cysteine (●) and 5-methyltryptophan (○) and somatic hybrids H3 (□) and H1 (△) of the resistant parental cells. The values represent the increase in fresh weight as a percentage of the value of the same cell line grown in the absence of the inhibitors. (White and Vasil 1979)

the resistant and sensitive parental forms was also found (White and Vasil 1979; Fig. 57).

Similar hybridizations utilizing mutants with 5-methyltryptophan resistance were performed by Harms et al. (1981) to screen somatic hybrids of carrot. In contrast to the situation in *Nicotiana*, the degree of resistance to the tryptophan analog in carrot hybrids did not differ from the resistance of the initial parental line. This dominant type of expression of 5-methyltryptophan resistance in carrot hybrids is not contradictory to the codominance observed in *Nicotiana*, since the mutants in question are of independent origin and were obtained not only from different parental mutants, but also from different plant species. Kameya et al. (1981) recovered somatic hybrid plants after fusion of *Daucus carota* cells, which were resistant to 5-methyltryptophan and were defective in morphogenesis, with callus cells of *Daucus capillifolius* and found that the resistance to the tryptophan analog in these hybrids was expressed as a semidominant character.

6.1.3.2 Resistance to S-(2-Aminoethyl)-L-Cysteine

White and Vasil (1979) produced hybrid cell lines of *Nicotiana sylvestris* by fusion of protoplasts isolated from cell variants resistant to S-(2-aminoethyl)-L-cysteine (a lysine analog) and 5-methyltryptophan. The hybrid cells possessed double resistance and it was found that the degree of resistance to the lysine analog in the hybrids was the same as in the initial resistant parental strain – thus, this character displays dominance upon hybridization (Fig. 57). A similar study was performed with carrot cells (Harms et al. 1981). In this case, the fusion of carrot cells resistant to S-(2-aminoethyl)-L-cysteine and 5-methyltryptophan resulted in hybrid cells with a double resistance and the level of resistance to the lysine analog indicated a dominant type of phenotype expression. Similar inferences have been made after studies with animal cell hybrids (Harris 1973).

6.1.3.3. Resistance to Cycloheximide

The experiments described in the previous sections have employed antimetabolites as an active, selective component of a system of somatic hybrid screening. Lazer et al. (1981) utilized the resistance to cycloheximide as a nonselective marker in analyzing somatic hybrids. Cycloheximide resistant carrot cells were pretreated with iodoacetate and then fused with a cycloheximide sensitive, albino strain of carrot. Green hybrid plants were selected and callus tissues were produced from them. These tissues were found to be sensitive to cycloheximide, with the character of resistance to the drug sorting out at a frequency of 10^{-4}, which is three orders of magnitude higher than the spontaneous frequency of mutation. This observation is in good agreement with the inference of the recessive nature of the resistance to cycloheximide.

6.1.3.4 Resistance to Azetidine-2-Carboxylate

Kameya et al. (1981) analyzed somatic hybrid plants produced by fusion of *Daucus capillifolius* cells with cells of a carrot line that was resistant to azetidine-2-carboxylate and was also defective in morphogenesis. The hybrids recovered demonstrated resistance to the drug comparable to resistant parental cells. Thus, the resistance to this drug behaved as a dominant character in this experiment.

6.1.3.5 Additional Complementation Studies

Conclusions concerning the recessiveness or codominance of various genes and their ability to complement have also been drawn by a number of other workers who employed parasexual hybridization of previously unstudied chlorophyll deficient mutants (Melchers 1977a, Dudits et al. 1977; Cocking et al. 1977; Melchers et al. 1978; Schieder 1977a, 1978a, 1980a; Gleba 1979; Sidorov and Maliga 1982). This type of investigation is still not performed as a matter of routine, if for no other reason, then because of the technical difficulties. A second limitation on the possible use of parasexual techniques for the analysis of nuclear genes is the aneuploidy of most of the well-studied cell mutants and, thus, their genetic irregularity, instability, and in some cases, the sterility of the hybrid plants obtained (Glimelius et al. 1978a; Müller 1978; Gleba and Berlin 1979).

6.1.4 Cosegregation of Extranuclear Genes

If heterozygotes for extranuclear genes can be produced, they can of course be used to study the process of mitotic segregation of extranuclear genes in the progeny cells. If the characters coded for by extranuclear genes segregate independently in the mitotic process, they are controlled by different genophores and vice versa; the invariant cosegregation of different characters coded for by extranuclear genes demonstrates the linkage of their corresponding genes (the possible nature of this linkage is beyond the scope of this book). This cosegregation analysis is an elegant way to determine gene linkage groups of extranuclear genes. These investigations seem to be especially suitable in the case of higher plants, whose cells contain at least two extranuclear genophores.

In our experiments (Gleba et al. 1978a; Gleba 1980), cytoplasmic hybrids of *Nicotiana tabacum* (×) *N. debneyi* that were heterozygous for the genes coding for plastome chlorophyll deficiency, the large subunits of Fraction I protein (see Chap. 4) and cytoplasmic male sterility (see Chap. 4) were utilized for studies of the patterns of extranuclear gene segregation. Callus tissues were derived from the leaves of variegated forms and plants were regenerated after 2 months of unorganized growth. Biochemical analysis of Fraction I protein was performed on several dozen green and white plants, which were segregants for the gene for chlorophyll deficiency. In all cases, the white plants possessed only *N. tabacum*-type large subunits of Fraction I protein and all of the green plants possessed only *N. debneyi*-type large subunits (Table 10). Thus, the sorting out of a species specific plastome was accompanied in all cases by the sorting out of the large subunit polypeptides of the appropriate parent. This same phenomenon was also observed in a similar analysis of green and white sexual progeny of variegated hybrids. The obvious cosegregation of these two characters confirms the premise that the polypeptides of the large subunit of Fraction I protein are coded for by the plastome. To the best of our knowledge, these data were the first direct evidence for the genetic linkage of extranuclear genes in higher plants.

A second important observation made in these experiments was that the heterozygosity for these characters was not connected perforce with the heterozygosity for those genes coding for the cytoplasmic male sterility and that the heterozygotes for male sterility (fertility) were frequently homoplastidic and thus were complete segregants for plastome encoded characters (see Chap. 4; Figs. 16, 17). These results practically exclude the possibility that the genes coding for the cytoplasmic male sterility might be located in the plastome and have also confirmed the presence of at least two independently segregating genophores outside the nucleus (Gleba et al. 1978a; Gleba 1980).

Belliard et al. (1978, 1979) performed extensive studies of plasmagene cosegregation. Using restriction endonuclease mapping, they analyzed chloroplast DNA (chl DNA) and mitochondrial DNA (mt DNA) of parasexual hybrid plants bearing novel morphological types of cytoplasmic male sterility (cms) as a result of the hybridization. As initial material, tobacco varieties Samsun and Xanthi were fused with a tobacco cms analog, Techne variety, with the cytoplasm of *Nicotiana debneyi* (Belliard et al. 1977a, b). The cytoplasmic male sterility of this cms analog was expressed morphologically as underdevelopment and twisting of the anthers, which were also completely sterile, and cleft petals. The initial material was also marked by petiolated leaves in the Samsun and Xanthi varieties and sessile leaves in the Techne variety. The leaf morphology of *Nicotiana* is a polygenic Mendelian character. Biochemical analysis was carried out for both sterile and fertile (cybrid) progeny possessing leaf morphology of one of the parents and flowers of a morphologically intermediate form. Electrophoretic analysis of Eco RI fragments of chl DNA from nine plants demonstrated the absence of cosegregation between the type of chl DNA and the characters of sterility (fertility) and modified flower shape. In all cases, the chl DNA of the hybrids was identical to one of the parental forms and the type of parental chl DNA was not correlated with the form of male sterility (fertility) observed (Figs. 21, 22).

The mitochondrial DNA of nine cybrids was analyzed in a similar way. The interpretation of these results is complicated by the fact that mt DNA, although similar in molecular weight to chl DNA, is usually digested by restriction enzymes into a greater number of fragments than chl DNA. This phenomenon has been explained as a result of the heterogeneity of mt DNA molecules in higher plant cells (see, for example, the review of Leving and Pring, 1979). Nevertheless, the results of the mt DNA analyses were striking. Each of the nine cybrids possessed mt DNA that differed not only from the parental forms, but also from the sum of the parental forms. Moreover, the mt DNA of all but one cybrid yielded new fragments. The number of restriction fragments in these cybrids varied from 35–41 and could not have arisen from the parental forms of mt DNA. The sums of the molecular weights of the restriction fragments of the cybrid mt DNA is closer to the value for the Techne parental form. At the same time, a correlation has been observed between the male sterility (fertility) and flower morphology on the one hand and the ratio of specific parental restriction fragments in the cybrid mt DNA on the other hand. A transition towards sterility and an increase in morphological abnormalities were accompanied by a quantitative increase in the number of specific fragments of *N. debneyi* mt DNA (Tables 14 and 15). On the whole, these results clearly demonstrate that the cytoplasmic male sterility studied in these experiments is most probably associated with the mitochondrion.

Another group of investigators has studied the segregation of plasmagenes in a somewhat more complex system (Zelcer et al. 1978; Aviv et al. 1980; Galun and Aviv 1979; Aviv and Galun 1980; Galun et al. 1982). Protoplasts of a cms analog of *Nicotiana tabacum,* possessing the cytoplasm of *N. suaveolens,* were irradiated with X-rays and fused with *N. sylvestris* protoplasts. The fusion products were cultured in mannitol-containing medium in which the protoplasts of *N. sylvestris* do not divide. Three different types of plants were found among the regenerants of seven callus colonies. They were identified as (1) plants of *N. sylvestris* bearing the plastids of *N. suaveolens;* (2) nuclear hybrids *N. sylvestris* (×) *N. tabacum* containing the plastids of *N. suaveolens;* and (3) plants of *N. sylvestris* with *N. sylvestris* plastids. Since the authors did not specify the exact number of plants studied, one can only state that in at least three segregants the cosegregation of chl DNA and the other two characters (resistance to tentoxin and the large subunit polypeptides of ribulose-1,5-bisphosphate carboxylase) was observed. The cosegregation of the last two characters was shown in at least six different plants. In the same experiments, the cytoplasmic male sterility character segregated independently of both nuclear and plastome markers. Independent segregation of cytoplasmic male sterility and resistance to tentoxin was also demonstrated for other hybrid plants (Aviv and Galun 1980). These cybrid plants were obtained by fusion of *N. sylvestris* protoplasts bearing the cytoplasmic male sterility character (produced in the foregoing somatic hybridization experiments) with protoplasts of *N. tabacum*. A detailed study of the mitochondrial DNA of seven segregants obtained in the second experiment was conducted using the restriction endonucleases Sal I and Xho I and revealed differences in the 33 Sal I and 36 Xho I fragments (Galun et al. 1982). In all but one cybrid, the mt DNA typical for one of the parents plus one to four bands typical for the other parent plus one or several new bands were found. In one plant, a pure parental type of mt DNA was found. An excellent correlation was observed between the sterility/fertility characters and the presence of four Sal I and eight Xho I fragments of mt DNA. No correlation was observed between the sterility/fertility characters and the type of chloroplast DNA and/or the resistance to tentoxin.

The research group of Maliga at the Biological Research Center of the Hungarian Academy of Sciences in Szeged, Hungary has also studied the segregation of extranuclear genes in interspecific hybrids of *Nicotiana*. In one of these experiments, callus protoplasts of a streptomycin resistant mutant of *Nicotiana tabacum* were hybridized with mesophyll cells of *N. knightiana*. Hybrids were selected by mechanical isolation and cloning. A total of 43 subclones were obtained by joint culture of three fusion products. Restriction endonuclease Eco RI digestion of chl DNA from ten plants regenerated from different subclones and electrophoretic analysis of the fragments demonstrated an absolute correlation between the resistance to streptomycin and the *N. tabacum* type chl DNA (Menczel et al. 1981).

In other experiments, different techniques for the transfer of plastids from tobacco cells of plastome mutants (Sidorov et al. 1981a) or with streptomycin resistance (Menczel et al. 1981, 1982; Maliga et al. 1982) into the cells of *N. plumbaginifolia* were studied. The fusions were performed with either tobacco cytoplasts or tobacco cells inactivated by irradiation. The transfer of plastids was confirmed by the acquisition of streptomycin resistance or plastome deficiency by the cells of *N. plumbaginifolia,* as well as by the presence of tobacco specific Eco RI restriction fragments of chl DNA in the hybrid cells. The analysis of chl DNA in a total of 20 plants confirmed the cosegregation of the streptomycin resistance character with the type of chloroplast DNA and the analysis of two plants has shown the cosegregation of the plastome chlorophyll deficiency with the chl DNA type.

Additional studies have been performed in Szeged to investigate the fate of the parental mitochondiral DNA upon cell hybridization. Nagy et al. (1982) studied the mt DNA of *Nicotiana tabacum* (×) *N. knightiana* hybrids, produced previously by joint cloning of three individual heteroplasmic protoplast fusion products (Menczel et al. 1981). The digestion of the parental mt DNA with the restriction endonuclease Sal I resulted in the formation of 31 *N. tabacum* and 26 *N. knightiana* fragments, of which 21 fragments were unique. The analysis of eight nuclear hybrids (five streptomycin resistant forms with *N. tabacum* plastids and three sensitive forms) revealed unique types of mt DNA in each of them. The number of fragments varied from 29–35 and the mt DNA of seven of these plants contained one to three new fragments which were not present in the parents. A total of five novel fragments were observed. Similar results were also obained in the analysis of six plants using the restriction endonuclease Bam HI. In this case, all of the hybrid plant mt DNAs also differed from the parental forms and from each other. These analyses were repeated at least four times during the course of a year and no changes or anomalies were observed in the restriction spectra.

These workers have also studied the location of sequences in mt DNAs homologous to the genes for ribosomal RNA of *Escherichia coli*. For this purpose, mt DNA restriction fragments were hybridized with radioactive plasmid DNA containing inserts of the *E. coli* DNA coding for 16 S and 23 S ribosomal RNA. The six types of hybrid plant mt DNA studied differed from the parental forms and from each other for both the extent or degree of hybridization and for the specific localization of these DNA sequences (Fig. 23).

Recently, Poulsen et al. (1980) and Schiller et al. (1982) compared the data on the large subunits of ribulose-1,5-bisphosphate carboxylase and of restriction patterns of chloroplast DNA in three lines of potato (×) tomato somatic hybrids (Melchers et al. 1978) and have also found cosegregation of these characters.

In summary, most of the characters coded for by plasmagenes demonstrate true mitotic segregation and sorting out of pure parental types, with the exceptions of the mt DNAs and the characters of cytoplasmic male sterility and abnormal flower morphology. During the process of mitotic segregation, the characters coded for by the plasmagenes segregate in distinct groups (i.e., cosegregate). The higher plant cell contains at least two cosegregation or linkage groups of plasmagenes and their segregation is completely independent. The data of cosegregation analyses are summarized in Table 19. Independent experiments have provided unequivocal results: the characters of resistance to tentoxin and streptomycin, the plastome chlorophyll deficiency and the type of polypeptides of the large subunit of ribulose-1,5-bisphosphate carboxylase cosegregate with the corresponding chloroplast DNA and are thus coded for by this genophore. This conclusion for some of these characters had been previously drawn on the basis of genetic and biochemical analyses. For example, a mutation of plastome chlorophyll deficiency was correlated with an alteration of chloroplast DNA (Wong-Staal and Wildman 1973); the large subunit of ribulose-1,5-bisphosphate carboxylase was found to be coded for by extranuclear genes and synthesized in chloroplasts, and its genes had been mapped in chloroplast DNA (Chan and

Table 19. Linkage groups of extranuclear characters demonstrated by parasexual hybridization. Cosegregation is indicated by + and no correlation is indicated by –

Character	Character			
	Type of chl DNA[a]	Type of RBPC[b] large subunit	Plastome chlorophyll deficiency	Resistance to tentoxin
Type of chl DNA[a]		+[h, j, o]	+[l]	+[h]
Type of RBPC[b] large subunit	+[h, j, o]		+[e]	+[h, p]
Plastome chlorophyll deficiency	+[l]	+[e]		
Resistance to tentoxin	+[h]	+[h, p]		
Resistance to streptomycin	+[k, n]			
Type of mt DNA[c]	–[g, m, q]			
Cytoplasmic male sterility	–[d, h, q]	–[e, h, p, r]		–[h, i, p, q]
Flower morphology	–[d, h, q]	–[e, h, r]		–[h, i, q]

Table 19 (continued)

Character	Character			
	Resistance to streptomycin	Type of mt DNA[c]	Cytoplasmic male sterility	Flower morphology
Type of chl DNA[a]	+[k, n]	–[d, h, q]	[d, h, q]	–[d, h, q]
Type of RBPC[b] large subunit			–[e, h, p, r]	–[e, h, r]
Plastome chlorophyll deficiency				
Resistance to tentoxin			–[h, i, p, q]	–[h, i, p]
Resistance to streptomycin				
Type of mt DNA[c]			+[g, q]	+[g, q]
Cytoplasmic male sterility		+[g, q]		
Flower morphology		+[g, q]		

[a]chloroplast DNA; [b]ribulose-1,5-bisphosphate carboxylase (Fraction I protein); [c]mitochondrial DNA; [d]Belliard et al. (1978); [e]Gleba et al. (1978); Gleba (1978); [f]Aviv and Galun (1979); [g]Belliard et al. (1979); [h]Aviv et al. (1980); [i]Aviv and Galun (1980); [j]Poulsen et al. (1980); [k]Menczel et al. (1981); [l]Sidorov et al. (1981); [m]Komarnitsky et al. (1982); [n]Maliga et al. (1982); [o]Schiller et al. (1982); [p]Bonnett and Glimelius, personal communication; [q]Galun et al., 1982; [r]Glimelius et al., 1981

Wildman 1972; Blair and Ellis 1973); the resistance to tentoxin was associated with changes in the coupling factor I of chloroplasts and appeared to be coded for by plasmagenes (Steele et al. 1976, 1977; Durbin and Uchytil 1977). However, the studies using hybridization of somatic cells have for the first time permitted the confirmation of these statements by strict formal genetic analysis.

The two characters of cytoplasmic male sterility and abnormal flower morphology segregate independently of the chloroplast DNA and from the characters coded for by chl DNA. On the other hand, they cosegregate with the type of mitochondrial DNA. In this case, however, cosegregation must be discussed cautiously, since the types of restriction spectra of mt DNA as well as the types of modifications in flower morphology arising during mitotic segregation are not identical to the initial parental types and thus the question is not that of simple alternative states of mt DNA type. Nevertheless, a clear quantitative correlation exists in the hybrids between the relative content of fragments specific for each parental form and the prevalence of characters of this parent is observed.

The next logical step in this line of investigation consists of "passing through the double heterozygotes" all of the available extranuclear characters (such as resistance to streptomycin, kanamycin, and tentoxin; polypeptide composition of ribulose-1,5-bisophosphate carboxylase (Fraction I protein) and coupling factor I; protein structure of 70 S ribosomes; and restriction spectra of mitochondrial and chloroplast DNA, etc.) and to attribute these extranuclear characters to definite linkage groups based on the pattern of segregation observed in the mitotic process. At present, two separate linkage groups of higher plant extranuclear genes are known. One consists of plastome chlorophyll deficiency, resistance to streptomycin, tentoxin, and the polypeptide composition of the large subunit of Fraction I protein and the other group consists of the characters of cytoplasmic male sterility. This fact is important in itself since in no other type of eukaryotic organism has more than one linkage group of extranuclear genes ever been revealed by genetic methods.

Recombined mitochondrial DNA molecules have been found in somatic hybrids of animal cells (Dawid et al. 1974). Since both genophore DNAs of higher plants are of much higher molecular weight than the mt DNA of animals, recombination in a plant system need not necessarily be so frequent. The tremendous opportunities for genetic analysis which would be opened up provided a technique of extranuclear gene recombination were available are self-evident. It should be mentioned that practically all of the investigations of chl DNA in somatic hybrids have revealed only pure parental types of chl DNA. Doubts have been expressed in this connection about the possibility that the hybridization of cells may result in recombination of chl DNA and whether the experimental systems are sensitive enough to detect this process. Interestingly, an abstract has recently been communicated by Conde (1981), who analyzed the chloroplast DNA of *Nicotiana glauca* (×) *N. langsdorffii* parasexual hybrids using the restriction endonuclease Xba I. A novel fragment with a molecular weight of 1.46×10^6 was detected in one of the hybrids. Since the number of studies of chl DNA in somatic hybrids is still rather limited and since in some cases the variability of chl DNA has been observed (for example, in studies of tobacco cytoplasmic sterile analogs obtained by substitution of cytoplasm; Kung et al. 1982), it seems reasonable to predict that recombinant forms of chloroplast DNA are not that rare and more examples will be found as more studies are completed.

In contradistinction to chloroplast DNA, molecules of plant mitochondrial DNA undergo frequent and dramatic changes in parasexual hybridization (Belliard et al. 1979; Nagy et al. 1981; Galun et al. 1982). These changes may arise as a result of the interaction of molecules of two parents resulting in the formation of recombinant forms of mt DNA molecules containing nucleotide sequences from both parents (an extremely interesting possibility), from intraspecific intermolecular recombination, from intramolecular rearrangements, and/or from selective replication and dominance in the hybrid cytoplasm of certain molecular forms from a heterogeneous mt DNA population, which constituted a minor fraction in the parental cells and, thus, were not detected by the methods used. Although it is scarcely feasible to definitively answer the question about the recombination of mitochondrial DNA at this stage, it is nevertheless clear that the hybridization of somatic cells, unlike sexual crossing, leads to profound changes in the mitochondrial genome and that exciting perspectives are thereby opened for the genetic analysis of the mitochondrial genome in higher plants and for its reconstruction using hybridization of somatic cells.

6.1.5 Analysis of Mitotic Cycle Mechanisms

It is conceivable that this sort of investigation in higher plant cells might be less informative than similar studies of hybrids and heterokaryons of animal cells since plant cells seem to be considerably less stringent in their cell cycle programming and their cell division machinery can be influenced by a variety of external physiological factors.

In particular, there has been no efficient cell cycle synchronization of plant cell populations and it is quite possible that just the very nature of the cells of higher plants might be an unsurmountable cause of the difficulties associated with synchronization. Interestingly, Szabadas et al. (1981) has demonstrated a relatively efficient synchronization (about 80%) of cultured cells and thus points the way toward further studies. Nonetheless, it is obvious that the study of these mechanisms is very complicated.

Several studies have shown that mitoses in heterokaryons may be synchronous after hybridization of mesophyll cells in G_1 phase with protoplasts isolated from actively growing callus tissues, which are heterogeneous for cell cycle phases (Kao et al. 1974; Constabel et al. 1975a, b; Gosch and Reinert 1976, 1978; Constabel et al. 1977; Constabel 1978). This lends support to the theory of cytoplasmic signals for the initiation of mitosis. The cytoplasmic signals responsible for the initiation of mitosis are not species specific since the synchronization, though not always complete, was also observed in heterokaryons obtained by cell fusion of distantly related plant species (Kao et al. 1974; Gosch and Reinert 1976; Kao 1977). A specific elimination of chromosomes of the parental mesophyll cells takes place in the fusion of mesophyll and callus cells to form the distantly related hybrids *Glycine* (×) *Nicotiana* (Kao 1977), *Arabidopsis* (×) *Brassica* (Gleba and

Hoffmann 1978), and *Atropa* (×) *Nicotiana* (Gleba et al. 1982c). This elimination might be connected with events similar to the premature chromatin condensation observed in animal hybrid cells that results in its preferential elimination (Ringertz and Savage 1976).

Sexual hybridization results in the formation of a common, "hybrid" nucleus in the zygote, but the chromosomes of the two parents are not necessarily randomly "mixed" within the common nucleus. On the contrary, there are indications that the parental genomes remain spatially separated even after many successive mitotic divisions (one of the interesting studies on flowering plants is Finch et al. 1981). Possible mechanisms of this phenomenon are discussed by Bostock and Sumner (1978). Studies of primary cell divisions in somatic hybrid cells of different species (Kao 1977; Constabel et al. 1977) show that the chromosomes of both parents are spatially arranged as separate groups during the first two to four successive divisions and form clearly delineated regions in metaphase and anaphase (Fig. 31). This indicates the preservation of the strict spatial arrangement of chromosomes in the interphase nucleus and probably the transmission of this arrangement via mitosis. It is evident that such regulated organization of chromosomal spatial arrangement in the nucleus of hybrid cells might be a significant factor affecting the process of interspecific chromosomal rearrangements. Therefore, we made an attempt to elucidate to what degree the specific arrangement of chromosomes of two species in a common nucleus is inherited and preserved in long-term cultures of cell lines of *Atropa belladonna* (×) *Nicotiana chinensis*, after 12 months of unorganized in vitro propagation (Gleba et al. 1983c). The chromosomes in the cells of long-term cultures are "mixed" to a large extent, as reflected by their apparently random distribution in mitotic metaphase plates (Fig. 50). At the same time, this analysis showed rather large regions with chromosomes of only one species to be present in a significant number of metaphases. This is quite logical from a statistical point of view, taking into consideration the fact that the process of the production of the squashed preparations results in a projection of chromosomes situated in different planes onto only one plane. It is rather difficult to rationalize this phenomenon by random fluctuations in the distribution of chromosomes. For example, 35 chromosomes of tobacco are situated in a sector where *Atropa* chromosomes are absent in Fig. 50. Moreover, the statistical analysis of a number of preparations by calculation of Mahalonobis' generalized distance between two chromosome types in metaphases has demonstrated that the distribution of the two types of chromosomes are different at a high confidence level ($p = 0.95$). This confirms the theory that the ordered chromosomal arrangement observed in the squashed cell preparations is probably not an artifact arising as a result of different hydrodynamic properties of the two types of chromosomes. The spatial separation of the two genomes was preserved at least partially in these cells after 12 months of culture. Further study of this question should consist of analysis of the inheritance of the spatial arrangement of specific, individual chromosomes from both parents. An excellent example would be the study of chromosomes with nucleolar organizers since, in this way, it should be possible to ascertain if hybrid nuleoli are formed in the hybrid nuclei. Spatial separation of the chromosomes of both parents has also been observed in metaphases of 6-month-old cultures of the interfamily cell hybrid *Pisum sativum* (×) *Nicotiana chinensis* (Gleba et al. 1983b), as well as in metaphases of the somatic hybrid plants *Nicotiana tabacum* (×) *N. glauca* (Evans, personal communication).

As was mentioned in Chap. 4, hybrids with chromosome instability were produced after fusion of normal euploid somatic plant cells (tobacco mesophyll protoplasts) with cells of the same species isolated from long-term callus cultures characterized by chromosome instability. In these experiments, the character of chromosome instability was dominant (Gleba and Berlin 1979; Gleba 1980). Since these were intraspecific hybridizations, this chromosome instability seems not to be a consequence of some sort of incompatibility within the hybrid cell and thus be a consequence of its hybrid nature only, but appears to be the result of inheritance of a bona fide parental instability character. Further study of dominant and codominant characters of chromosome instability in hybrid cells may elucidate the mechanisms of this instability. The chromosome instability observed in hybrid cells may be due to a defect in chromosome function of the chromosomes from one of the parents. If so, then the instability should affect only the chromosomes of that parent with the instability character. Alternatively, the instability may be caused by a defect of the spindle apparatus and all of the chromosomes of the hybrid should be subject to irregular distribution to a more or less equal extent after mitosis. This problem may be studied with the aid of a sufficient number of genetic markers for the individual chromosomes of both parents. Such studies have already been performed for animal hybrid cells and have demonstrated that the instability affects all of the chromosomes of the hybrid cells and not just those of the parent with the original character of instability (Bengtsson et al. 1975).

6.1.6 Analysis of the Mechanisms of Differentiation and Morphogenesis

The fate of epigenetic programs in cells can be modified by parasexual hybridization. The fusion of an undifferentiated callus cell with a differentiated leaf mesophyll cell results in the return of the mesophyll nucleus to the cell cycle and in a prompt disappearance of all visible signs of differentiation in the dividing hybrid cell. However, this phenomenon is not unique to parasexual hybrids, since the mesophyll cells of many species go through this de-differentiation in standard conditions of in vitro culture, without hybridization. Perhaps the only exception are the leaf mesophyll cells of Graminae species, which apparently can not reverse their specialization (Potrykus et al. 1977a). It should be mentioned that the hybridization of mesophyll protoplasts of Graminae species with callus cells nevertheless results in the destruction of their epigenetic determinants (Kao et al. 1974) (Fig. 29). It is not unlikely that the annihilation of most of the epigenetic programs of parental cells takes place in hybrid plant cells like in animal hybrid cells (Ringertz and Savage 1976).

Four papers concerning the inheritance of the ability of cells to undergo morphogenesis upon induction will be discussed here. It had been demonstrated (Gleba and Berlin 1979; Gleba 1980) that a hybrid cell obtained by fusion of a tobacco callus cell defective for morphogenesis with a morphologically potent tobacco mesophyll cell is capable of morphogenesis. That is, the character of morphogenetic ability behaves as a dominant character during hybridization. In this case, the parental line probably lost its morphogenetic ability as a result of long-term in vitro culture. Similar conclusions were reached in the study of hybridization of tobacco cell mutants for nitrate reductase which were also defective for morphogenesis (Glimelius et al. 1978a). These hybrids were morphogenetically active, in contradistinction to both parents. This work illustrates the complexity of the interpretation of the mechanisms of this phenomenon, since the deficiency for morphogenesis in these nitrate reductase deficient mutants of tobacco is obviously connected with the absence of one enzyme (one gene) that has no apparent connection with the mechanisms of morphogenesis as such (Müller 1978). The restoration of morphogenetic potential upon interspecific hybridization was also described by Hungarian scientists (Maliga et al. 1977). *Nicotiana knightiana* cells incapable of morphogenesis in the culture conditions employed were fused with *N. sylvestris* mutant cells that were resistant to kanamycin and had lost their morphogenetic potential (Dix et al. 1977). The restoration of morphogenesis found in these hybrids is somewhat difficult to interpret, since an interspecific hybrid had been produced in this case and since the physiological conditions for the hybrids' morphogenetic induction may differ considerably from those conditions optimal for the parental cells. Also, the morphogenetic block of one of the parents is due to a mutation that is not likely to be associated with the mechanisms of morphogenesis. In our opinion, these studies say more about the complicated dependence of morphogenesis upon primary metabolism rather than anything about the nature of this process as such.

The restoration of morphogenetic potential is also observed in distantly related (intertribal) combinations of species, as demonstrated by our efforts to produce *Arabidobrassica* hybrids (Gleba and Hoffmann 1979; 1980). In these hybrids, it is unlikely that the restoration of morphogenetic potential is the result of simple complementation of a pair of genes. The *Brassica* parent was incapable of complete morphogenesis and the *Arabidopsis* parent had lost its capacity for morphogenesis during the process of in vitro culture. In direct contrast to the results in closely related species combinations, the restoration of morphogenetic potential in *Arabidobrassica* may be associated with an increase in genetic and perhaps even epigenetic heterogeneity within the hybrid cell population after hybridization.

Wullems et al. (1980) studied the parasexual hybridization of a tobacco strain with hormone independent growth, which was derived from tumorous cells produced following infection with *Agrobacterium tumefaciens,* fused with mesophyll protoplasts of a streptomycin resistant tobacco mutant. The selection of hybrids was based on hormone independent growth and resistance to streptomycin. Four hybrid cell colonie capable of shoot regeneration were recovered. Analysis demonstrated the activity of lysopine dehydrogenase and therefore the presence and expression of the T-region of Ti plasmid DNA from *Agrobacterium.* Whereas the initial hormone independent strain was defective for shoot formation, the hybrids were capable of shoot formation, although they were unable to from roots. A similar lack of rhizogenesis has been observed in shoots obtained after in vitro transformation of tobacco protoplasts by *Agrobacterium* (Marton et al. 1979).

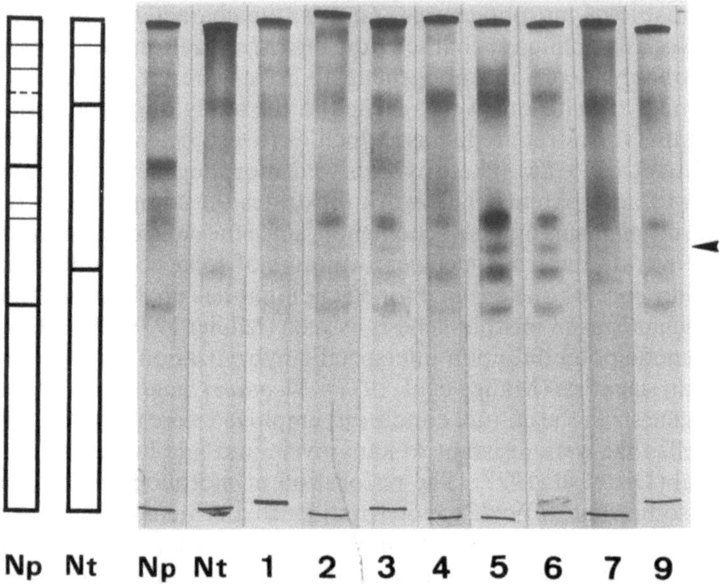

Fig. 58. Electrophoretic analysis of esterase isoenzymes in callus cells of somatic hybrids between tobacco cells, transformed with *Agrobacterium tumefaciens* (×) *Nicotiana plumbaginifolia*. The schematic diagram and zymograms are of *N. plumbaginifolia* (*Np*), *N. tabacum* (*Nt*), and several different hybrid cell lines. The presence of a new band, apparently the result of formation of hybrid esterase molecules, is indicated in the hybrids by an *arrow*. (Kanevsky and Gleba, in preparation)

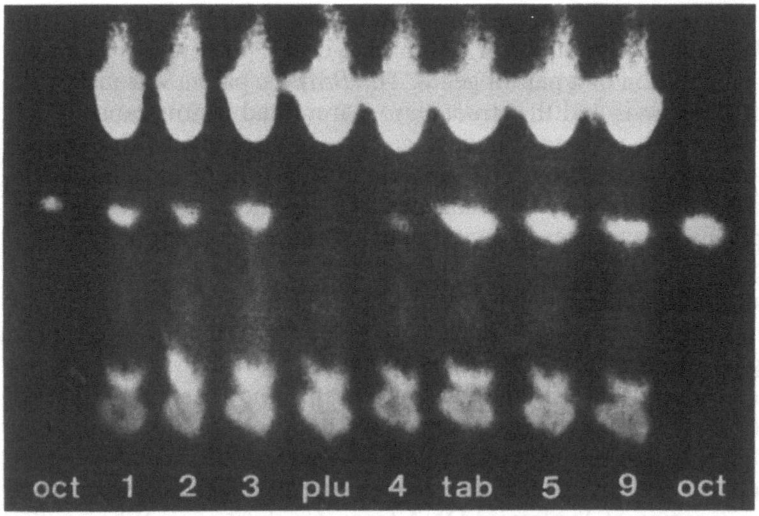

Fig. 59. Electropherogram of the products present in the reaction mixture of assays for lysopine dehydrogenase activity in extracts of different hybrid calli of *Nicotiana tabacum* crown gall tumor cells (×) *N. plumbaginifolia*. The different lanes represent: authentic octopine *oct*, *N. plumgaginifolia* extracts *plu*, tobacco crown gall calli extracts *tab*, and extracts from six different somatic hybrids *1-9*. Note the presence of octopine in the hybrid cells. (Kanevsky and Gleba, in preparation)

These results illustrate that the T-DNA of the Ti plasmid of *Agrobacterium* is transferred to parasexual progeny and the suppression of rhizogenesis in transformed cells seems to be a dominant character. Similar observations have recently been made by Kanevsky and Gleba utilizing fusion of hormone independent, *Agrobacterium* transformed tobacco cells with protoplasts of *Nicotiana plumbaginifolia* (Figs. 58, 59).

Recent findings by Chinese workers (Xianghui et al. 1982) seem to indicate that the T-region DNA can also be transferred from one cell to another via cell fusion. After fusion of tumorous tobacco cells with mesophyll petunia protoplasts, plants were recovered that had the morphology of petunia as well as the activity of lysopine dehydrogenase. These results also indicate that the T-DNA is inherited or transferred as a dominant character upon somatic cell fusion. However, the data obtained are insufficient for an unequivocal interpretation of this phenomenon.

Additional applications of DNA and gene transfer utilizing the Ti plasmid, while not directly connected with protoplast fusion, are nonetheless potentially powerful tools for genetic engineering of higher plant cells and have been recently reviewed (van Montagu and Schell 1981; Bevan and Chilton 1982).

6.2 Practical Use of Somatic Cell Hybridization

6.2.1 Introduction

As is abundantly clear from Chap. 3–5, somatic cell hybridization permits the hybridization of plants forms and species which are impossible to cross sexually. Moreover, owing to the pecularities of behavior of parental genes in hybridization by fusion of somatic and cultured cells, the hybrid plants produced by the parasexual method differ in their genetic constitution from the sexual hybrids between the same parents. This phenomenon represents the value of this novel method of hybridization for practical plant breeding. Since the perspectives for the involvement of somatic hybridization in breeding have already been broadly discussed, we shall limit ourselves here to a short enumeration of a few experimental works which resulted in the production of specific hybrids which represent valuable initial material for plant breeding. For example, cell fusion can be used for the rescue of mutations obtained in morphogenetically defective cell lines and/or production of fertile plants carrying these new characters. Direct crosses involving haploid or aneuploid plant forms (which are otherwise sterile) are also possible, permitting direct plant breeding work on the low ploidy forms of natural polyploids, such as potato. Somatic cell fusion also permits rapid hybridization experiments of perennial plants, such as trees. Of course, the possibilities offered by somatic hybridization are much more diverse than these few examples.

6.2.2 Bypassing of Incompatibility in Interspecific Hybridization

Intrageneric interspecific hybridization by sexual crossing is often difficult or impossible and is a major problem for plant breeders. In some cases, the limits of

Fig. 60. Heterotic effect (hybrid vigor) in somatic hybrid plants of *Datura innoxia* (×) *Datura discolor*. From left to right: *Datura innoxia*, somatic hybrid *Datura innoxia* (×) *Datura discolor*, and *Datura discolor*. (Schieder and Krumbiegel 1979)

sexual incompatibility can be bypassed by affecting the process of fertilization and/or the development of the embryos and sometimes the use of in vitro embryo culture is successful. However, all of the attempts to produce sexual hybrids have been unsuccessful for many species combinations which are of great practical interest. Several of the initial experiments performed by using fusion of isolated protoplasts with the aim of production of interspecific hybrids valuable for practical breeding illustrate that somatic hybridization represents a true panacea in these cases.

Schieder (1978a) obtained the fertile amphidiploids *Datura innoxia* (×) *D. discolor* and *D. innoxia* (×) *D. stramonium* by fusing diploid mesophyll protoplasts of a chlorophyll defective mutant of *D. innoxia* with protoplasts of the two other *Datura* species. According to numerous investigators, conventional hybridization by sexual crossing between these species is impossible. These *Datura* hybrids produced by protoplast fusion can be used for the breeding of industrial thornapple species, presently used for the industrial production of scopolamine, since heterosis (hybrid vigor) is observed in these combinations and the content of alkaloids, is 20%–25% higher than in the parental forms (Fig. 60).

Another practical success consists of the production of interspecific *Nicotiana* hybrids. This genus encompasses more than 60 species and most of them can be easily crossed with each other. However, there are some important exceptions. The majority of investigations in hybridization of *Nicotiana* species involve commercial varieties of tobacco, *Nicotiana tabacum* L., as one of the parents. Some of the wild *Nicotiana* species, particularly species of the *Repandae* section, *N. repanda, N. nesophila,* and *N. stocktonii,* are of interest for industrial breeding due to their resistance to a number of diseases, but they are unfortunately not sexually crossable with tobacco. Evans et al. (1981, 1982) have fused mesophyll protoplasts from albino tobacco plants homozygous for Su gene with mesophyll protoplasts of these wild species and subsequently selected hybrids as heterozygotes for Su gene. Fertile hybrids were obtained in the combinations *N. tabacum* (×) *N. nesophila* and *N. tabacum* (×) *N. stocktonii*. No hybrids were obtained upon protoplast fusion of tobacco with *N. repanda*. It should be mentioned that diploid somatic cells were used for fusion in these experiments and the hybridization resulted in the production of fertile amphidiploids which could be directly utilized in plant breeding. Nagao (1982) recovered three somatic hybrid plants, one of which was fertile, after the fusion of mesophyll protoplasts of a tobacco aurea mutant with those of *Nicotiana repanda*. This report is the first successful direct hybridization of these two species. It should be stressed again that the hybrids *N. tabacum* (×) *N. repanda* could not be produced by sexual means, but have been obtained only by the parasexual technique of protoplast fusion.

Schenck (1982) has been successful in producing somatic hybrids between *Brassica oleracea* and *B. campestris*, thus artificially resynthesizing the natural amphidiploid *Brassica napus* for the first time. These results are of great importance, since the genetic basis of the rape cultivars is quite narrow.

Power et al. (1980) have obtained fertile amphidiploid hybrids via protoplast fusion of the sexually incompatible species *Petunia parodii* and *P. parviflora*. These hybrids permit the valuable characters of *P. parviflora*, such as branched, prostrate growth, to be introduced into the breeding of ornamental *Petunia* forms.

Somatic hybridization has also permitted the crossing of distantly related plant species belonging to different genera, tribes, and even families, but in all cases to date, only sterile plants or cells incapable of morphogenesis have been recovered (see Chap. 5). Though in this case we apparently deal with the genetic noncoordination within distantly related hybrids, there are several applications for distantly related hybrids in practical work. One of them is the use of distantly related plant hybridization for the selection of novel cell lines producing pharmacologically important substances – in this case, the reconstruction of fertile plants is not necessary and genetically stable in vitro cell lines might be the desired result of breeding. Other potentials include the breeding of hybrid plant forms for which the sexual process is not obligatory or commercially important, such as in the vegetative propagation of potato. As a result of parasexual hybridization, novel and important sterile plant forms might be produced. Finally, one promising approach for practical investigations is the production of highly asymmetric forms bearing only a few genes from one parent along with a complete gene set from the other parent. This might permit the production of fertile plant forms with chromosome fragments from distantly related species "implanted" by cell fusion.

6.2.3 Reconstruction of Cytoplasmic Genomes

The genophores of the cytoplasm code for a number of practically important characters such as the rate of photosynthesis, resistance to various diseases, and cytoplasmic male sterility. The routine sexual crossing permits the substitution of cytoplasm via numerous recurrent crosses in the great majority of cultured plant species (reviewed by Kirk and Tilney-Basset 1978), but as a result of the uniparental maternal inheritance of plasmagenes, this substitution simultaneously affects all of the plasmagenes. The possibility to produce plants bearing only some of the cytoplasmic genes from one species (plastome, for example) as well as some of the cytoplasmic genes from another species (mitochondrial genes, for example) has been realized only as a result of the elaboration of the techniques of somatic hybridization (see Chap. 4).

Biparental inheritance and independent segregation of plasmagenes in protoplast fusion products were first described by Belliard et al. (1978), Gleba (1978c), and Gleba et al. (1978a). However, rigorous analysis of cytoplasmic segregants was performed only after the advent of restriction endonuclease mapping of chloroplast and mitochondrial DNA. Three groups of investigators conducted such studies on interspecific somatic hybrids of tobacco, *Nicotiana tabacum* (×) *N. tabacum* (with the cytoplasm of *N. debneyi*) (Belliard et al. 1978, 1979), *N. tabacum* (with the cytoplasm of *N. suaveolens*) (×) *N. sylvestris* (Aviv and Galun 1980; Galun et al. 1982), and *N. tabacum* (×) *N. knightiana* (Menczel et al. 1981a; Nagy et al. 1981). Komarnitsky et al. (1981, 1983) performed similar analyses on interspecific hybrids of potato.

The main conclusion of the results obtained by all of the groups working with tobacco is that there are forms bearing the plastome of one parental species and the mitochondrial genome of another, as well as forms with the plastome of one species and a reconstructed mitochondrial genome among the great diversity of hybrid plant forms obtained from fused protoplasts (see Sect. 4.2, 4.5, and 6.1). It is not out of place to state here that the novel tobacco forms produced by parasexual techniques represent valuable material for plant breeders and underscore the relevance and influence of the two cytoplasmic genophores upon agronomically important traits of this plant.

Since plant geneticists and breeders have not been able to produce plants with heterozygous cytoplasm, there have been no attempts to evaluate the contribution of plasmagenes to the control of various characters, including economically important ones. Evans et al. (1982) compared clones of sexually and parasexually synthesized hybrids between (*Nicotiana tabacum* and *N. nesophila* and found unexpectedly high and statistically significant variation between the clones of the somatic hybrids, but not between the sexual hybrids, for pollen viability, flower and leaf morphology, and trichome density. Since all of the somatic hybrid clones had 96 chromosomes, the variability is best explained as a result of cytoplasmic segregation or recombination, although mitotic recombination of nuclear genes or chromosomal rearrangements can not be excluded. This study stresses once again the urgent necessity to analyze the influence of plasmagenes in higher plants. Several *Nicotiana* hybrids and cybrids with euploid chromosome numbers and well-characterized, new, reconstructed cytoplasmic genomes are now available for such an analysis.

Heteroplasmic forms of potato were obtained via protoplast fusion by Butenko and Kuchko (1979). Biochemical analysis of chloroplast and mitochondrial DNA of these plants via restriction endonuclease mapping, performed in our laboratory by I. Komarnitsky (Komarnitsky et al. 1981, 1983), has shown that the plastome of the hybrid originated from the wild species *Solanum chacoense* and the mitochondrial DNA originated from *S. tuberosum*. This hybrid was obtained without any intentional selective pressure at the level of the cytoplasmic genophores and seems to be a consequence of random segregation and sorting out during the process of propagation of the cytoplasmic heterozygote fusion products.

6.2.4 Transfer of Cytoplasmic Genomes and Cytoplasmic Male Sterility

In hybridization by protoplast fusion, the plasmagenes are inherited biparentally, the nuclear genes are inherited both biparentally and uniparentally, and mitotic segregation of plasmagenes takes place in the hybrid products (see Chap. 4). Thus, it is obvious that some cybrid forms that bear the nucleus of one of the parents and the cytoplasmic genophore(s) of the other parent must be present within the spectrum of plant forms obtained after regeneration of the fusion products. This means that hybridization by protoplast fusion may result in production of hybrid plants with substituted cytoplasm within one crossing cycle, which is a very rapid transfer of cytoplasm. The need to transfer the cytoplasmic male sterility character is often encountered in practical plant breeding. This need has traditionally been meet by applications of sexual crossing: a hybrid between a female donor and the recipient of the cytoplasm is produced and then all the chromosomes of the plasma donor are substituted by recurrent crossing of the hybrid with the recipient. Such a procedure is often rather protracted and requires several recurrent crosses (four or five crosses for tobacco with $2n=48$ chromosomes) must be performed so as to eliminate the chromosomes of the species which was the donor of the cytoplasm. Even then there is no guarantee that none of the donor nuclear material has been integrated.

The problem of cytoplasmic gene transfer for those genes coding for cytoplasmic male sterility has been solved by somatic hybridization for some cultured plants such as tobacco (Belliard et al. 1977a, b; Zelcer et al. 1978; Aviv et al. 1980; Aviv and Galun 1980; Gleba 1980; Glimelius et al. 1981) and petunia (Izhar and Power 1979; Izhar and Tabib 1980; Bergounioux-Bunisset and Perennes 1980).

Although cybrid plant forms are naturally produced with a rather high frequency during protoplast fusion, there are several techniques available to selectively produce cybrids in some cases. In order to efficiently effect the transfer of organelles, different authors have employed the irradiation of the cells of the cytoplasm donor species (Aviv and Galun 1980) or the enucleation of the cells of the cytoplasm donor species (Maliga et al. 1982). The irradiation with X-rays and gamma rays in doses of 50–300 J kg^{-1} was found to be efficient in partial or complete inactivation of the nuclei of the donor cells. Moreover, this treatment completely arrests the divisions of the nonfused donor cells and thus the irradiation also serves as a screening factor (Aviv and Galun 1980; Sidorov et al. 1981a; Menczel et al. 1982). In spite of the fact that these doses of irradiation are a strong mutagenic factor, no mutant or aneuploid forms containing some of the donor species chromosomes have been found among the cybrid forms produced (Menczel et al. 1982). Inactivation with iodoacetate may also be employed in those cases where other methods of suppression of the propagation of the nonfused recipient cells are not available (Sidorov et al. 1981a). A practical example

Fig. 61. Chloroplast transfer by protoplast fusion in *Nicotiana*. From left to right: *N. tabacum* plant with a plastome chlorophyll deficiency, *N. plumbaginifolia*, and a somatic hybrid (cybrid) of *N. tabacum* plastome mutant (×) *N. plumbaginifolia*. The cybrid plant on the *right* is of *N. plumbaginifolia* type with *N. tabacum* plastids. (Sidorov et al. 1981 a)

of chloroplast transfer in *Nicotiana* somatic hybrids is shown in Fig. 61. Of course, these techniques have been realistically elaborated for and shown practical success only in *Nicotiana* species.

7 Conclusion

The experiments reported in this book clearly demonstrate that hybrids of higher plants can really be obtained by protoplast fusion and subsequent regeneration of the hybrid cells, thereby avoiding sexual hybridization. The techniques of parasexual hybridization have been elaborated primarily for Solanaceae species, in particular for the production of parasexual hybrid plants of *Nicotiana* (including tobacco), *Datura, Petunia, Solanum* (potato), *Lycopersicon* (tomato), and *Atropa*. Parasexual hybrids have also been produced among members of the Cruciferae (*Arabidopsis* and *Brassica*) and Umbelliflorae (*Daucus, Petroselinum,* and *Angelica*) families. Hybrids of soybean and pea (Fabaceae) have been obtained at the cell level. Parasexual hybridization can now be exercised as a tool for genetic engineering and production of novel plants for these plant taxa.

Hybridization by protoplast fusion permits the production of hybrids between parental forms which can not be crossed sexually because of their deficiency for morphogenesis and/or gametogenesis. For example, hybrids between haploid plants as well as between cultured cells deficient for morphogenesis have been produced using this method.

Either protoplasts from leaf mesophyll cells or protoplasts from in vitro cultured callus tissues are now generally used as one or both of the initial parental materials for parasexual hybridization. The method of cell agglutination by polyethylene glycol with subsequent treatment with a high pH and high Ca^{+2} solution is most frequently employed to induce protoplast fusion. The frequency of fusion of protoplasts in this procedure is about 10^{-1}, as determined by cytological analyses. A variety of different techniques have been elaborated for the screening of parasexual hybrids. The most efficient screening methods are based on the use of genetic or "physiological" complementation or on mechanical isolation and individual culture of the heterokaryons.

The techniques of parasexual hybridization are relatively simple and in contradistinction to the other techniques of genetic engineering, such as recombinant molecule technology, do not require a lot of expensive instruments and materials. Laboratories for somatic cell engineering can be easily organized in any institute or breeding center. The minimum experimental base for such a laboratory includes the facilities for aseptic work, incubators or culture rooms for growing the plants and cells, and of course, the chemicals, cellulolytic enzymes, phytohormones, and other components of the nutritive media.

Parasexual hybridization by protoplast fusion is confronted with some serious unsolved methedological problems. One of the major obstacles to wide application of somatic cell hybridization is the lack of a generally applicable and efficient technique for hybrid screening. However, the progress in this area has been so rapid and impressive in the last few years (see Chap. 3) that this problem may be regarded as generally solved. The most promising approaches to this problem are based on individual culturing of single fusion products and the use of genetic and physiological complementation.

Another serious and much more complicated problem hampering the use of parasexual hybridization is the absence of techniques for protoplast and cell culture and/or plant regeneration for a number of theoretically and practically important plant species. First and foremost, techniques must be developed for Gramineae species. The solution of this problem may take many years since the development of culture techniques and especially the formulation of appropriate nutritive media are chiefly empirical undertakings. The intense studies in this direction over the last years have met with only very limited success.

The genetic irregularity of most parasexual hybrids (which may be caused by the very conditions of the experimental work) is also an obstacle to the wide application of parasexual hybridization, especially for practical purposes. To avoid this problem, genetically regular initial plant material has to be utilized, rather than the cells of habituated callus cultures which are frequently employed now. Thus, not only mesophyll protoplasts, but also protoplasts from other plant tissues and organs need to be tried in attempts to produce genetically regular hybrids.

Parasexual hybridization by protoplast fusion permits the production of novel combinations of parental genes in higher plants. In particular, as illustrated by concrete examples in published experiments, one can produce via parasexual hybridization:

1. plant which are heterozygous for extranuclear genes;
2. plants whose nucleus originates from one parent and cytoplasm from the other;
3. plants bearing some extranuclear genes from one parent and the remaining extranuclear genes from the other.

It is either very difficult or impossible to obtain these gene combinations via conventional sexual crossing. Parasexual hybridization permits the independent and direct manipulation of the extranuclear genes of plant cells and is therefore of both theoretical and practical interest. Theoretically, the potential to produce cytoplasmic heterozygotes by protoplast fusion means the availability of a sophisticated technique for formal genetic analysis of extranuclear genes in the plant cell. At present, the most promising perspectives are the further study of linkage groups of extranuclear genes, based on the cosegregation of extranuclear gene determinants in progeny of double cytoplasmic heterozygotes and the potential to reveal and apply genetic recombination of extranuclear genes. The transfer of cytoplasm and particularly the transfer of genes for cytoplasmic male sterility from one species or variety to another in just one cycle of hybridization is the most obvious application of protoplast fusion for practical purposes. A similar procedure may conceivably be employed to generate new varieties bearing the character of cytoplasmic male sterility for the purpose of industrial production of hybrid seeds. Such an approach might replace the long-term recurrent hybridization by sexual crossing which is exercised today. The genetic diversity of the extranuclear genomes of plant varieties may be increased by the transfer and utilization of alien cytoplasm, as well as by the "hybridization" of the cytoplasm, which has become practical thanks to the method of protoplast fusion. The increase in genetic diversity is one of the principal goals of plant breeding. The chloroplast and mitochondria are both responsible for supplying energy in the plant cell and their genetic reconstruction or manipulation has to be a part of any active program of research in agricultural plant breeding.

Parasexual hybridization by protoplast fusion permits the crossing of distantly related plant species which can not be crossed sexually. Unfortunately, these hybrid cells, although viable, regenerate abnormal hybrid plants which are sterile and incapable of sexual reproduction. Thus, the parasexual hybrids of distantly related plant species produced to date are useless for practical plant breeding. Nevertheless, there are some evidences of theoretical value of these "monsters." They demonstrate that the cell hybridization of distantly related plant species is possible and that the genetic "machinery" of these plant cells are capable of working together in a rather coordinated fashion. This cooperation is most evident at the level of unorganized growth and reproduction of the hybrid cells and is also manifested in the extremely great capacity for coordinated morphogenesis in these hybrids. Somewhat of a surprise was the extent of the noncoordination of cellular and supercellular processes in hybrid cells of distantly related species, which is illustrated by directed elimination and/or reconstruction of chromosomes resulting from imperfections in the cell division process and from defects in morphology resulting from noncoordination of morphogenetic processes. However, this limited coordination in plant cells constructed by joining different genetic programs may very well become a valuable tool for the analysis of the mechanisms of cell inheritance and morphogenesis, comparable to the role played by mutants in a number of other genetic studies. The analysis of deficiencies in physiology of cells constructed in different ways embodies the "synthetic" spirit and purpose of genetic engineering as we see it and is in contrast to the passive "analytic" method of investigation of living organisms.

The hybridization of distantly related species is of interest in practical breeding of agricultural plants as a means for transfer of small amounts of genetic material from one species to another. The parasexual hybrids of distantly related species produced to date are morphologically abnormal and it is not likely that the near future will bring the production of fertile amphieuploid hybrids. However, asymmetric hybrids, in which the genome of one of the parents is represented by an incomplete set of chromosomes, are of some interest. Some of these asymmetric hybrids are morphologically more normal than symmetric hybrids and it may be supposed that the further reduction of genetic material from the second parent will permit the synthesis of fertile hybrid plants which retain a few certain genes of the second, donor parent. Methods for the induction of asymmetry in hybrids of plant species and especially methods of screening for the desired genes might transform this procedure into an efficient way of expansion of species or variety gene pools via transfer of small amounts of genetic material between distantly related plant species by protoplast fusion. Certainly, this topic is of interest for future plant breeding.

To conclude, we would like to recall a phrase from Bulgakav's novel "*The Master and Margarita*", where Satan persuades the Master:

"Dont't you want to sit, like the Faust, over a retort, in the hope of sculpting a new homunculus? On this way, Master, on this way." [1]

The results of the experiments reported in this book force us to believe more and more that the way chosen by their authors for *sculpting a novel plant* is the efficient one.

1 Bulgakov MA (1975) Novels. Chudozhestvennaya Literatura Publishing House Moscow

References

Abo El-Nil MM, Hildebrandt AC (1976) Cell wall regeneration and colony formation from isolated single geranium protoplasts in microculture. Can J Bot 54, 1530–1534

Abstracts of the Fourth International Symposium on Yeast and Other Protoplasts (1975) Nottingham, U K 8–12 September

Akazawa T (1975) The structure and function of Fraction 1 protein. Regulatory aspects of photosynthetic CO_2 fixation in chloroplasts. In: Reinhold L, Liwschitz Y (eds) Progress in phytochemistry, Vol II, pp 107–141. London: Interscience

Amerkhanova MB, Fedotov AF, Shiyan AN, Vinetsky YP (1976) The split of DNA in higher plants by restriction nuclease Eco RI. Dokl Akad Nauk USSR 227:746–749

Anonymous (1975) Isolation and culture of rice protoplasts. Sci 18:779–784

Arcioni S, Davey MR, Santos AVP dos, Cocking EC (1982) Somatic embryogenesis in tissues from mesophyll and cell suspension protoplasts of Medicago coerula and M. glutinosa. Z Pflanzenphysiol 106:105–110

Ashton NW, Cove DJ (1977) The isolation and preliminary characterization of auxotrophic and analogue resistant mutants of the moss, Physcomitrella patens. Mol Gen Genet 154:87–95

Atchison BA, Whitfeld PR, Bottomley W (1976) Comparison of chloroplast DNAs by specific fragmentation with Eco RI endonuclease. Mol Gen Genet 148:263–269

Aviv D, Fluhr R, Edelman M, Galun E (1980) Progeny analysis of the interspecific somatic hybrids: Nicotiana tabacum (CMS) + N. sylvestris with respect to nuclear and chloroplast markers. Theor Appl Genet 56:145–150

Aviv D, Galun E (1980) Restoration of fertility in cytoplasmic male sterile Nicotiana sylvestris by fusion with X-irradiated N. tabacum protoplasts. Theor Appl Genet 58:121–128

Bajaj YPS (1977) Protoplast isolation, culture and somatic hybridization. In: Reinert J, Bajaj YPS (eds) Applied and fundamental aspects of plant cell, tissue, and organ culture. Springer, Berlin Heidelberg New York, pp 467–496

Bajaj YPS, Gosch G, Ottma M, Weber A, Gröbler A (1978) Production of polyploid and aneuploid plants from anthers and mesophyll protoplasts of Atropa belladonna and Nicotiana tabacum. Ind J Exp Biol 16:947–953

Banks MS, Evans PK (1976) A comparison of the isolation and culture of mesophyll protoplasts from several Nicotiana species and their hybrids. Plant Sci Lett 7, 409–416

Bapat VA, Schieder O (1981) Protoplast culture of several members of the genus Physalis. Plant Cell Rep 1:69–70

Batra A, Abel WO (1981) Development of the moss plants from isolated and regenerated protoplasts. Plant Sci Lett 20:183–189

Baur E (1909) Das Wesen und die Erblichkeitsverhältnisse der „Varietates albomarginate hort“ von Pelargonium zonale. Z Vererbungsl 1:330–351

Bayliss MW (1980) Chromosomal variation in plant tissue culture. Int Rev Cytol 11A:113–144

Bedbrook JR, Bogorad L (1976) Endonuclease recognition sites mapped on Zea mays chloroplast DNA. Proc Natl Acad Sci USA 73:4309–4313

Belitser NV, Sidorov VA, Tarasenko VA, Gleba YY, Sytnik KM (1977) Electron microscopic study of tobacco mesophyll protoplasts. I. Ultrastructural and cytochemical study of tobacco protoplasts on early stages of cultivation. Sov Cytol Genet 11:259–267

Belliard G, Pelletier G, Ferault M (1977a) Fusion de protoplasts de Nicotiana tabacum à différents: étude des hybrides cytoplasmiques néo-formés. CR Acad Sci D 284:749–752

Belliard G, Pelletier G, Ferault M (1977b) Cytoplasmic hybridization in Nicotiana by protoplast fusion and transfer of cytoplasmic male sterility and fertility factors. In: Interspecific hybridization in plant breeding, pp 237–242. Proc. 8th Congress, Madrid

Belliard G, Pelletier G, Vedel F, Quetier F (1978) Morphological characteristics and chloroplast DNA distribution in different cytoplasmic parasexual hybrids of Nicotiana tabacum. Mol Gen Genet 165:231–238

Belliard G, Vedel F, Pelletier G (1979) Mitochondrial recombination in cytoplasmic hybrids of Nicotiana tabacum by protoplast fusion. Nature 281:401–402

Bendich AJ, McCarthy BG (1970) DNA comparisons among barleys, oats, rye and wheat. Genetics 65:545–565

Bengtsson BO, Nabholz M, Kennett R, Bodmer WF, Povey S, Swallow D (1975) Human intraspecific somatic cell hybrids: A genetic and karyotypic analysis of crosses between lymphocytes and D98/AH-2. Somatic Cell Genet 1:41–64

Bennici A, D' Amato FD (1978) In vitro regeneration of durum wheat plants. I. Chromosome numbers of regenerated plantlets. Z Pflanzenzücht 81:305–311

Bennici A, Buiatti M, D' Amato F (1968) Nuclear conditions in haploid Pelargonium in vivo and in vitro. Chromosoma 24:194–201

Bergmann, L (1963) Growth and division of single cells of higher plants in vitro. J Gen Physiol 43:841–851

Bergounioux-Bunisset C, Perennes C (1980) Transfert des facteurs cytoplasmiques de la fertilité mâle entre 2 lignées de Petunia hybrida par fusion de protoplastes. Plant Sci Lett 19:143–149

Berlin J, Widholm JM (1978) Amino acid uptake by amino acid analog resistant tobacco cell lines. Z Naturforsch C33:634–640

Bevan MW, Chilton MD (1982) T-DNA of the Agrobacterium TI and RI plasmids. Annu Rev Genet 16:357–384

Bhojwani SS, Power JB, Cocking EC (1977) Isolation, culture and division of cotton callus protoplasts. Plant Sci Lett 8:85–89

Bhojwani SS, White DWR (1982) Mesophyll protoplasts of white clover: isolation, culture and organogenesis. Plant Sci Lett 26:265–271

Binding H (1966) Regeneration und Verschmelzung nackter Laubmoosprotoplasten. Z Pflanzenphysiol 55:305–321

Binding H (1974) Cell cluster formation by leaf protoplasts from axenic cultures of haploid Petunia hybrida L. Plant Sci Lett 2:185–188

Binding H (1975 a) Regeneration von haploiden und diploiden Pflanzen aus Protoplasten von Petunia hybrida L. Z Pflanzenphysiol 74:327–356

Binding H (1975b) Reproducibly high plating efficiencies of isolated mesophyll protoplasts from shoot cultures of tobacco. Physiol Plant 35:225–227

Binding H (1979) Subprotoplasts and organelle transplantation. In: Sharp WR, Larsen PO, Paddock EF, Raghavan V (eds) Plant cell and tissue culture. Principles and applications. Ohio State University Press, Columbus, pp 789–806

Binding H, Nehls R (1977) Regeneration of isolated protoplasts to plants in Solanum dulcamara L. Z Pflanzenphysiol 85:279–282

Binding H, Nehls R (1978a) Somatic cell hybridization of Vicia faba + Petunia hybrida. Mol Gen Genet 164:137–143

Binding H, Nehls R (1978b) Regeneration of isolated protoplasts of Vicia faba L. Z Pflanzenphysiol 88:327–332

Binding H, Nehls R (1980) Protoplast regeneration to plants in Senecio vulgaris L. Z Pflanzenphysiol 99:183–186

Binding H, Nehls R, Schieder O, Sopory SK, Wenzel G (1978) Regeneration of mesophyll protoplasts isolated from dihaploid clones of Solanum tuberosum. Physiol Plant 43:52–54

Binding H, Nehls R, Kock R, Finger J, Mordhorst G (1981) Comparative studies on protoplast regeneration in herbaceous species of the Dicotyledoneae class. Z Pflanzenphysiol 101: 119–130

Binding H, Jain SM, Finger J, Mordhorst G, Nehls R, Gressel J (1982) Somatic hybridization of an atrazine resistant biotype of Solanum nigrum with Solanum tuberosum. Theor Appl Genet 63:273–277

Binding, H et al. (1984) Somatic cell hybrids and plant formation. In: Reinert J (ed) Results and problems in cell differentiation, Vol 12. Springer, Berlin Heidelberg New York (in press)

Birky CW, Perlman PS, Byers TJ (eds) (1974) Genetics and biogenesis of mitochondria and chloroplasts. Ohio State University Press, Columbus

Bobrow M, Cross J (1974) Differential staining of human and mouse chromosomes in interspecific cell hybrids. Nature 251:77–79
Bobrow M, Madan K, Pearson L (1972) Staining of some specific regions of human chromosomes, particularly the secondary constriction of No 9. Nature New Biol 238:124
Bonnett HT, Glimelius K (1983) Somatic hybridization in Nicotiana: Behavior of organelles after fusion of protoplasts from male-fertile and male-sterile cultivars. (submitted)
Blair GE, Ellis RJ (1973) Protein synthesis in chloroplasts. I. Light-driven synthesis of the large subunit of Fraction I protein by isolated pea chloroplasts. Biochim Biophys Acta 319:223–234
Blaschek W, Hess D, Hoffmann F (1974) Transcription in aus Protoplasten isolierten Zellkernen von Nicotiana und Petunia. Z Pflanzenphysiol 72:262–271
Bogorad L (1977) Genes for chloroplast ribosomal RNAs and ribosomal proteins: gene dispersal in eukaryotic genomes. In: Brinkley BR, Porter KR (eds) International cell biology 1976–1977. Rockefeller University Press, New York, pp 175–182
Bogorad L (1981) Chloroplasts. J Cell Biol 91:256s–270s
Bonnett HT (1976) On the mechanism of the uptake of Vaucheria chloroplasts by carrot protoplasts treated with polyethylene glycol. Planta 131:229–233
Bonnett HT, Banks MS (1977) Chloroplast incorporation, survival and replication in foreign cytoplasm. In: Brinkley BR, Porter KR (eds) International cell biology 1976–1977. Rockefeller University Press, New York, pp 225–231
Bonnett HT, Eriksson T (1974) Transfer of algal chloroplasts into protoplasts of higher plants. Planta 120:71–79
Bottomley W (1980) Fraction 1 protein. In: Reinert J (ed) Results and problems in cell differentiation, Vol X. Chloroplasts. Springer, Berlin Heidelberg New York
Bourgin JP, Missonier C (1978a) Culture of haploid mesophyll protoplasts from Nicotiana alata. Z Pflanzenphysiol 87:55–64
Bourgin JP, Missonier C (1978b) Culture of mesophyll protoplasts of haploid Nicotiana alata Zink and Otto. Incompatibility Newsl 9:25–30
Bourgin JP, Missionier C, Chupeau Y (1976) Culture de protoplastes de mésophylle de Nicotiana sylvestris Spegazzini et Comes haploide et diploide. CR Acad Sci D 282:1853–1856
Bourgin JP, Chupeau Y, Missonier C (1979) Plant regeneration from mesophyll protoplasts of several Nicotiana species. Physiol Plantarum 45:288–292
Boyd YL, Harris H (1973) Correction of genetic defects in mammalian cells by the input of small amount of foreign genetic material. J Cell Sci 13:841–861
Boyes CJ, Zapata FJ, Sink KC (1980) Isolation, culture and regeneration to plants of callus protoplasts of Salpiglossis simuta L. Z Pflanzenphysiol 99:471–474
Brar DS, Rambold S, Gamborg O, Constabel F (1979) Tissue culture of corn and sorghum. Z Pflanzenphysiol 95:377–388
Brar DS, Rambold S, Constabel F, Gamborg OL (1980) Isolation, fusion and culture of sorghum and corn protoplasts. Z Pflanzenphysiol 96:269–276
Braun A (1959) A demonstration of the recovery of the crown-gall tumor cell with the use of complex tumors of single cell origin. Proc Natl. Acad Sci USA 45:932–938
Bui-Dang-Ha D, Mackenzie IA (1973) The division of protoplasts from Asparagus officinalis L and their growth and differentiation. Protoplasma 78:215–221
Bui-Dang-Ha D, Norreel B, Masset A (1975) Regeneration of Asparagus officinalis L through callus cultures derived from protoplasts. J Exp Bot 26:263–270
Burgess J, Linstead PJ (1981) Studies of the growth and development of protoplasts of the moss Physcomitrella patens, and its control by light. Planta 151:331–338
Burgutin AB, Butenko RG, Frolova LV, Gleba YY (1977) Some peculiarities of plants produced by protoplast fusion of Nicotiana tabacum L mutants. Sov Physiol Biochem Cult Plants 9:291–299
Burk LG (1960) Male-sterile flower anomalies in interspecific tobacco hybrids. J Hered 51:27–31
Burk LG, Mann TJ (1970) Onset, prevention and restoration of male-sterile flower anomalies in tobacco. Interaction of the Nicotiana repanda plasmon and N. tabacum genome. J Hered 61:143–146
Butenko RG (1964) Culture of isolated tissues and physiology of plant morphogenesis. Moscow, Nauka
Butenko RG (1979) Cultivation of isolated protoplasts and hybridization of somatic plant cells. Int Rev Cytol 59:323–392

Butenko RG, Kuchko AA (1979) Physiological aspects of procurement, cultivation and hybridization of isolated potato protoplasts. Sov Plant Physiol 26:901
Butenko RG, Kuchko AA (1980) Somatic hybridization of Solanum tuberosum L and Solanum chacoense Bitt. by protoplast fusion. In: Ferenczy L, Farkas GL (eds) Advances in Protoplast Research, Akad Kiado, Budapest, pp 293–300
Butenko RG, Kuchko AA, Vitenko VA, Avetisov VA (1977) Production and cultivation of isolated protoplasts from leaf mesophyll of Solanum tuberosum L and Solanum chacoense Bitt. Sov Plant Physiology 24:660–664
Caboche M (1980) Nutritional requirements of protoplast-derived haploid tobacco cells grown at low cell densities in liquid medium. Planta 149:7–18
Carlson PS (1973a) Toward a parasexual cycle in higher plants. Colloq Int Cent Nat Rech Sci 212:497–505
Carlson PS (1973b) The use of protoplasts for genetic research. Proc Natl Acad Sci USA 70(2):598–602
Carlson PS (1973c) Methionine sulfoximine-resistant mutants of tobacco. Science 180:1366–1368
Carlson PS, Smith HH, Dearing RD (1972) Parasexual interspecific plant hybridization. Proc Natl Acad Sci USA 69:2292–2294
Cell Genetics in Higher Plants see Ref. Dudits et al. 1976a
Chaleff RS (1981) Genetics of higher plants. Applications of cell culture. Cambridge University Press. Cambridge
Chaleff RS, Carlson PS (1974) Somatic cell genetics of higher plants. Annu Rev Genet 8:267–278
Chan PH, Wildman SG (1972) Chloroplast DNA codes for the primary structure of the large subunit of Fraction I protein. Biochem Biophys Acta 277:677–680
Chan PH, Sakano K, Singh S, Wildman SG (1972) Crystalline Fraction I protein: preparation in large yield. Science 176:1145–1146
Chellappan KP, Seeni S, Gnanam A (1980) Photosynthetic studies with mesophyll protoplasts from Notonia-Grandiflora, a Crassulacean acid metabolism plant. Physiol Plant 48:403–410
Chen K, Kung SD, Gray JC, Wildman SG (1976) Subunit polypeptide composition of Fraction I protein from various plant species. Plant Sci Lett 7:429–434
Chen K, Wildman SG, Smith HH (1977) Chloroplast DNA distribution in parasexual hybrids as shown by polypeptide composition of Fraction I protein. Proc Natl Acad Sci USA 74:5109–5112
Chien, YC, Kao KN, Wetter LR (1982) Chromosome and isozyme studies of Nicotiana tabacum – Glycine max hybrid cell lines. Theor Appl Genet 62:301–304
Chi-kuei T, Ying-ch'ien C, Yün-lo C, Su-hsuen W (1977) Regeneration of plants from tobacco protoplasts and some factors affecting the plant regeneration. Sci Sin 20:458–468
Chourey PS, Zurawski DB (1981) Callus formation from protoplasts of a maize cell culture. Theor Appl Genet 59:341–344
Chupeau Y, Bourgin JP, Missonier C, Morel G (1973) Les protoplastes de plantes supérieures. Etat actuel et perspectives. Bull Soc Bot Fr 120:175–188
Chupeau Y, Bourgin JP, Missonier C, Dorion N, Morel G (1974) Préparation et culture de protoplastes de divers Nicotiana. C R Acad Sci D 278:1565–1568
Chupeau Y, Missonier C, Hommel MC, Goujaud J (1978) Somatic hybrids of plants by fusion of protoplasts. – Observation on model system Nicotiana glauca-Nicotiana langsdorffii. Mol Gen Genet 165:239–246
Clayton EE (1950) Male sterile tobacco. J Hered 41:171–175
Cocking EC (1960) A method for the isolation of plant protoplasts and vacuoles. Nature 187:962–963
Cocking EC (1965) Plant protoplasts. In: McCarthy JD, Duddington CL (eds) Viewpoints in biology Vol 4, Butterworth, London, pp 170–203
Cocking EC (1971) Fusion of isolated protoplasts: a first step towards the somatic hybridization of plants. Colloq Int Cent Nat Rech Sci 193:303–317
Cocking EC (1973) Plant cell modification: problems and perspectives. Colloq Int Cent Nat Rech Sci 212:327–341
Cocking EC (1976a) Fusion and somatic hybridization of higher plant protoplasts. In: Peberdy JF et al. (eds) Fourth int. symp. Yeast and other protoplasts: microbial and plant protoplasts. Academic Press, London, pp 81–86
Cocking EC (1976b) A new procedure for the selection of somatic hybrids in plants. In: Dudits D et al. (eds) Cell genetics in higher plants. Akad Kiado, Budapest, pp 141–148
Cocking EC (1978) Selection and somatic hybridization. In: Thorpe T (ed) Frontiers of plant tissue culture 1978. IAPTC, Calgary, pp 151–158

Cocking EC (1979) Somatic hybridization by the fusion of isolated protoplasts – an alternative to sex. In: Sharp WR, Larsen PO, Paddock EF, Raghavan V (eds) Plant cell and tissue culture. Principles and applications. Ohio State University Press, Columbus, pp 353–370

Cocking EC (1981) Opportunities from the use of protoplasts. Philos Trans R Soc Lond B 292:557–568

Cocking EC, Power JB, Evans PK, Safwat F, Frearson EM, Hayward C, Berry SF, George D (1974) Naturally occurring differential drug sensitivities of cultured plant protoplasts. Plant Sci Lett 3:341–350

Cocking EC, George D, Price-Jones MJ, Power JB (1977) Selection procedures for the production of interspecies somatic hybrids of Petunia hybrida and Petunia parodii. II. Albino complementation selection. Plant Sci Lett 10:7–12

Cocking EC, Davey MR, Pental D, Power JB (1981) Aspects of plant genetic manipulation (a review). Nature 293:265–269

Conde MF (1981) Chloroplast DNA recombination in Nicotiana somatic parasexual hybrids. Genetics 97:26

Constabel F (1978) Development of protoplast fusion products, heterokaryocytes. In: Thorpe TA (ed) Frontiers of plant tissue culture 1978. IAPTC, Calgary, pp 141–150

Constabel F, Kirkpatrick JW, Gamborg OL (1973) Callus formation from mesophyll protoplasts of Pisum sativum. Can J Bot 51:2105–2106

Constabel F, Dudits D, Gamborg OL, Kao KN (1975a) Nuclear fusion in intergenetic heterokaryons. A note. Can J Bot 53:2092–2095

Constabel F, Kirkpatrick JW, Kao KN, Kartha KK (1975b) The effect of canavanine on the growth of cells from suspension cultures and on intergeneric heterokaryocytes of canavanine sensitive and tolerant plants. Biochem Physiol Pflanz 168:319–325

Constabel F, Weber G, Kirkpatrick JW, Pahl K (1976) Cell division of intergeneric protoplast fusion products. Z Pflanzenphysiol 79:1–7

Constabel F, Weber G, Kirkpatrick JW (1977) Sur la compatibilité des chromosomes dans les hybrides intergénériques de cellules de Glycine max x Vicia haljastana. CR Acad Sci D 285:319–322

Constabel F, Rambold S, Shyluk JP, Letouneau D, Kurz WGW, Kutney JP (1981) Alkaloid production in Catharanthas roseus cell cultures. 10. Mitoxic effect of 3′, 4′-dehydrovinblastine. Z Pflanzenphysiol 105:53–58

Correns C (1909) Vererbungsversuche mit blaß (gelb) grünen und buntblättrigen Sippen bei Mirabilis jalapa, Urtica pilulifera und Lumaria annua. Z Vererbungsl 1:291–329

Coutts RHA, Wood KR (1975) The isolation and culture of cucumber mesophyll protoplasts. Plant Sci Lett 4:189–193

Cove DJ (1979) Genetic studies of nitrate assimilation in Aspergillus nidulans. Biol Dev 54:291–337

Crepy L, Chupeau MC, Chupeau Y (1982) The isolation and culture of leaf protoplasts of Cichorium intybus and their regeneration into plants. Z Pflanzenphysiol 107:123–132

D'Amato F (1978) Chromosome number variation in cultured cells and regenerated plants. In: Thorpe T (ed) Frontiers of plant tissue culture 1978. IAPTC, Calgary, pp 287–295

Das NK, Patau K, Skoog F (1954) Initiation of mitosis and cell division by kinetin and IAA in excised tobacco pith tissue. Physiol Plant 9:640–659

Davey MR, Mathias RJ (1979) Close-packing of plasma membrane particles during wall regeneration by isolated higher plant protoplasts. Fact or artefact. Protoplasma 100:85–100

Davey MR, Frearson EM, Withers LA, Power JB (1974a) Observations on the morphology, ultrastructure and regeneration of tobacco leaf epidermal protoplasts. Plant Sci Lett 2(1):23–27

Davey MR, Bush E, Power JB (1974b) Cultural studies of a dividing legume leaf protoplast system. Plant Sci Lett 3:127–133

Davey MR, Frearson EM, Power JB (1976) Polyethylene glycoinduced transplantation of chloroplasts into protoplasts: an ultrastructural assessment. Plant Sci Lett 7:7–16

Davey MR, Cocking EC, Freeman J, Draper J, Pearce N, Tudor I, Hernalsteens JP, Beuckeleer M de, Montagu M van, Shell J (1980a) The use of plant protoplasts for transformation by Agrobacterium and isolated plasmids. In: Ferenczy L, Farkas G, Lazar G (eds) Adv Protoplast Res Proc 5th Int. Protoplast Symp. Akad Kiado, Budapest

Davey MR, Cocking EC, Freeman J, Pearce N, Tudor I (1980b) Transformation of Petunia protoplasts by isolated Agrobacterium plasmids. Plant Sci Lett 18:307–313

Davey MR, Pearce N, Cocking EC (1980c) Fusion of legume root nodule protoplasts with non-legume protoplasts: ultrastructural evidence for the functional activity of Rhizobium bacteroids in a heterokaryotic cytoplasm. Z Pflanzenphysiol 99:435–448

David A, David H (1979) Isolation and callus formation from cotyledon protoplasts of pine (Pinus pinaster). Z Pflanzenphysiol 94:173–178

Dawid IB, Horak I, Coon HG (1974) The use of hybrid somatic cells as an approach to mitochondrial genetics in animals. Genetics 78:459–471

Deka PC, Sen SK (1976) Differentiation in calli originated from isolated protoplasts of rice (Oryza sativa L) through plating technique. Mol Gen Genet 145:239–244

Diettrich B, Neumann D, Luckner M (1980) Protoplast-derived clones from cell cultures of Digitalis purpurea. Planta Med 38:375–382

Diettrich B, Neumann D, Luckner M (1982) Clonation of protoplast derived cells of Digitalis lanata suspension cultures. BPP 177:176–183

Dix PJ, Street HE (1974) Effects of p-fluorophenylalanine (PFP) on the growth of cell lines differing in ploidy and derived from Nicotiana sylvestris. Plant Sci Lett 3:283–288

Dix PJ, Joo F, Maliga P (1977) Cell line of Nicotianasylvestris with resistance to kanamycin and streptomycin. Mol Gen Genet 157:285–290

Donn G (1978) Cell division and callus regeneration from leaf protoplasts of Vicia narbenensis. Z Pflanzenphysiol 86:65–76

Dos Santos AVP, Outka DE, Cocking EC, Davey MR (1980) Organogenesis and somatic embryogenesis in tissues derived from leaf protoplasts and leaf explants of Medicago sativa. Z Pflanzenphysiol 99:261–270

Douglas GC, Keller WA, Setterfield G (1981a) Somatic hybridization between Nicotiana rustica and N. tabacum. 1. Isolation and culture of protoplasts and regeneration of plants from cell cultures of wild-type and chlorophyll-deficient strains. 2. Protoplast fusion and selection and regeneration of hybrid plants. Can J Bot 59:208–219, 220–227

Douglas GC, Wetter LR, Keller WA, Setterfield G (1981b) Somatic hybridization between Nicotiana rustica and N. tabacum IY. Analysis of nuclear and chloroplast genome expression in somatic hybrids. Can J Bot 59:1509–1513

Douglas GC, Wetter LR, Nakamura C, Keller WA, Setterfield G (1981c) Somatic hybridization between N. rustica and N. tabacum. 3. Biochemical, morphological and cytological analysis of somatic hybrids. Can J Bot 59:228–237

Dudits P, Farkas GL, Maliga P (1976a) Cell genetics in higher plants. Akad Kiado, Budapest

Dudits D, Kao KN, Constabel F, Gamborg OL (1976b) Embryogenesis and formation of tetraploid and hexaploid plants from carrot protoplasts. Can J Bot 54:1063–1067

Dudits D, Hadlaczky G, Levi E, Fejer O, Haydu Z, Lazar G (1977) Somatic hybridization of Daucus carota and D. capillifolius by protoplast fusion. Theor Appl Genet 51:127–132

Dudits D, Hadlaczky G, Bajszar GY, Koncz C, Lazar G, Horvath G (1979) Plant regeneration from intergeneric cell hybrids. Plant Sci Lett 15:101–112

Dudits D, Fejer O, Hadlaczky G, Koncz C, Lazar GB, Horvath G (1980a) Intergeneric gene transfer mediated by plant protoplast fusion. Mol Gen Genet 179:283–288

Dudits D, Hadlaczky G, Lazar G, Haydu Z (1980b) Increase in genetic variability through somatic cell hybridization of distantly related plant species. In: Sala F, Parisi B, Cella R, Cifferi O (eds) Plant cell cultures. Elsevier/North Holland, Amsterdam, pp 207–214

Dulieu H (1972) The combination of cell and tissue culture with mutagenesis for the induction and isolation of morphological or developmental mutants. Phytomorphology 22:283–295

Durand J (1979) High and reproducible planting efficiencies of protoplasts isolated from in vitro grown haploid N. sylvestris. Z Pflanzenphysiol 93:283–296

Durand J, Potrykus I, Donn G (1973) Plantes issues protoplastes de Pétunia. Z Pflanzenphysiol 69:26–34

Durbin RD, Uchytill TF (1977) Cytoplasmic inheritance of protoplast coupling factor 1 subunits. Biochem Genet 15:1143–1146

Edwards GE, Robinson SP, Tyler NJC, Walker DA (1978) Photosynthesis by isolated protoplasts, protoplast extracts and chloroplasts of wheat. Influence of orthophosphate, pyrophosphate, and adenylates. Plant Physiol 62:313

Edwardson JR (1970) Cytoplasmic male sterility. Bot Rev 36:341–420

Ephrussi B (1972) Hybridization of somatic cells. Princeton University Press, Princeton, NJ

Eriksson T (1965) Studies on the growth requirements and growth measurements of cell cultures of Haplopappus gracilis. Physiol Plant 18:976–993

Eriksson T (1967) Effects of UV and X-ray radiation on in vitro cultivated cells of Haplopappus gracilis. Physiol Plant 20:507–558

Eriksson T (1971) Isolation and fusion of plant protoplasts. Colloq Int Cent Nat Rech Sci 193:297–302
Eriksson T, Jonasson K (1969) Nuclear division in isolated protoplasts from cells of higher plants grown in vitro. Planta 89:85–89
Eriksson T, Bonnett H, Glimelius K, Wallin A (1974) Technical advances in protoplast isolation, culture and fusion. In: Street HE (ed) Tissue culture and plant science 1974. Academic Press, London
Evans DA (1979) Chromosome stability of plants regenerated from mesophyll protoplasts of Nicotiana species. Z Pflanzenphysiol 95:459–463
Evans DA (1982) Plant regeneration and genetic analysis of somatic hybrid plants. In: Earle ED (ed) Plant regeneration and genetic variability. Praeger, Eastbourne
Evans DA, Gamborg OL (1982) Chromosome stability of cell suspension cultures of Nicotiana spp. Plant Cell Rep 1:104–107
Evans DA, Wetter LR, Gamborg OL (1980a) Somatic hybrid plants of Nicotiana glauca and Nicotiana tabacum obtained by protoplast fusion. Physiol Plant 48:225–230
Evans DA, Crocomo OJ, Carvalho MTV de (1980b) Protoplast isolation and subsequent callus regeneration in sugarcane. Z Pflanzenphysiol 98:355–358
Evans DA, Flick CE, Jensen RA (1981) Disease resistance: incorporation into sexually incompatible somatic hybrids of the genus Nicotiana. Science 213:907–909
Evans DA, Flick CE, Kut SA, Reed SM (1982) Comparison of Nicotiana tabacum and Nicotiana nesophila hybrids produced by ovule culture and protoplast fusion. Theor Appl Genet 62:193–198
Facciotti D, Pilet PE (1979) Plants and embryoids from haploid Nicotiana sylvestris protoplasts. Plant Sci Lett 15:1–6
Finch RA, Smith JB, Bennett MD (1981) Hordeum and Secale mitotic genomes lie apart in a hybrid. J Cell Sci 52:391–403
Finer JJ, Smith RH (1982) Isolation and culture of protoplasts from cotton (Gossypium klotzchianum Anderss.) callus cultures. Plant Sci Lett 26:147–151
Flavell RB, Rimpau J, Smith DB (1977) Repeated sequence DNA relationships in four cereal genomes. Chromosoma 63:205–222
Flick CE, Evans DA (1982) Evaluation of cytoplasmic segregation in somatic hybrids of Nicotiana: tentoxin sensitivity. J Hered 73:264–266
Flick CE, Evans DA (1983) Isolation, culture and plant regeneration from protoplasts isolated from flower petals of ornamental Nicotiana species. Z Pflanzenphysiol (accepted)
Flick CE, Jensen RA, Evans DA (1981) Isolation, protoplast culture and plant regeneration of PFP-resistant variants of N. tabacum Su-Su. Z Pflanzenphysiol 103:239–246
Fowke LC, Rennie PJ, Kirkpatrick JW, Constabel F (1975a) Ultrastructural characterisrics of intergenesic protoplast fusion. Can J Bot 53:272–278
Fowke LC, Bech-Hansen CW, Gamborg OL, Constabel F (1975b) Electron-microscope observations of mitosis and cytokinesis in multinucleate protoplasts of soybean. J Cell Sci 18:491–508
Fowke LC, Rennie PJ, Kirkpatrick JW, Constabel F (1976) Ultrastructure of fusion products from soybean cell culture and sweet clover leaf protoplasts. Planta 130:39–45
Fowke LC, Constabel F, Gamborg OL (1977) Fine structure of fusion products from soybean cell culture and pea leaf protoplasts. Planta 135:257–266
Frearson EM, Power JB, Cocking EC (1973) The isolation, culture and regeneration of Petunia leaf protoplasts. Dev Biol 33:130–137
Friend KK, Chen S, Ruddle FH (1976) Differential staining of interspecific chromosomes in somatic cell hybrids by alkaline Giemsa stain. Somatic Cell Genet 2:183–188
Furner IJ, King J, Gamborg OL (1978) Plant regeneration from protoplasts isolated from a predominantly haploid suspension culture of Datura innoxia. Plant Sci Lett 11:169
Galbraith DW, Galbraith JEC (1979) A method for identification of fusion of plant protoplasts derived from tissue cultures. Z Pflanzenphysiol 93:149–158
Galun E, Arzee-Gonen P, Fluhr R, Edelman M, Aviv D (1982) Cytoplasmic hybridization in Nicotiana: mitochondrial DNA analysis in progenies resulting from fusion between protoplasts having different organelle constitutions. Mol Gen Genet 186:50–56
Gamborg OL, Shyluk JP (1976) Tissue culture, protoplasts, and morphogenesis in flax. Bot Gaz 137:301–306
Gamborg OL, Miller RA, Ojima K (1968) Nutrient requirements of suspension cultures of soybean root cells. Exp Cell Res 50:151–158
Gamborg OL, Constabel F, Fowke L, Kao KN, Ohyama K, Pelcher L (1974) Protoplast and cell culture methods in somatic hybridization in higher plants. Can J Genet Cytol 16:737–750

Gamborg OL, Constabel F, Kao K, Ohyama K (1975) Plant protoplasts in genetic modifications and production of intergeneric hybrids. In: Markham R, Davies DR, Hopwood DA, Horne RW (eds) Modification of the information content of plant cells. Elsevier/North Holland, Amsterdam Oxford

Gamborg OL, Shyluk JP, Fowke LC, Wetter LR, Evans D (1979) Plant regeneration from protoplasts and cell cultures of Nicotiana tabacum sulfer mutant (Su/Su). Z Pflanzenphysiol 95:255–264

Gatenby AA, Cocking EC (1977) Callus formation from protoplasts of marrow stem kale. Plant Sci Lett 8:275–280

Gay L (1976) The development of leafy gametophytes from isolated protoplasts of Polytrichum juniperinum wild. Z Pflanzenphysiol 79:33–39

Giles KL (1977) Chloroplast uptake and genetic complementation. In: Reinert J, Bajaj YPS (eds) Applied and fundamental aspects of plant cell, tissue, and organ culture. Springer, Berlin Heidelberg New York, pp 536–550

Gill R, Rashid A, Maheshwari SC (1979) Isolation of mesophyll protoplasts of Nicotiana rustica and their regeneration into plants flowering in vitro. Physiol Plant 47:7–10

Gleba DM, Gleba YY (1978a) Mutations and epigenetic changes in culture of cells and protoplasts of higher plants. Sov Cytol Genet 12:458–469

Gleba DM, Gleba YY (1978b) Effect of chemical mutagenesis on survival and plating efficiency of isolated mesophyll protoplasts of tobacco, Nicotiana tabacum. Sov Genet 14:1935–1945

Gleba DM, Gleba YY, Piven NM, Slavenas IY (1978b) Cytophotometric studies of DNA content in tobacco mesophyll protoplasts during cultivation and the effect of treatment with N-nitroso-N-methylurea. Sov Genet 14:1946–1954

Gleba YY (1974) Culture of isolated protoplasts as model for genetic engineering. Thesis Kiev

Gleba YY (1978a) Microdroplet culture. Tobacco plants from mesophyll protoplasts. Naturwissenschaften 65:158–159

Gleba YY (1978b) Cell divisions in hybrid products obtained by fusion of protoplasts from distantly related species. In: Alferman AW, Reinhard E (eds) Production of natural compounds by cell culture methods. Gesellschaft für Strahlen- und Umweltforschung, München, pp 258–265

Gleba YY (1978c) Extranuclear inheritance investigated by somatic hybridization. In: Thorpe TA (ed) Frontiers of plant tissue culture 1978, University of Calgary, Calgary, pp 95–102

Gleba YY (1979) Non-chromosomal inheritance in higher plants as studied by somatic cell hybridization. In: Sharp WR et al. (eds) Plant cell and tissue culture – principles and applications. Ohio State University Press, Columbus, pp 775–788

Gleba YY (1980) Protoplast fusion and genetic engineering of higher plants. Habil. Thesis, Leningrad

Gleba YY, Berlin J (1979) Somatic hybridization by protoplast fusion in Nicotiana: fate of nuclear genetic determinants. Abstr Fifth Int Protopl Symp, Szeged, Hungary, p 73

Gleba YY, Evans DA (1982) Transmission genetics of somatic hybrid plants: In: Sharp WR, Evans DA, Ammirato PV, Yamada Y (eds) Applications of plant tissue culture methods for crop improvement. Macmillan, Chicago (in press)

Gleba YY, Hoffmann F (1978) Hybrid cell lines Arabidopsis thaliana + Brassica campestris: no evidence for specific chromosome elimination. Mol Gen Genet 165:257–264

Gleba YY, Hoffmann F (1979) Arabidobrassica: Plant-genome engineering by protoplast fusion. Naturwissenschaften 66:547–554

Gleba YY, Hoffmann F (1980) Arabidobrassica: A novel plant obtained by protoplast fusion. Planta 149:112–117

Gleba YY, Sytnik KM (1982) Protoplast fusion and genetic engineering of higher plants. Naukova Dumka, Kiev

Gleba YY, Shvydkaya LG, Butenko RG, Sytnik KM (1974a) Culture of isolated protoplasts. Sov Plant Physiol 21:598–605

Gleba YY, Khasanov MM, Slusarenko AG, Butenko RG, Vynetsky YP (1974b) Uptake of Bacillus subtilis ^{3}H-DNA by isolated protoplasts of tobacco, Nicotiana tabacum L. Dokl Akad Nauk USSR 219:1478–1481

Gleba YY, Butenko RG, Sytnik KM (1975) Fusion of protoplasts and parasexual hybridization in Nicotiana tabacum L. Dokl Akad Nauk USSR 221.1196–1198

Gleba YY, Piven NM, Komarnitsky IK, Sytnik KM (1978a) Parasexual cytoplasmic hybrids (cybrids) Nicotiana tabacum + N. debneyi obtained by protoplast fusion. Dokl Akad Nauk USSR 240:1223–1226

Gleba YY, Sidorov VA, Artemenko VS (1978b) Cultivation and fusion of protoplasts isolated from callus tissue of Arabidopsis thaliana. Sov Plant Physiol 25:161–165

Gleba YY, Kohlenbach HW, Hoffmann F (1978c) Root morphogenesis in somatic hybrid cell lines Arabidopsis thaliana + Brassica campestris. Naturwissenschaften 65:655–656

Gleba YY, Momot VP, Cherep NN, Skarzhynskaya MV (1982a) Intertribal hybrid cell lines of Atropa belladonna (×) Nicotiana chinensis obtained by cloning individual protoplast fusion products. Theor Appl Genet 62:75–79

Gleba YY, Okolot A, Momot VP, Cherep NN, Skarzhynskaya MV (1982b) Somatic hybrids Atropa belladonna (×) Nicotiana tabacum obtained by cloning individual heterokaryocytes. Plant Cell Rep (in press)

Gleba YY, Momot VP, Okolot A, Cherep NN, Skarzhynskaya MV (1982c) Genetic processes in intergeneric hybrid cells Atropa (×) Nicotiana. I. Analysis of cell clones grown in vitro. Theor Appl Genet (submitted)

Gleba YY, Kanevsky I, Cherep NN (1983a) Transmission genetics of hybridization by protoplast fusion in Nicotiana. II. Genetic constitution of N. tabacum (×) N. plumbaginifolia hybrids obtained using selective and non-selective screening methods (submitted)

Gleba YY, Momot VP, Okolot AN, Cherep NN (1983b) Isolation of somatic cell hybrid Pisum sativum (×) Nicotiana chinensis (submitted)

Gleba YY, Momot VP, Okolot AN, Cherep NN, Skarzhynskaya MV, Kotov V (1983c) Genetic processes in intergeneric cell hybrids Atropa (×) Nicotiana. I. Genetic constitution of cells of different clonal origin grown in vitro (submitted)

Glimelius K, Bonnett HT (1981) Somatic hybridization in Nicotiana: Restoration of photoautotrophy to an albino mutant with defective plastids. Planta 153:497–503

Glimelius K, Eriksson T, Grafe R, Müller AJ (1978a) Somatic hybridization of nitrate reductase-deficient mutants of Nicotiana tabacum by protoplast fusion. Physiol Plant 44:273–277

Glimelius K, Wallin A, Eriksson T (1978b) Concanavalin A improves the polyethylene glycol method for fusing plant protoplasts. Physiol Plant 44:92–96

Glimelius K, Chen K, Bonnett HT (1981) Somatic hybridization in Nicotiana: Segregation of organellar traits among hybrid and cybrid plants. Planta 153:504–510

Gosch G, Reinert J (1976) Nuclear fusion in intergeneric heterokaryocytes and subsequent mitosis of hybrid nuclei. Naturwissenschaften 63:534

Gosch G, Reinert J (1978) Cytological identification of colony formation of intergeneric somatic hybrid cells. Protoplasma 96:23–38

Gosch G, Bajaj YPS, Reinert J (1975a) Isolation, culture, and fusion studies on protoplasts from different species. Protoplasma 85:327–336

Gosch G, Bajaj YPS, Reinert J (1975b) Isolation, culture, and induction of embryogenesis in protoplasts from cell-suspensions of Atropa belladonna. Protoplasma 86:405–410

Goss SJ, Harris H (1975) New method for mapping genes in human chromosomes. Nature 255:680–684

Goss SJ, Harris H (1977) Gene transfer by means of cell fusion. I. Statistical mapping of the human X-chromosomes by analysis of radiation-induced gene segregation. J Cell Sci 25:17–37

Grambow HJ, Kao KN, Miller RA, Gamborg OL (1972) Cell division and plant development from protoplasts of carrot cell suspension cultures. Planta 103:348–355

Gresshoff PM (1978) Interspecific hybridization of Arabidopsis thaliana and Trifolium repens (white clover). Arabidopsis Inf Serv 15:24–27

Gresshoff PM (1980) In vitro culture of white clover: callus suspension, protoplast culture and plant regeneration. Bot Gaz 141:157–164

Griesbach RJ, Malmberg RL, Carlson PS (1982) An improved technique for the isolation of higher plant chromosomes. Plant Sci Lett 24:55–60

Grimsley NH, Ashton NW, Cove DJ (1977a) The production of somatic hybrids by protoplast fusion in the moss, Pyscomitrella patens. Mol Gen Genet 154:97–100

Grimsley NH, Ashton NW, Cove DJ (1977b) Complementation analysis of auxotrophic mutants of the moss, Physcomitrella patens, using protoplast fusion. Mol Gen Genet 155:101–107

Grimsley NH, Featherstone DR, Courtice GRM, Ashton NW, Cove DJ (1980) Somatic hybridization following protoplast fusion as a tool for the genetic analysis of development in the moss Physcomitrella patens. In: Ferenczy L, Farkas GL (eds) Advances in protoplast research. Akad Kiado, Budapest, pp 363–376

Gunn RE, Shepard JF (1981) Regeneration of plants from mesophyll derived protoplasts of British potato (Solanum tuberosum L) cultivars. Plant Sci Lett 22:97–101

Gupta N, Carlson PS (1972) Preferential growth of haploid plant cells in vitro. Nature New Biol 239:86

Gwozdz EA, Waliszewska B (1979) Regeneration of enzymatically isolated protoplasts of the moss Funaria hydrometrica L Sibth. Plant Sci Lett 15:41–46

Hagemann R (1959) Plasmatische Vererbung. Ziemsen, Wittenberg

Harms CT, Potrykus I (1978a) Enrichment for heterokaryocytes by the use of iso-osmotic density gradients after plant protoplast fusion. Theor Appl Genet 53:49–56

Harms CT, Potrykus I (1978b) Fractionation of plant protoplast types by iso-osmotic density gradient centrifugation. Theor Appl Genet 53:57–64

Harms CT, Potrykus I, Widholm JM (1981) Complementation and dominant expression of amino acid analogue resistance markers in somatic hybrid clones from Daucus carota after protoplast fusion. Z Pflanzenphysiol 101:377–390

Harms CT, Oertli JJ, Widholm JM (1982) Characterization of amino acid analogue resistant somatic hybrid cell lines of Daucus carota L. Z Pflanzenphysiol 106:239–249

Harris H (1974) Nucleus and cytoplasm. Oxford University Press, London New York

Harris M (1973) Phenotypic expression of drug resistance in hybrid cells. J Natl Cancer Inst 50:423–429

Hartmann JX, Kao KN, Gamborg OL, Miller RA (1973) Immunological methods for the agglutination of protoplasts from cell suspension cultures from different genera. Planta 112:45–56

Hastie AC (1970) Benlate induced instability of Aspergillus diploids. Nature 226:771

Hayward C, Power JB (1975) Plant production from leaf protoplasts of Petunia parodii. Plant Sci Lett 4:407–410

Heimer YM, Filner P (1970) Regulation of the nitrate assimilation pathway of cultured tobacco cells. II. Properties of a variant cell line. Biochim Biophys Acta 215:152–165

Herrmann RG, Bohnert HJ, Kowallik KV, Schmitt JM (1975) Size, conformation and purity of chloroplast DNA of some higher plants. Biochim Biophys Acta 378:305–317

Hess D (1973) Die Aufnahme von FITC und FITC-serumalbumin in isolierte Protoplasten von Petunia hybrida. Z Pflanzenphysiol 69:280–286

Heyn RF, Rörsch A, Schilperoort RA (1974) Prospects in genetic engineering of plants. Quart Rev Biophys 7:35–73

Hoffmann F, Adachi T (1981) "Arabidobrassica": chromosomal recombination in asymmetric intergeneric hybrid cells. Planta 153:586–593

Hopwood DA (1981) Genetic studies with bacterial protoplasts. Annu Rev Microbiol 35:237–272

Hopwood DA, Wright HM (1981) Protoplast fusion in streptomyces – fusions involving ultraviolet-irradiated protoplasts. J Gen Microbiol 126:21–28

Ivantsov AI, Akhmetov RR (1978) Cultivation of isolated corn protoplasts. Sov Plant Physiol 25:897–899

Iwai S, Nagao T, Nakata K, Kawashima N, Matsuyama S (1980) Expression of nuclear and chloroplastic genes coding for fraction I protein in somatic hybrids of Nicotiana tabacum + rustica. Planta 147:414–417

Iwai S, Nakata K, Nagao T, Kawashima N, Matsuyama S (1981) Detection of the Nicotiana rustica chloroplast genome coding for the large subunit of Fraction I protein in a somatic hybrid which only the N. tabacum chloroplast genome appeared to have been expressed. Planta 152:478

Izhar S, Power JB (1979) Somatic hybridization in Petunia male sterile cytoplasmic hybrid. Plant Sci Lett 14:49–56

Izhar S, Tabib Y (1980) Somatic hybridization in Petunia. Part 2: heteroplasmic state in somatic hybrids followed by cytoplasmic segregation into male sterile and male fertile lines. Theor Appl Genet 57:241–246

Jensen RG, Bahr JT (1977) Ribulose 1,5 – bisphosphate carboxylase-oxygenase. Annu Rev Plant Physiol 28:379–400

Jing-fen Jia, Potrykus I (1981) Mesophyll protoplasts from Solanum melongena var depressum Bailey regenerate to fertile plants. Plant Cell Rep 1:71–72

Jones L, Hildebrandt A, Riker A, Wu J (1960) Growth of somatic tobacco cells in microculture. Am J Bot 47:468–475

Johnson LB, Stuteville DL, Higgins RK, Skinner DZ (1981) Regeneration of alfalfa plants from protoplasts of selected S clones. Plant Sci Lett 20:297–304

Kajita S, Matsui C, Syono K, Suzuki M, Nagata T (1980) Fine structure of fusion bodies formed between pea root nodule and tobacco mesophyll protoplasts. Z Pflanzenphysiol 97:233–240
Kameya T (1973) The effects of gelatin on aggregation of protoplasts from higher plants. Planta 115:77–82
Kameya T (1975a) Induction of hybrids through somatic cell fusion with dextran sulfate and gelatin. J J Genet 50:236–246
Kameya T (1975b) Culture of protoplasts from chimeral plant tissue of nature. J J Genet 50:417–420
Kameya T (1979) Studies on plant cell fusion: effects of dextran und pronase E on fusion. Cytologia 44:449–456
Kameya T, Takahashi N (1972) The effects of inorganic salts on fusion of protoplasts from roots and leaves of Brassica species. Jpn J Genet 47:215–217
Kameya T, Horn ME, Widholm JM (1981) Hybrid shoot formation from fused Daucus carota and D. capillifolius protoplasts. Z Pflanzenphysiol 104:459–466
Kao KN (1977) Chromosomal behaviour in somatic hybrids of soybean-Nicotiana glauca. Mol Gen Genet 150:225–230
Kao KN, Michayluk MR (1974) A method for high-frequency intergeneric fusion of plant protoplasts. Planta 115:355–367
Kao KN, Michayluk MR (1975) Nutritional requirement for growth of Vicia hajastana cells and protoplasts at a very low population density in liquid media. Planta 126:105–110
Kao KN, Michayluk MR (1980) Plant regeneration from mesophyll protoplasts of alfalfa. Z Pflanzenphysiol 96:135–142
Kao KN, Wetter RL (1977) Advances in techniques of plant protoplast fusion and culture of heterokaryocytes. In: Brinkley BR, Porter KR (eds) International cell biology 1976–1977. Rockefeller University Press, New York, pp 216–224
Kao, KN, Keller WA, Miller RA (1970a) Cell divisions in newly formed cells from protoplasts of soybean. Exp. Cell Res. 62:338–340
Kao KN, Miller RA, Gamborg OL, Harvey BL (1970b) Variations in chromosome number and structure in plant cells grown in suspension cultures. Can J Genet Cytol 12:297–301
Kao KN, Gamborg OL, Miller RA, Keller WA (1971) Cell divisions in cells regenerated from protoplasts of soybean and Haplopappus gracilis. Nature New Biol 232:124
Kao KN, Gamborg OL, Michayluk MR, Keller WA, Miller RA (1973) The effects of sugars and inorganic salts on cell regeneration and sustained division in plant protoplasts. Colloq Int Cent Nat Rech Sci 212:207–213
Kao KN, Constabel F, Michayluk MR, Gamborg OL (1974) Plant protoplast fusion and growth of intergeneric hybrid cells. Planta 120:215–227
Kartha KK, Michayluk MR, Kao KN, Gamborg OL, Constabel F (1974a) Callus formation and plant regeneration from mesophyll protoplasts of rape plants (Brassica napus L. cv. Zephyr). Plant Sci Lett 3:265–271
Kartha KK, Gamborg OL, Constabel F, Kao KN (1974b) Fusion of rapeseed and soybean protoplasts and subsequent division of heterokaryocytes. Can J Bot 52:2435–2436
Kawashima N, Wildman SG (1970) Fraction I protein. Annu Rev Biochem 21:325–358
Keller WA, Melchers G (1973) The effect of high pH and calcium on tobacco leaf protoplasts fusion. Z Pflanzenphysiol 28c:737–741
Kemble RJ, Flavell RB, Bretell RIS (1982) Mitochondrial DNA analysis of fertile and sterile maize plants derived from tissue cultures with the Texas male sterile cytoplasm. Theor Appl Genet 62:213–218
Kevei F, Peberdy JF (1979) Induced segregation in interspecific hybrids of Aspergillus nidulans and Aspergillus regulosus obtained by protoplast fusion. Mol Gen Genet 170:213–218
Khasanov MM, Butenko RG (1979) Cultivation of isolated protoplasts from cotyledons of cotton. Fiziologiya Rastenii 26:77–81
King PJ (1980) Plant tissue culture and the cell cycle. Adv Biochem Eng 18:1–38
Kirk JTO, Tilney-Bassett RAE (1978) The plastids. Elsevier/North Holland, Amsterdam
Klercker J von (1892) Eine Methode zur Isolierung lebender Protoplasten. Ofvers Vetensk Akad Forhandl 49:463–474
Koblitz H (1975) Isolierung und Kultivierung von Protoplasten aus Calluskulturen von Catharanthus roseus. BPP 167:489–499

Koblitz H (1976) Isolierung und Kultivierung von Protoplasten aus Calluskulturen der Gerste. BPP 170:287–293

Kohlenbach HW, Bonke E (1975) Isolation and culture of mesophyll protoplasts of Hyoscyamus niger L. var. annuus Sims. Experientia 31:1281–1283

Kohlenbach HW, Wenzel G, Hoffmann F (1982) Regeneration of Brassica-napus plantlets in cultures from isolated protoplasts of haploid stem embryos as compared with leaf protoplasts. Z Pflanzenphysiol 105:131–142

Kolodner R, Tewari KK (1972a) Molecular size and conformation of chloroplast deoxyribonucleic acid from pea leaves. J Biol Chem 247:6355–6364

Kolodner R, Tewari KK (1972b) Physicochemical characterization of mitochondrial DNA from pea leaves. Proc Natl Acad Sci USA 69:1830–1834

Komarnitsky IK, Gleba YY (1981) Fraction I protein analysis of parasexual hybrids Arabidopsis thaliana + Brassica campestris. Plant Cell Rep 1:67–68

Komarnitsky IK, Kuchko AA, Butenko RG, Vitenko VA (1980) Ribulose diphophate carboxylase of somatic hybrid of potato. Dokl Akad Nauk USSR 4:75–78

Komarnitsky IK, Kuchko AA, Shlumukov LR, Butenko RG (1981) Analysis of chl-DNA of interspecific somatic hybrid of potato. Dokl Akad Nauk USSR 256:495–496

Komarnitsky IK, Kuchko AA, Butenko RG (1983) Analysis of mitochondrial DNA of interspecific somatic hybrid of potato. Dokl Akad Nauk USSR (in press)

Kostoff D (1943) Cytogenetics of the genus Nicotiana. Sofia, States Printing House

Krishnamurthy KV, Hendre RR, Godbole DA, Kulkarni VM, Mascarenhas AF, Jagannathan V (1979) Isolation and regeneration of rose bud callus protoplasts. Plant Sci Lett 15:135–138

Krumbiegel G, Schieder O (1979) Selection of somatic hybrids after fusion of protoplasts from Datura innoxia Mill and Atropa belladonna L. Planta 145:371–376

Krumbiegel G, Schieder O (1981) Comparison of somatic and sexual incompatibility between Datura innoxia and Atropa belladonna. Planta 153:466–470

Kumar A, Wilson D, Cocking EC (1981a) Polypeptide composition of fraction I protein of the somatic hybrid between Petunia parodii and Petunia parviflora. Biochem Genet 19:255–262

Kumar A, Cocking EC, Bovenberg WA, Kool AJ (1982) Restriction endonuclease analysis of chloroplast DNA in interspecies somatic hybrids of Petunia. Theor Appl Genet 62:377–383

Kumar PM, Bhattacharya P, Bhattacharya NM, Sen SK (1981b) Induction of callus from the mesophyll protoplasts of Brassica campestris, an oilseed crop. Ind J Exp Biol 19:656–657

Kunakh VA (1978) Variability of chromosome numbers in ontogenesis of higher plants. Sov Cytol Genet 2:160–173

Kung SD (1976) Tobacco Fraction I protein: a unique genetic marker. Science 191:429–434

Kung SD (1977) Expression of chloroplast genomes in higher plants. Annu Rev Plant Physiol 28:401–437

Kung SD, Gray JC, Wildman SG, Carlson PS (1975) Polypeptide composition of Fraction I protein from parasexual hybrid plants in the genus Nicotiana. Science 187:353–355

Kung SD, Zhu YS, Shen GF (1982) Nicotiana chloroplast genome III. Chloroplast DNA evolution. Theor Appl Genet 61:73–79

Kuster E (1909) Über die Verschmelzung nackter Protoplasten. Ber Dtsch Bot Ges 27:589–598

Landgren CR (1976a) Patterns of mitosis and differentiation in cells derived from pea root protoplasts. Am J Bot 63:473–480

Landgren CR (1976b) Influence of culture conditions on mitotic activity in protoplasts derived from Pisum root cortical explants. Protoplasma 87:49–70

Landgren CR, Bonnett HT (1979) Culture of albino tobacco protoplasts treated with polyethylene glycol to induce chloroplast incorporation. Plant Sci Lett 16:15–22

Landgren CR, Torrey JG (1973a) Mitotic activity in cells derived from protoplasts of seedling pea roots. Am J Bot [Suppl] 60:8

Landgren CR, Torrey JG (1973b) The culture of protoplasts derived from explants of seedling pea roots. Colloq Int Cent Nat Res Sci 212:281–289

Larkin PJ, Scowcroft WR (1981) Somaclonal variation – a novel source of variability from cell cultures for plant improvement. Theor Appl Genet 60:197–214

Larsen PO, Paddock EF, Raghavan V (eds) (1979) Plant cell and tissue culture – principles and applications. Ohio State University Press, Columbus

Lazar GB, Dudits D, Sung ZR (1981) Expression of cycloheximide resistance in carrot somatic hybrids and their segregants. Genetics 98:347–356

Leible MB, Shoeman RL, Schweiger HG (1982) Ribulose-1,5-bisphosphate carboxylase, a marker for chloroplast species specificity in Acetabularia. Biochim Biophys Acta 699:60–66

Levings CS, Pring DR (1976) Restriction endonuclease analysis of mitochondrial DNA from normal and Texas cytoplasmic male-sterile maize. Science 193:158–160

Levings CS, Pring DR (1979) Mitochondrial DNA of higher plants and genetic-engineering. In: Setlow JK, Hollaender A (eds) Genetic engineering, Vol 1. Principles and methods. Plenum Press, New York, pp 205–222

Lhoas P (1961) Mitotic haploidization by treatment of Aspergillus niger diploids with para-fluorophenylalanine. Nature 190:744

Lin Zhon-ping, MC (1981) Enucleation of garlic protoplasts by means of cytochalasin B and centrifugation. In: Schweiger HG (ed) International cell biology 1980–1981. Springer, Berlin Heidelberg New York, pp 879

Lonnendonker N, Schieder O (1980) Amylase isoenzymes of the genus Datura as a simple method for an early identification of somatic hybrids. Plant Sci Lett 17:135–139

Lorz H, Potrykus I (1978) Investigations on the transfer of isolated nuclei into plant protoplasts. Theor Appl Genet 53:251–256

Lorz H, Potrykus I (1979) Regeneration of plants from mesophyll protoplasts of Atropa belladonna. Experientia 35:313–314

Lorz H, Wernicke W, Potrykus I (1979) Culture and plant regeneration of Hyoscyamus protoplasts. Planta Med 36:21–29

Lorz H, Paszkowski J, Dierksventling C, Potrykus I (1981) Isolation and characterization of cytoplasts and miniprotoplasts derived from protoplasts of cultured cells. Physiol Plantar 53:385–391

Lu CY, Vasil V, Vasil IK (1981) Isolation and culture of protoplasts of Panicun maximum Jacq. (Guinea Grass): Somatic embryogenesis and plantlet formation. Z Pflanzenphysiol 104:311–318

Lu CY, Pental D, Cocking EC (1982) Plant regeneration from seedling cotyledon protoplasts. Z Pflanzenphysiol 107:59–64

Lucy JA (1970) The fusion of biological membranes. Nature 277:814–817

Mackenzie IA, Bui-Dang-Ha D, Davey MR (1973) Some aspects of isolation, fine structure and growth of protoplasts from Asparagus officinalis L. Coloq Int Cent Nat Res Sci 212:291–299

Maeda M (1979) Isolation and culture of protoplasts from moss protonemata. Bot Mag 92:105–110

Maeda M, Ito M (1981) Isolation of protoplasts from fern prothallia and their regeneration to gemetophytes. Bot Mag 94:35–40

Magnien E, Dalschaert X, Devreux M (1980) Different radiosensitivities of Nicotiana plumbaginifolia leaves and regeneration protoplasts. Plant Sci Lett 19:231–241

Maliga P, Sz.-Breznovits A, Marton L (1973) Streptomycin resistant plants from callus culture of haploid tobacco. Nature New Biol 244:29–30

Maliga P, Sz.-Breznovits A, Marton L, Joo F (1975) Non-Mendelian streptomycin-resistant tobacco mutant with altered chloroplasts and mitochondria. Nature 255:401–402

Maliga P, Lazar G, Joo F, Nagy AH, Menczel L (1977) Restoration of morphogenic potential in Nicotiana by somatic hybridization. Mol Gen Genet 157:291–296

Maliga P, Kiss ZR, Nagy AH, Lazar G (1978) Genetic instability in somatic hybrids of Nicotiana tabacum and Nicotiana knightiana. Mol Gen Genet 163:145–151

Maliga P, Lorz H, Lazar G, Nagy F (1982) Cytoplast-protoplast fusion for interspecific chloroplast transfer in Nicotiana. Mol Gen Genet 185:211–215

Malmberg R, Griesbach RJ (1980) The isolation of mitotic and meiotic chromosomes from plant protoplasts. Plant Sci Lett 17:141–147

Mareno V, Roing LA, Garcia B (1980) Isolation and culture of protoplasts of melon Cucumis melo L. In: Application des cultures in vitro pour l'amelioration des plantes potageres. Reunion Eucarpia, sect. Legumes, Versailles 1980. EUCARPIA, Versailles, pp 166–171

Maretzki A, Nickell LJ (1973) Formation of protoplasts from sugarcane cell suspensions and the regeneration of cell cultures from protoplasts. Colloq Int Cent Nat Res Sci 212:51–63

Marton L, Wullems GJ, Molendijk L, Schilperoort RA (1979) In vitro transformation of cultured cells from N. tabacum by Agrobacterium tumefaciens. Nature 277:129–130

Marton L, Sidorov V, Biassini G, Maliga P (1982) Complementation in somatic hybrids indicates four types of nitrate reductase deficient lines in Nicotiana plumbaginifolia. Mol Gen Genet 187:1–3

Marzluf GA (1981) Regulation of nitrogen metabolism and gene expression in fungi. Microbiol Rev 45:437–461

Meadows MG, Potrykus I (1981) Hoechst 33258 as a vital stain for plant cell protoplasts. Plant Cell Rep 1:77–79

Medgyesy P, Menczel L, Maliga P (1980) The use of cytoplasmic streptomycin resistance – Chloroplast transfer from Nicotiana tabacum into Nicotiana sylvestris and isolation of their somatic hybrids. Mol Gen Genet 179:693–698

Melchers G (1977) Microbial techniques in somatic hybridization by fusion of protoplasts. In: Brinkley B, Porter KR (eds) International cell biology 1976–1977. Rockefeller University Press, New York, pp 207–215

Melchers G (1980) The future. Int Rev Cytol [Suppl] 11B:241–253

Melchers G (1981) Somatic and sexual genetics combined in higher plants. In: Proc XIV Intern Congr Genet., Moscow, Vol 1, pp 208–220

Melchers G (1982) The first decennium of somatic hybridization. In: Materials of the Vth Int Congr of plant tissue cell cult, Tokyo

Melchers G, Labib G (1973) Plants from protoplasts. Significance for genetics and breeding. Colloq Int Cent Nat Res Sci 212:367–372

Melchers G, Labib G (1974) Somatic hybridization of plants by fusion of protoplasts. I. Selection of light resistant hybrids of "haploid" light sensitive varieties of tobacco. Mol Gen Genet 135:277–294

Melchers G, Sacristan MD (1977) Somatic hybridization of plants by fusion of protoplasts. II. The chromosome numbers of somatic hybrid plants of 4 different fusion experiments. In: La culture des tissues et des cellules des vegetaux. Masson, Paris, pp 169–177

Melchers G, Sacristan MD, Holder AA (1978) Somatic hybrid plants of potato and tomato regenerated from fused protoplasts. Carlsberg Res Commun 43:203–218

Menczel L, Lazar G, Maliga P (1978) Isolation of somatic hybrids by cloning Nicotiana heterokaryons in nurse cultures. Planta 143:29–32

Menczel L, Nagy F, Kiss Z, Maliga P (1981) Streptomycin resistant and sensitive somatic hybrids of N. tabacum × N. knightiana: correlation and resistance to N. tabacum plastids. Theor Appl Genet 59:191–198

Menczel L, Galiba G, Nagy F, Maliga P (1982) Effect of radiation dosage on the efficiency of chloroplast transfer by protoplast fusion in Nicotiana. Genetics 100:487–495

Mendel RR, Muller AJ (1978) Reconstitution of Nadh-nitrate reductase in vitro from nitrate reductase-deficient Nicotiana tabacum mutants. Mol Gen Genet 161:77–80

Mendel RR, Muller AJ (1979) Nitrate reductase-deficient mutant cell lines of Nicotiana tabacum. Mol Gen Genet 177:145–153

Messerschmidt M (1974) Callus formation and differentiation of protoplasts isolated from cotyledons of Pharbitis nil. Z Pflanzenphysiol 74:175–178

Microbial and Plant Protoplasts (1976) In: Peberdy JF, Rose AH, Rogers HJ, Cocking EC (eds) Proc fourth int symp. Yeast and other protoplasts. Academic Press, London, pp 370

Mitra J, Mapes MO, Steward FC (1960) Growth and organised development of cultured cells. IV. The behaviour of the nucleus. Am J Bot 47:357–368

Montagu M van, Schell J (1982) The Ti Plasmids of Agrobacterium. Curr Top Microbiol Immunol 96:237–254

Muhlbach HP (1980) Different regeneration potentials of mesophyll protoplasts from cultivated and wild species of tomato. Planta 148:89–95

Muller AG, Grafe R (1978) Isolation and characterization of cell lines of Nicotiana tabacum lacking nitrate reductase. Mol Gen Genet 161:67–76

Murashige T, Nakano R (1966) Tissue culture as a potential tool in obtaining polyploid plants. J Hered 57:114–118

Murashige T, Nakano R (1967) Chromosome complement as a determinant of the morphogenic potential of tobacco cells. Am J Bot 54:963–970

Murashige T, Skoog F (1962) A revised medium for rapid growth and bioassays with tobacco tissue cultures. Physiol Plant 15:473–497

Murty BR, Swaminathan MS (1957) Cytogenetic studies in derivatives of Nicotiana rustica × N. tabacum. Euphytica 6:227–236

Nagao T (1978) Somatic hybridization by fusion of protoplast. I. The combination of Nicotiana tabacum and N. rustica. Jpn J Crop Sci 47:491–498

Nagao T (1979) Somatic hybridization by fusion of protoplasts. II. Combinations of Nicotiana tabacum and N. glutinosa and of N. tabacum and N. alata. Jpn J Crop Sci 48:385–392

Nagata T, Takebe I (1971) Plating of isolated tobacco mesophyll protoplasts on agar medium. Planta 99:12–20

Nagata T, Eibl H, Melchers G (1979) Fusion of plant protoplasts induced by a positively charged synthetic phospholipid. Z Naturforsch Sect C 34:460–462

Nagl W (1979) Differential DNA replication in plants. A critical review. Z Pflanzenphysiol 95:283–314

Nagl W, Hoffmann F (1980) Arabidobrassica: Evidence for intergeneric somatic hybrid nature from electron microscopic morphometry of chromatin. Eur J Cell Biol 21:227–228

Nagy F, Torok I, Maliga P (1981) Extensive rearrangements in the mitochondrial DNA in somatic hybrids of Nicotiana tabacum and N. knightiana. Mol Gen Genet 183:437–439

Nagy JI, Maliga P (1976) Callus induction and plant regeneration from mesophyll protoplasts of Nicotiana sylvestris. Z Pflanzenphysiol 78:453–455

Nakata K, Oshima H (1982) Cytoplasmic chimaericity in the somatic hybrids of tobacco. Abstr. Vth int congr plant tissue cell cult, 11–16 July 1982, Tokyo p 110

Negrutiu I (1981) Improved conditions for large-scale culture, mutagenesis and selection of haploid protoplasts of Nicotiana plumbaginifolia Viviani. Z Pflanzenphysiol 104:431–442

Negrutiu I, Mousseau J (1980) Protoplast culture from in vitro grown plants of Nicotiana sylvestris Speg. and Comes. Z Pflanzenphysiol 100:373–376

Negrutiu I, Mousseau J (1981) Large-scale culture of protoplasts of Nicotiana sylvestris Speg. Comes isolated from in vitro propagated plants. Z Pflanzenphysiol 103:367–372

Negrutiu I, Muller JF (1981) Culture conditions of protoplast-derived cells of Nicotiana sylvestris for mutant selection. Plant Cell Rep 1:14–17

Nehls R (1978a) Use of metabolic inhibitors for selection of fusion products of higher plant protoplasts. Mol Gen Genet 166:117–118

Nehls R (1978b) Isolation and regeneration of protoplasts from Solanum nigrum L. Plant Sci Lett 12:183–188

Ninnemann H, Juttner F (1981) Volatile substances from tissue cultures of potato, tomato and their somatic fusion products. – Comparison of gas chromatographic patterns for identification of hybrids. Z Pflanzenphysiol 103:95–108

Nitsch JP, Ohyama K (1971) Obtention de plantes à partir de protoplastes haploides cultivés in vitro. CR Acad Sci D 273:801–804

Nitzsche W (1973) Mitotische Chromosomenreduction in höheren Pflanzen durch 3-Fluor-Phenylalanin. Naturwissenschaften 60:390

Nitzsche W, Wenzel G (1977) Haploids in plant breeding. P Parey, Berlin Hamburg

O'Connell, PJ, Brady CJ (1981) Multiple forms of the large subunit of wheat ribulose bisphosphate generated by excess iodoacetamide. Biochim Biophys Acta 670:355–361

Ohyama K, Nitsch JP (1972) Flowering haploid plants obtained from protoplasts of tobacco leaves. Plant Cell Physiol 13:229–236

Ono K, Ohyama K, Gamborg OL (1979) Regeneration of the Liverwort Marchantia polymorpha L. from protoplasts isolated from cell suspension culture. Plant Sci Lett 14:225–230

Pandey KK (1975) Sexual transfer of specific genes without gemetic fusion. Nature 256:310–313

Pandey KK (1979) Gametic gene transfer in Nicotiana by means of irradiated pollen. Genetica 49:53–69

Partanen CR, Power JB, Cocking EC (1980) Isolation and division of protoplasts of Pteridium aquilinumm. Plant Sci Lett 17:333–338

Passiatore JE, Sink KC (1981) Plant regeneration from leaf mesophyll protoplasts of selected ornamental Nicotiana species. J Am Soc Hort Sci 106:799–803

Patnaik G, Cocking EC, Hamill J, Pental D (1982) A simple procedure for the manual isolation and identification of plant heterokaryons. Plant Sci Lett 24:105–110

Peberdy JF, Rose AH, Rogers HJ, Cocking EC (eds) (1976) Microbial and plant protoplasts. Academic Press, New York London

Pelcher LE, Gamborg OL, Kao KN (1974) Bean mesophyll protoplasts: production, culture and callus formation. Plant Sci Lett 3:107–111

Piven NM (1980) Fusion of isolated protoplasts and cytoplasmic inheritance in Nicotiana. Thesis Kiev

Piven NM (1981) Regeneration of intact plants from isolated leaf mesophyl protoplasts of Nicotiana debneyi Domin. Sov Cytol Genet 15:35–39

Plant Cell and Tissue Culture, see Ref. Larsen PO, Paddock EF, Raghavan V 1979, Ohio State University Press, Columbus

Poirier-Hamon S, Rao PS, Harada H (1974) Culture of mesophyll protoplasts and stem segments of Antirrhinum majus/Snapdragon. J Exp Bot 25:752–760
Pontecorvo G (1958) Trends in genetic analysis. Columbia University Press, New York
Pontecorvo G (1971) Induction of directional chromosome elimination in somatic cell hybrids. Nature 230:367–369
Potrykus I (1973) Transplantation of chloroplasts into protoplasts of Petunia. Z Pflanzenphysiol 70:364–366
Potrykus I (1980) The old problem of protoplast culture: Cereals. In: Ferenczy L, Farkas GL, Lazar G (eds) Advances in protoplast research. Akad Kiado, Budapest, pp 243–254
Potrykus I, Hoffmann F (1973) Transplantation of nuclei into protoplasts of higher plants. Z Pflanzenphysiol 69:287–289
Potrykus I, Harms CT, Lorz H (1976) In: Dudits D, Farkas GL, Maliga P (eds) Cell genetics in higher plants. Akad Kiado, Budapest, pp 129–140
Potrykus I, Lorz H, Harms CT (1977a) On some selected problems and results concerning culture and genetic modification of higher plant protoplasts. In: Barz W, Reinhard E, Zenk MH (eds) Plant tissue culture and its bio-technological application. Springer, Berlin Heidelberg New York, pp 323–333
Potrykus I, Harms CT, Lorz H, Thomas E (1977b) Callus formation from stem protoplasts of corn (Zea mays L) Mol Gen Genet 156:347–350
Potrykus I, Harms CT, Lorz H (1979 a) Multiple-drop-array/MDA/technique for the large-scale testing of culture media variations in hanging microdrop cultures of single cell systems. Technique. Plant Sci Lett 14:231–236
Potrykus I, Harms CT, Lorz H (1979 b) Determination of phytohormone combinations for optimal division response in Nicotiana tabacum protoplast cultures. Plant Sci Lett 14:237–244
Potrykus I, Harms CT, Lorz H (1979c) Callus formation from cell culture protoplasts of Corn (Zea mays L.) Theor Appl Genet 54:209–214
Poulson C, Porath D, Sacristan M, Melchers G (1980) Peptide mapping of the ribulose biphosphate carboxylase small subunit from the somatic hybrid of tomato and potato. Carlsberg Res Commun 45:249–268
Povilaitis B, Cameron DR (1963) A mutation causing chlorophyll deficiency in Nicotiana tabacum. Can J Genet Cytol 5:233–238
Power JB, Berry SF (1979) Plant regeneration from protoplasts of Browallia viscosa. Z Pflanzenphysiol 94:469–474
Power JB, Cocking EC (1970) Isolation of leaf protoplasts: Macromolecule uptake and growth substance response. J Exp Bot 21:64–70
Power JB, Cocking EC (1971) Fusion of plant protoplasts. Sci Prog 59:181–198
Power JB, Cocking EC (1977) Selection systems for somatic hybrids. In: Reinert J, Bajaj YPS (eds) Applied and fundamental aspects of plant cell, tissue, and organ culture. Springer, Berlin Heidelberg New York, pp 456–497
Power JB, Cummins SE, Cocking EC (1970) Fusion of isolated plant protoplasts. Nature 255:1016–1018
Power JB, Frearson EM, Hayward C, Cocking EC (1975) Some consequences of the fusion and selective culture of petunia and Parthenocissus protoplasts. Plant Sci Lett 5:197–207
Power JB, Frearson EM, Hayward C, George D, Evans PK, Berry SF, Cocking EC (1976a) Somatic hybridization of Petunia hybrida and P. parodii. Nature 263:500–502
Power JB, Frearson EM, George AA (1976b) The isolation, culture and regeneration of leaf protoplasts in the genus Petunia. Plant Sci Lett 7:51–55
Power JB, Berry SF, Frearson EM, Cocking EC (1977) Selection procedures for the production of inter-species somatic hybrids of Petunia hybrida and Petunia parodii. I. Nutrient media and drug sensitivity complementation selection. Plant Sci Lett 10:1–6
Power JB, Berry SF, Chapman JV, Cocking EC, Sink KC (1979) Somatic hybrids between unilateral cross-incompatible Petunia species. Theor Appl Genet 55:97–100
Power JB, Berry SF, Chapman JV, Cocking EC (1980) Somatic hybridization of sexually incompatible Petunias: P. parodii, P. parviflora. Theor Appl Genet 57:1–4
Raveh D, Galun E (1975) Rapid regeneration of plants from tobacco protoplasts plated at low densities. Z Pflanzenphysiol 76:76–79
Razdan MK, Cocking EC, Power JB (1980) Callus regeneration from mesophyll protoplasts of sweet pea (Lathyrus odoratus L.) Z Pflanzenphysiol 96:181–184

Redenbaugh K, Ruzin S, Bartholomew J, Bassham JA (1982) Characterization and separation of plant protoplasts via flow cytometry and cell sorting. Z Pflanzenphysiol 107:65–80
Reinert J (1963) Growth of single cells from higher plants on synthetic media. Nature 200:90–91
Reinert J, Ardenne R von (1964) Abhängigkeit des Teilungswachstums isolierter Einzelzellen aus Vitis vinifera-Kulturen von benachbarten wachsenden Geweben. Z Naturforsch Teil B:1150–1156
Reinert J, Bajaj YPS (eds) (1977) Applied and fundamental aspects of plant cell tissue, and organ culture. Springer, Berlin Heidelberg New York, pp 803
Reinert J, Gosch G (1976) Continuous division of heterokaryons from Daucus carota and Petunia hybrida protoplasts. Naturwissenschaften 63:534
Rieger R, Michaelis A, Green MM (1976) A glossary of genetics and cytogenetics, 4th edn. Springer, Berlin Heidelberg New York
Ringertz NS, Savage RE (1976) Cell hybrids. Academic press, New York, pp 366
Roberts RJ (1982) Restriction and modification enzymes and their recognition sequences. Nucl Acid Res 10:r117–r143
Rodgers A (1979) Detection of small amounts of human DNA in human-rodent hybrids. J Cell Sci 38:391–403
Roper W (1981) Callus formation from protoplasts derived from cell suspension cultures of Vicia faba L. Z Pflanzenphysiol 101:75–78
Sacristan MD, Melchers G (1969) The caryological analysis of plants regenerated from tumorous and other callus cultures of tobacco. Mol Gen Genet 105:317–333
Sacristan MD, Melchers G (1977) Regeneration of plants from „habituated" and „Agrobacterium-transformed" single cell clones of tobacco. Mol Gen Genet 152:111–117
Sager R (1972) Cytoplasmic genes and organelles. Academic Press, New York
Sager R (1975) Patterns of inheritance of organelle genomes: molecular basis and evolutionary significance. In: Birky CW Jr, Perlman PS, Byers TJ (eds) Genet. Biogen. Mitochondria Chloroplasts. Ohio State University Press, Columbus, pp 252–267
Sand SA (1968) Genetic modification of cytoplasmic male sterility in tobacco. J Hered 59:175–177
Sand SA, Christoff GT (1973) Cytoplasmic-chromosomar interactions and altered differentiation in tobacco. J Hered 64:24–30
Sangwan RS, Harada H (1978) Abstr. fourth Int Congr plant tissue, cell culture, Calgary, Canada, p 59
Saxena PK, Gill R, Rashid A, Maheshwari SC (1981a) Plantlet formation from isolated protoplasts of Solanum melongena L. Protoplasma 108:355–356
Saxena PK, Gill R, Rashid A, Maheswari SC (1981b) Isolation and culture of protoplasts of Capsicum annuum L. and their regeneration into plants flowering in vitro. Protoplasma 108:357–360
Scandalios JG (1974) Isozymes in development and differentiation. Annu Rev Plant Physiol 25:225–258
Schenck H (1982) Brassica napus: successful resynthesis by protoplast fusion between B. oleracea and B. campestris. Abstr. Vth Int. Congr. Plant Tissue Cell Cult, Tokyo, p 109
Schenck H, Hoffmann F (1979) Callus and root regeneration from mesophyll protoplasts of basis Brassica species: B. campestris, B. oleracea, and B. nigra. Z Pflanzenzuecht 82:354–362
Schieder O (1974) Fusionen zwischen Protoplasten aus Mutanten von Sphaerocarpes donnellii Aust. BPP 165:433–435
Schieder O (1975a) Selektion einer somatischen Hybride nach Fusion von Protoplasten auxotropher Mutanten von Sphaerocarpos donnellii Aust. Z Pflanzenphysiol 74:357–365
Schieder O (1975b) Regeneration von haploiden und diploiden Datura innoxia Mill. mesophyll-Protoplasten zu Pflanzen. Z Pflanzenphysiol 76:462–466
Schieder O (1976) Isolation of mutants with altered pigments after irradiating haploid protoplasts from Datura innoxia Mill. with X-rays. Mol Gen Genet 149:251–254
Schieder O (1977a) Hybridization experiments with protoplasts from chlorophyll-deficient mutants of some Solanaceae species. Planta 137:253–258
Schieder O (1977b) Attempts in regeneration of mesophyll protoplasts of haploid and diploid wild type lines, and those of chlorophyll-deficient strains from different Solanaceae. Z Pflanzenphysiol 84:275–281
Schieder O (1978a) Somatic hybrids of Datura innoxia Mill. + Datura discolor Bernh. and of Datura innoxia Mill. + Datura stramonium L. var. tatula L. I. Selection and characterisation. Mol Gen Genet 162:113–119
Schieder O (1978b) Genetic evidence for the hybrid nature of somatic hybrids from Datura innoxia Mill. Planta 141:333–334

Schieder O (1980a) Somatic hybrids between a herbaceous and two tree Datura species. Z Pflanzenphysiol 98:119–127

Schieder O (1980 b) Somatic hybrids of Datura innoxia Mill. + Datura discolor Bernh. and of Datura innoxia Mill. + Datura stramonium L. var. tatula L. II. Analysis of progenies of three sexual generations. Mol Gen Genet 179:387–390

Schieder O, Krumbiegel G (1979) Höhere Erträge bei Arzneipflanzen durch künstliche Zellfusion. Umschau 79:545–546

Schieder O, Vasil IK (1980) Protoplast fusion and somatic hybridization. Int Rev Cytol 11B:21–46

Schiller B, Herrmann RG, Melchers G (1982) Restriction endonuclease analysis of plastid DNA from tomato, potato and some of their somatic hybrids. Mol Gen Genet 186:453–459

Schum A, Priel W (1981) Regeneration of callus from mesophyll protoplasts of Chrysanthemum morifolium Ramat. Gartenbauwissenschaft 46:91–92

Schwartz AG, Cook PR, Harris H (1971) Correction of a genetic defect in a mammalian cell. Nature New Biol 230:5–8

Scowcroft WR, Larkin PJ (1980) Isolation, culture and plant regeneration from protoplasts of Nicotiana debneyi. Aust J Plant Physiol 7:635–644

Scowcroft WR, Larkin PJ (1981) Cloroplast DNA assorts randomly in intraspecific somatic hybrids of Nicotiana debneyi. Theor Appl Genet 60:179–184

Secor GA, Shepard JF (1981) Variability of protoplast-derived potato clones. Crop Sci 21:102–105

Shahin EA, Shepard JF (1980) Cassava mesophyll protoplasts: isolation, proliferation and shoot formation. Plant Sci Lett 17:459–465

Shepard JF (1982) Cultivar dependent cultural requirements in potato protoplast regeneration. Plant Sci Lett 26:127–132

Shepard JF, Totten RE (1977) Mesophyll cell protoplasts of potato: Isolation, proliferation, and plant regeneration. Plant Physiol 60:313–316

Shepard JF, Bidney D, Shahin E (1980) Potato protoplasts in crop improvement. Science 208:17–24

Shimada T, Tabata M (1967) Chromosome numbers in cultured pith tissue of tobacco. Jpn J Genet 42:195–201

Sidorov VA, Maliga P (1982) Fusion-complementation analysis of auxotrophic and chlorophyll-deficient lines isolated in haploid Nicotiana plumbaginifolia protoplast cultures. Mol Gen Genet 186:328–332

Sidorov VA, Belitser NV, Artemenko VS, Gleba YY, Sytnik, KM (1977) Electron microscopic study of tobacco mesophyll protoplasts. II. Ultrastructure of cultured protoplasts on the stage of primary cell colonies. Sov Cytol Genet 11:291–297

Sidorov VA, Gleba Y, Sytnik KM (1978) Ultrastructural study of Arabidopsis thaliana (L.) Heynh. cultured protoplasts and arabidopsis + tobacco fusion products. Arabidopsis Inf Serv 15:70–77

Sidorov VA, Menczel L, Nagy F, Maliga P (1981a) Chloroplast transfer in Nicotiana based on metabolic complementation between irradiated and iodoacetate treated protoplasts. Planta 152:341–345

Sidorov VA, Menczel, L, Maliga P (1981b) Isoleucine-requiring Nicotiana plant deficient in threonine deaminase. Nature 294:87

Simmonds JA, Simmonds DH, Cumming BG (1979) Isolation and cultivation of protoplasts from morphogenetic callus cultures in Lilium. Can J Bot 57:512–516

Singh M, Sinhar U (1976) Chloral hydrate induced haploidization in Aspergillus nidulans. Experientia 32:1144–1145

Sink KC, Power JB (1977) Isolation, culture and regeneration of leaf protoplasts of Petunia parviflora Juss. Plant Sci Lett 10:335–340

Skarzhynskaya MV (1980) Production of hybrid cell lines by individual cloning of single fusion products of isolated protoplasts. Ukr Bot Zh 36:58–61

Skarzhynskaya MV, Cherep NN, Gleba Y (1982) Potato and tobacco hybrid cell lines and plants obtained by cloning individual protoplast fusion products. Sov Cytol Genet 6:42–48

Skene KGM (1975) Production of callus from protoplasts of cultured grape pericarp. Vitis 14:177–180

Smillie RM, Gibbons GC (1979) Chilling and heat sensitivity in chloroplast membranes of somatic hybrid plants of tomato and potato. Proc Aust Biochem Soc 12:101

Smillie RM, Melchers F, Wettstein D von (1979) Chilling resistance of somatic hybrids of tomato and potato. Carlsberg Res Commun 44:127–132

Smith HH (1974) Model systems for somatic cell plant genetics. Bioscience 24:269–276

Smith HH (1979) The genus as a genetic resource. In: Durbin RD (ed) Nicotiana. Procedures for experimental use. USDA Bull 1586:1–16

Smith HH, Mastrangelo-Hough IA (1979) Genetic variability available through cell fusion. In: Sharp WR et al. (eds) Plant cell and tissue culture. Principles and applications. Ohio State University, Columbus, pp 265–286

Smith HH, Kao KN, Combatti NC (1976) Interspecific hybridization by protoplast fusion in Nicotiana: confirmation and extension. J Hered 67:123–128

Smolenskaya IN, Raldugina GN (1981) Protoplast culture from cell suspension of sugar beet. Sov Plant Physiol 28:1022–1029

Sondahl MR, Chapman MS, Sharp WR (1980) Protoplast liberation, cell wall reconstitution, and callus proliferation in Coffea arabica L. callus tissues. Turrialba 30:161–165

Steele JA, Uchytil TF, Durbin RD, Bhatnagar P, Rich DH (1976) Chloroplast coupling factor I: a species-specific receptor for tentoxin. Proc Natl Acad Sci USA 73:2245–2248

Steele JA, Uchytil TF, Durbin RD (1977) The binding of tentoxin to a tryptic digest of chloroplast coupling factor I. Biochim Biophys Acta 459:347–355

Stewart RN, Burk LG (1970) Independence of tissues derived from apical layers in ontogeny of the tobacco leaf and ovary. Am J Bot 57:1010–1016

Stumm I, Meyer Y, Abel WO (1975) Regeneration of the moss Physcomitrella patens (Hedw.) from isolated protoplasts. Plant Sci Lett 5:113–118

Sunderland N (1973) Nuclear cytology. In: Street HE (ed) Plant tissue and cell culture. Blackwell, Oxford, pp 161–190

Szabados L, Hadlaczky G, Dudits D (1981) Uptake of isolated plant chromosomes by plant protoplasts. Plasma 151:141–145

Tabata M, Yamamoto H, Hiraoka N (1968) Chromosome constitution and nicotine formation in mature plants derived from cultured pith of tobacco. Jpn J Genet 43:319–322

Takebe I (1975) The use of protoplasts in plant virology. Annu Rev Phytopathol 13:105–126

Takebe I, Labib G, Melchers G (1971) Regeneration of whole plants from isolated mesophyll protoplasts of tobacco. Naturwissenschaften 58:318–320

Taylor DL (1970) Cloroplasts as symbiotic organelles. Int Rev Cytol 27:29–64

Ternovskiy MF, Grebenkin AP (1967) Cytoplasmic male sterility upon distantly related crosses in Nicotiana generus. Sov Genet 8:4–17

Thomas E, Hoffmann F, Potrykus I, Wenzel G (1976) Protoplast regeneration and stem embryogenesis of haploid androgenetic rape. Mol Gen Genet 145:245–248

Thomas E, King PJ, Potrykus I (1979) Improvement of crop plants via single cells in vitro assessment. Z Pflanzenzüchtung 82:1–30

Tsikov DK, Tsikova E (1971) Male sterility in backcrosses of reciprocal amphidiploids of Nicotiana debneyi Dom. and N. tabacum L. with N. tabacum L. CR Sci Agr Bulg 4:17–19

Uchimiya H (1978) Abstr. fourth Intern. Congr. plant tissue cult, Calgary, Canada, N 200, p 47

Uchimiya H (1982) Somatic hybridization between male sterile Nicotiana tabacum and N. glutinosa through protoplast fusion. Theor Appl Genet 61:69–72

Uchimiya H, Murashige T (1976) Influence of the nutrient medium on the recovery of dividing cells from tobacco protoplasts. Plant Physiol 57:424–429

Uhrig H (1981) Regeneration of protoplasts of diahaploid potato plants bleached by a herbicide (SAN 6706) Molec Gen Genet 181:403–405

Umiel N, Goldner R (1976) Effects of streptomycin on diploid tobacco callus cultures and the isolation of resistant mutants. Protoplasma 89:83–89

Vardi A, Spiegel-Roy P, Galun E (1975) Citrus cell culture: isolation of protoplasts, plating densities, effect of mutagens and regeneration of embryos. Plant Sci Lett 44:231–236

Vardi A, Spiegel-Roy P, Galun E (1982) Plant regeneration from Citrus protoplasts: variability in methodological requirements among cultivars and species. Theor Appl Genet 62:171–176

Vasil IK (1976) The progress, problems, and prospects of plant protoplasts research. Adv Agron 28:119–160

Vasil IK, Vasil V (1980) Isolation and culture of protoplasts. Int Rev Cytol 11A:1–20

Vasil IK, Ahuja MR, Vasil V (1979) Plant tissue cultures in genetic and plant breeding. Adv Genet 20:127–215

Vasil V, Hildebrandt AC (1965) Differentiation of tobacco plants from single, isolated cells in microcultures. Science 150:889–892

Vasil V, Vasil IK (1974) Regeneration of tobacco and petunia plants from protoplasts culture of corn protoplasts. In vitro 10:83–96

Vasil V, Vasil IK (1979) Isolation and culture of cereal protoplasts. I. Callus formation from pearl millet (Pennisetum mexicanum) protoplasts. Z Pflanzenphysiol 92:379–383

Vasil V, Vasil IK (1980) Isolation and culture of cereal protoplasts. II. Embryogenesis and plantlet formation from protoplasts of Pennisetum americanum. Theor Appl Genet 56:97–99

Vatsya B, Bhaskaran S (1981) Production of sub-protoplasts in Brassica oleracea var. capitata – a function of osmolarity of the media. Plant Sci Lett 23:277–282

Vaughan JG, Maclead AJ, Jones MMG (1976) The biology and chemistry of the Cruciferae, Academic Press, London

Vedel F, Quetier F, Bayen M (1976) Specific cleavage of chloroplast DNA from higher plants by Eco RI restriction nuclease. Nature 263:440–442

Veen JH van der (1957) Studies on the inheritance of leaf shape in Nicotiana tabacum L. Thesis, Wageningen

Wagner G, Hess D (1974) Haploide, diploide und triploide Pflanzen von Petunia hybrida aus Pollenkörnern. Z Pflanzenphysiol 73:273–276

Wallin A., Glimelius K, Eriksson T (1978) Enucleation of plant protoplasts by cytochalasin-B. Z Pflanzenphysiol 87:333–340

Wallin A, Glimelius K, Eriksson T (1979) Formation of hybrid cells by transfer of nuclei via fusion of miniprotoplasts from cell lines of nitrate reductase deficient tobacco. Z Pflanzenphysiol 91:89–94

Wenzel G, Schieder O (1973) Regeneration of isolated protoplasts from nicotinic acid-deficient mutants of the liverwort Sphaerocarpos donnellii Oust. Plant Sci Lett 1:421–423

Wernicke W, Thomas E (1980) Studies on morphogenesis from isolated protoplasts: shoot formation from mesophyll protoplasts of Hyoscyamus muticus and Nicotiana tabacum. Plant Sci Lett 17:401–407

Wernicke W, Lorz H, Thomas E (1979) Plant regeneration from leaf protoplasts of haploid Hyoscyamus-muticus L. produced via anther culture. Plant Sci Lett 15:239–250

Wetter LR (1977) Isoenzyme patterns in soybean-Nicotiana somatic hybrid cell lines. Mol Gen Genet 150:231–235

Wetter LR, Kao KN (1976) The use of isozymes in distinguishing the sexual and somatic hybrids in callus cultures derived from Nicotiana. Z Pflanzenphysiol 80:455–462

Wetter LR, Kao KN (1980) Chromosome and isoenzyme studies on cells derived from protoplast fusion of Nicotiana glauca with Glycine max – Nicotiana glauca cell hybrids. Theor Appl Genet 57:273–276

Wettstein D von, Poulson C, Holder AA (1978) Ribulose-1,5-bisphosphate carboxylase as a nuclear and chloroplast marker. Theor Appl Genet 53:193–198

White DWR, Bhojwani SS (1981) Callus formation from Trifolium arvense protoplast-derived cells plated at low densities. Z Pflanzenphysiol 102:257–262

White DWR, Vasil IK (1979) Use of amino acid analogue-resistant cell lines for selection of Nicotiana sylvestris somatic cell hybrids. Theor Appl Genet 55:107–112

Wildman SG, Bonner J (1947) The proteins of green leaves. I. Isolation, enzymatic properties and auxin content of spinach cytoplasmic proteins. Arch Biochem Biophys 14:381–413

Wildman SG, Lu-Liao C, Wong-Staal F (1973) Maternal inheritance, cytology and macromolecular composition of defective chloroplasts in a variegated mutant of Nicotiana tabacum. Planta 113:293–312

Willecke K (1978) Results and prospects of chromosomal gene transfer between cultured mammalian cells. Theor Appl Genet 52:97–104

Winkler H (1908) Solanum tubingense, ein echter Pfropfbastard zwischen Tomate und Nachtschatten. Ber Dtsch Bot Ges 26a:595–608

Wong-Staal F, Wildman SG (1973) Identification of a mutation in chloroplast DNA correlated with formation of defective chloroplasts in a variegated mutant of Nicotiana tabacum. Planta 113:313–326

Wright WE (1978) The isolation of heterokaryons and hybrids by a selective system using irreversible biochemical inhibitors. Exp Cell Res 112:395–407

Wright WE, Hayflick L (1975) Use of biochemical lesions for selection of human cells with hybrid cytoplasms. Proc Natl Acad Sci USA 72:1812–1816

Wullems GJ, Marton L, Molendijk L, Krens F, Ooms G, Wurzer-Figurelli L, Schilperoort RA (1979) Genetic modification of plant cells by transformation and somatic hybridization. In: Ferenczy L (ed) Advances of protoplast research, Proc. 5th Int. Protoplast Symp., 407–424. Akad Kiado, Budapest

Wullems GJ, Molendijk L, Schilperoort RA (1980) The expression of tumour markers in intraspecific somatic hybrids of normal and crown gall cells from Nicotiana tabacum. Theor Appl Genet 56:203–208

Xianghui L, Wenbin L, Meijuan H (1982) Somatic hybrid plants from intergeneric fusion between tobacco tumor B6S3 and Petunia hybrida W43 and expression of LpDH. Sci Si 25:611–619

Xu ZH, Davey MR, Cocking EC (1981) Isolation and sustained division of Phaseolus aureus root protoplasts. Z Pflanzenphysiol 104:289–298

Xu ZH, Davey MR, Cocking EC (1982a) Callus formation from root protoplasts of Glycine (soybean). Plant Sci Lett 24:111–115

Xu ZH, Davey MR, Cocking EC (1982b) Plant regeneration from root protoplasts of Brassica. Plant Sci Lett 24:117–121

Xu ZH, Davey MR, Cocking EC (1982c) Organogenesis from root protoplasts of the forage legumes Medicago sativa and Trigonella foenum-graecum. Z Pflanzenphysiol 107:231–236

Zagorska NA, Shamina ZB, Butenko RG (1971a) Peculiarities of morphogenesis in tissue culture of diploid and polyploid tobacco. Dokl Acad Nauk USSR 196:1209–1212

Zagorska NA, Shamina ZB, Butenko RG (1971b) Morphogenesis responses in tissue cultures of diploid and polyploid tobacco. Dokl Akad Nauk USSR 196:5–8

Zapata FJ, Sink KC (1981) Somatic embryogenesis from Lycopersicon peruvianum leaf mesophyll protoplasts. Theor Appl Genet 59:265–268

Zapata FJ, Evans PK, Power JB, Cocking EC (1977) Effect of temperature on division of leaf protoplasts of Lycopersicon exculentum and Lycopersicon peruvianum. Plant Sci Lett 8:119–124

Zapata FJ, Sink KC, Cocking EC (1981) Callus formation from leaf mesophyll protoplasts of three Lycopersicon species: L. esculentum, L. pimpinillifolium and L. hirsutum. Plant Sci Lett 23:41–46

Zelcer A, Aviv D, Galun E (1978) Interspecific transfer of cytoplasmic male sterility by fusion between protoplasts of normal Nicotiana sylvestris and X-ray irradiated protoplasts of male-sterile N. tabacum. Z Pflanzenphysiol 90:397–408

Zieg RG, Outka DE (1980) The isolation, culture and callus formation of soybean pod protoplasts. Plant Sci Lett 18:105–114

Zimmermann U (1982) Electric field-mediated related electrical phenomena. Biochim Biophys Acta (694:227–277)

Zimmermann U, Scheurich P (1981a) Fusion of Avena sativa mesophyll cell protoplasts by electrical breakdown. Biochim Biophys Acta 641:160–165

Zimmermann U, Scheurich P (1981b) High frequency fusion of plant protoplasts by electric field. Planta 151:26–32

Zimmermann U, Kuppers G, Salhani N (1982) Electric field-induced release of chloroplasts from plant protoplasts. Naturwissenschaften 69:451:452

Zosimovitch VP, Kunach VA (1975) Level, types and origin of chromosome aberration in culture of isolated plant tissues. Sov Genet 11:37–46

Subject Index

MIX
Papier aus verantwortungsvollen Quellen
Paper from responsible sources
FSC® C105338

If you have any concerns about our products,
you can contact us on
ProductSafety@springernature.com

In case Publisher is established outside the EU,
the EU authorized representative is:
Springer Nature Customer Service Center GmbH
Europaplatz 3, 69115 Heidelberg, Germany

Printed by Libri Plureos GmbH
in Hamburg, Germany